Value-Range Analysis of C Programs

Axel Simon

Value-Range Analysis of C Programs

Towards Proving the Absence of Buffer Overflow Vulnerabilities

Axel Simon

ISBN 978-1-84996-702-0 e-ISBN 978-1-84800-017-9
DOI: 10.1007/978-1-84800-017-9

British Library Cataloguing in Publication Data
A catalogue record for this book is available from the British Library

Softcover reprint of the hardcover 1st edition 2008

Printed on acid-free paper

Springer Science+Business Media
springer.com

To my parents.

Preface

A buffer overflow occurs when input is written into a memory buffer that is not large enough to hold the input. Buffer overflows may allow a malicious person to gain control over a computer system in that a crafted input can trick the defective program into executing code that is encoded in the input itself. They are recognised as one of the most widespread forms of security vulnerability, and many workarounds, including new processor features, have been proposed to contain the threat. This book describes a static analysis that aims to prove the absence of buffer overflows in C programs. The analysis is conservative in the sense that it locates every possible overflow. Furthermore, it is fully automatic in that it requires no user annotations in the input program.

The key idea of the analysis is to infer a symbolic state for each program point that describes the possible variable valuations that can arise at that point. The program is correct if the inferred values for array indices and pointer offsets lie within the bounds of the accessed buffer. The symbolic state consists of a finite set of linear inequalities whose feasible points induce a convex polyhedron that represents an approximation to possible variable valuations. The book formally describes how program operations are mapped to operations on polyhedra and details how to limit the analysis to those portions of structures and arrays that are relevant for verification. With respect to operations on string buffers, we demonstrate how to analyse C strings whose length is determined by a NUL character within the string.

We complement the analysis with a novel sub-class of general polyhedra that admits at most two variables in each inequality while allowing arbitrary coefficients. By providing polynomial algorithms for all operations necessary for program analysis, this sub-class of general polyhedra provides an efficient basis for the proposed static analysis. The polyhedral sub-domain presented is then refined to contain only integral states, which provides the basis for the combination of numeric analysis and points-to analysis. We also present a novel extrapolation technique that automatically inspects likely bounds on variables, thereby providing a way to infer precise loop invariants.

Target Audience

The material in this book is based on the author's doctoral thesis. As such it focusses on a single topic, namely the definition of a sound value-range analysis for C programs that is precise enough to verify non-trivial string buffer operations. Furthermore, it only applies one approach to pursue this goal, namely a fixpoint computation using convex polyhedra that approximate the state space of the program. Hence, it does not provide an overview of various static analysis methods but an in-depth treatment of a real-world analysis task. It should therefore be an interesting and motivating read, augmenting, say, a course on program analysis or formal methods.

The merit of this book lies in the formal definition of the analysis as well as the insight gained on particular aspects of analysing a real-world programming language. Most research papers that describe analyses of C programs lack a formal definition. Most work that is formal defines an analysis for toy languages, so it remains unclear if and how the concepts carry over to real languages. This book closes this gap by giving a formal definition of an analysis that handles full C. However, this book is more than an exercise in formalising a large static analysis. It addresses many facets of C that interact and that cannot be treated separately, ranging from the endianness of the machine, alignment of variables, overlapping accesses to memory, casts, and wrapping, to pointer arithmetic and mixing pointers with values.

As a result, the work presented is of interest not only to researchers and implementers of sound static analyses of C but to anyone who works in program analysis, transformation, semantics, or even run-time verification. Thus, even if the task at hand is not a polyhedral analysis, the first chapters, on the semantics of C, can save the reinvention of the wheel, whereas the latter chapters can serve in finding analogous solutions using the analysis techniques of choice. For researchers in static analysis, the book can serve as a basis to implement new abstraction ideas such as shape analyses that are combined with numeric analysis. In this context, it is also worth noting that the abstraction framework in this book shows which issues are solvable and which issues pose difficult research questions. This information is particularly valuable to researchers who are new to the field (e.g., Ph.D. students) and who therefore lack the intuition as to what constitutes a good research question.

Some techniques in this book are also applicable to languages that lack the full expressiveness of C. For instance, the Java language lacks pointer arithmetic, but the techniques to handle casting and wrapping are still applicable. At the other extreme, the analysis presented could be adapted to analyse raw machine code, which has many practical advantages.

The book presents a sound analysis; that is, an analysis that never misses a mistake. Since this ambition is likely to be jeopardised by human nature, we urge you to report any errors, omissions, and any other comments to us. To this end, we have set up a Website at `http://www.bufferoverflows.org`.

Acknowledgments

First and foremost, I would like to thank Andy King, who has become much more to me than a Ph.D. supervisor during these last years. He not only chose an interesting topic but also supported me with all his expertise and encouragement in a way that went far beyond his duties. Furthermore, my many friends at the Computing Laboratory at the University of Kent – who are too numerous to list here – deserve more credit than they might realise. I wish to thank them for their support and their ability to take my mind off work. My special thanks go to Paula Vaisey for her undivided support during the last months of preparing the manuscript, especially after I moved to Paris. I would also like to thank Carrie Jadud for her diligent proofreading.

Paris, *Axel Simon*
May 2008

Contents

Contributions

This section summarises the novelties presented in this book. Some of these contributions have already been published in refereed forums, such as our work on the principles of tracking NUL positions by observing pointer operations [167], the ideas behind the TVPI domain [172], a convex hull algorithm for planar polyhedra [168], the idea of widening with landmarks [170], the idea of an abstraction map that implicitly handles wrapping [171], and the use of Boolean flags to refine points-to analysis [166]. Overall, this book makes the following contributions to the field of static analysis:

1. Chapter 2: Defining the Core C intermediate language, which is concise yet able to express all operations of C.
2. Chapter 3: The observation of improved precision when implementing congruence analysis as a reduced product with $\mathbb{Z}$-polyhedra.
3. Chapters 4–6: A sound abstraction of C; in particular:
   a) Sound treatment of the wrapping behaviour of integer variables.
   b) Automatic inference of fields in structures that are relevant to the analysis. In particular, fields on which no information can be inferred are not tracked by the polyhedral domain and therefore incur no cost.
   c) Combining flow-sensitive points-to analysis with a polyhedral analysis of pointer offsets.
   d) Sound and precise approximation of pointer accesses when the pointer may have a range of offsets using access trees.
   e) A concise definition of an abstraction map between concrete and abstract semantics.
4. Chapter 7 presents a complete set of domain operations for planar polyhedra; in particular, a novel convex hull algorithm [168].
5. Chapter 8 presents the two-variables-per-inequality (TVPI) domain [172].
6. Chapter 9 describes how integral tightening techniques can be applied in the context of the TVPI domain.

7. Chapter 10 discusses techniques for adding polyhedral variables on-the-fly. Specifically, this chapter introduces the notion of typed polyhedral variables.
8. Chapter 11 details string buffer manipulation through pointers. The techniques presented in this book are a substantial refinement of [167].
9. Chapter 12 presents widening with landmarks [170], a novel extrapolation technique for polyhedra.
10. Chapter 13 discusses techniques for analysing a path of the program several times using a single polyhedron [166]. It uses the techniques developed to define a very precise points-to analysis.

The most important contribution of this book is a formal definition of a static analysis of a real-world programming language that is reasonably concise and – we hope – simple enough to be easily understood by other researchers in the field. We believe that the static analysis presented in this book will be useful as a basis for similar analyses and related projects.

List of Figures

1 Introduction

In 1988, Robert T. Morris exploited a so-called buffer-overflow bug in *finger* (a dæmon whose job it is to return information on local users) to mount a denial-of-service attack on hundreds of VAX and Sun-3 computers [159]. He created what is nowadays called a worm; that is, a crafted stream of bytes that, when sent to a computer over the network, utilises a buffer-overflow bug in the software of that computer to execute code encoded in the byte stream. In the case of a worm, this code will send the very same byte stream to other computers on the network, thereby creating an avalanche of network traffic that ultimately renders the network and all computers involved in replicating the worm inaccessible. Besides duplicating themselves, worms can alter data on the host that they are running on. The most famous example in recent years was the MSBlaster32 worm, which altered the configuration database on many Microsoft Windows machines, thereby forcing the computers to reboot incessantly. Although this worm was rather benign, it caused huge damage to businesses who were unable to use their IT infrastructure for hours or even days after the appearance of the worm. A more malicious worm is certainly conceivable [187] due to the fact that worms are executed as part of a dæmon (also known as "service" on Windows machines) and thereby run at a privileged level, allowing access to any data stored on the remote computer. While the deletion of data presents a looming threat to valuable information, even more serious uses are espionage and theft, in particular because worms do not have to affect the running system and hence may be impossible to detect.

Worms also incur high hidden costs in that software has to be upgraded whenever an exploitable buffer-overflow bug appears. A lot of effort on the part of the programmer is spent in confining intrusions by singling out those software components that need to run at the highest privilege level, with the aim of executing the majority of the (potentially erroneous) code at a lower privilege level. While this tactic reduces the potential damage of an attack, it does not prevent it. A laudable goal is therefore to rid programs of buffer-overflow bugs, which is the aim of numerous tools specifically created for this task. So far, no tool has been able to ensure the absence of exploitable buffer overflows without incurring either manual labour (program annotations) or performance losses (run-time checks). As a result, most security vulnerabilities today are still accredited to buffer-overflow errors in software [64, 126]. Interestingly, the US National Security Agency predicted a decade ago that buffer-overflow attacks would remain a problem for another ten years [173]. While many new projects part from C as the implementation language, most server software is legacy C code such that buffer overflows remain problematic. This book presents an analysis that has the potential to automatically detect all possible buffer overflows and thereby prove the absence of vulnerabilities if no overflow is found. This analysis is purely static; that is, it operates solely on the source code and neither modifies nor examines the program's behaviour at runtime. Furthermore, it works in a "push-button" style in that no annotations in the program are required in order to use the tool. The challenge in the pursuit of this fully automated, purely static analysis is threefold:

soundness: It must not miss any potential buffer overflows.

efficiency: It has to deliver the result in a reasonable amount of time.

completeness: It should not warn about overflows if the program is correct.

The question of whether a buffer overflow is possible is at least as difficult as the Halting Problem and therefore undecidable in general. Due to the nature of this problem, an effective analysis must necessarily compromise with respect to completeness. The key idea of a static analysis is to abstract a potentially infinite number of runs of a program (which stem from a potentially infinite number of inputs) into a finite representation that is able to express the property to be proved. The technical explanation of worms in the next section introduces the "property to be proved", namely that a program has no buffer overflows. The finite representation that we have chosen to express this property are sets of linear inequalities or, in their geometric interpretation, polyhedra. To motivate the choice of linear inequalities (rather than, say, finite automata as used in model checking [49]), we examine a small example program in Sect. 1.2. We then briefly comment on the three challenges of soundness, efficiency, and completeness of our analysis, a preview of the three parts that comprise this book. This chapter concludes with a comparison of related tools and a summary of our contributions.

1.1 Technical Background

In its simplest form, a program exploiting a buffer overflow manages to write beyond a fixed-sized memory region allocated on the stack. Consider, for example, a function that declares a local 2000-byte array `buffer` into which it copies parts of a byte stream that it receives from the network. The call

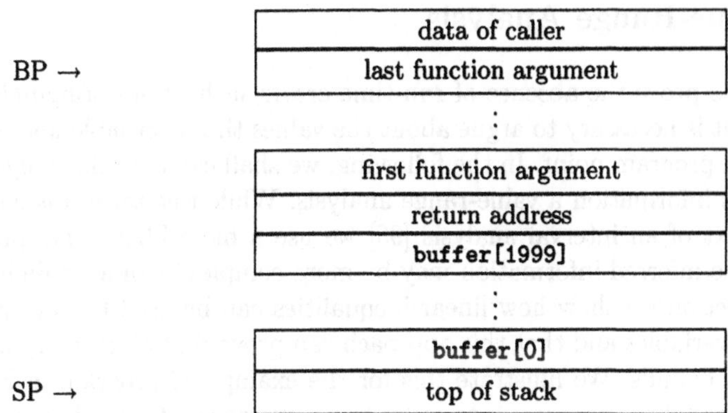

Fig. 1.1. A view of the stack after entering a function that declares a 2000-byte buffer. The pointers BP (base or frame pointer) and SP (stack pointer) manage the stack, which grows downwards (towards smaller addresses).

stack after invoking this function takes on a form that resembles the schematic representation in Fig. 1.1.

If a byte stream can be crafted such that more than 2000 bytes are copied to `buffer`, the memory beyond the end of the buffer will be overwritten, thereby altering the return address. A worm sets the return address to lie within `buffer` itself, with the effect that the byte stream from the network is run as a program when the function returns. It is the program encoded in the byte stream that determines the further action of the worm. A detailed description of how to craft one such input stream was given by a hacker known by the pseudonym of Aleph One, who presented a skeleton of a worm [141] that forms the basis of many known worms [159]. While the technical details are certainly interesting, the focus of this book lies in preventing such intrusions. Specifically, this work aims to prove the absence of buffer overflows, which is equivalent to showing that every memory access in a given program lies within a declared variable or dynamically allocated memory region. Detecting possible out-of-bounds accesses to variables is useful for any programming language with arrays (or plain memory buffers); however, only languages that do not check access bounds at run-time can create programs where buffer overflows create security vulnerabilities. The most prominent language in this category is C, a programming language that is widely used to implement networking software. Programmers chose C mostly for its ubiquity but also for the speed and flexibility that its low-level nature provides. However, it is exactly this low-level nature of C that makes program analysis challenging. Before Sect. 1.4 reviews the techniques to overcome the complexity of these low-level aspects, we detail what kinds of properties our analysis needs to extract from a program.

1.2 Value-Range Analysis

In order to prove the absence of run-time errors such as out-of-bounds array accesses, it is necessary to argue about the values that a variable may take on at a given program point. In the following, we shall call a static analysis that infers this information a value-range analysis. While this term was coined in the context of an interval analysis [95] we use a more liberal interpretation in that the inferred information may be more complex than a single interval. In this section we show how linear inequalities can be used to infer possible values of variables and that this approach can prove that all memory accesses lie within bounds. We illustrate this for the example C program in Fig. 1.2. The purpose of the program is to count the occurrences of each character in its first command-line argument. The idea is to define a table `dist`, where the ith entry stores the number of characters with the ASCII value i that have been observed in the input so far. Among the declared variables is the `dist` table containing 256 integers and a pointer to the input string `str`. In line 10, `str` is set to the beginning of the first command-line argument, namely `argv[1]`. This input string consists of a sequence of bytes that is terminated by a NUL character (a byte with the value zero). Note that the use of a NUL character to denote the length of the string is not enforced in C, even for arrays of bytes: The next line calls the function `memset`, which sets the bytes of a memory region to a given byte value, in this case zero. Here, the length of the buffer is passed explicitly as **`sizeof`**`(dist)` rather than being stored implicitly. The use of several conventions to store size information for memory regions is one of the idiosyncrasies of C that fosters incorrect memory management.

The **`while`** loop in lines 13–16 is the heart of the program. The loop iterates as long as the character currently pointed to by `str` is non-zero. Due to the `str++` statement in line 15, the loop will be executed for each character in the `argv[1]` buffer until the terminating zero character is encountered. The body of the loop increments the ith element of the `dist` array by one, assuming that the current character pointed to by `str` has the ASCII value i. Note that the character read by `*str` is converted to an integer, which ensures that the compiler does not emit a warning about automatic conversion from characters to an array index, which, according to the C standard [51], is of type **`int`**. The purpose of the last lines of the program is to print a fragment of the calculated character distribution to the screen.

Now consider the task of proving that all memory accesses are within bounds. While this task is trivial for variables such as `i` and `str`, expressing the correctness of the accesses to the memory regions `dist` and `*str` is complicated by the fact that the input string can be arbitrarily long.

In order to simplify the exposition, we assume that the program is run with exactly one command-line argument such that `argc` is equal to 2 and the return statement in line 9 is never executed. Under this assumption, the

```
1  #include<stdio.h>
2  #include<string.h>
3
4  int main(int argc, char* argv[]) {
5    int i;
6    char* str;
7    int dist[256];             /* Table of character counts.*/
8
9    if (argc!=2) return 1;         /* Expect one argument.*/
10   str = argv[1];               /* Let str point to input.*/
11   memset(dist, 0, sizeof(dist));        /* Clear table.*/
12
13   while (*str) {
14     dist[(int) *str]++;
15     str++;
16   };
17
18   for(i=32; i<128; i++)      /* Show dist for printable */
19     printf("'%c'␣:␣%i\n", i, dist[i]); /* characters.*/
20
21   return 0;
22 }
```

Fig. 1.2. Example C program that calculates the distribution of characters.

correctness of all memory accesses can be deduced with a few linear equalities and inequalities:

- The content of `argv[1]` is a pointer to a memory region of variable size x_s. Since we cannot explicitly represent an arbitrary number of array elements, we merely track the first known zero element of this memory region as x_n (the so-called NUL position), which indicates the end of the string. A conservative assumption is that the buffer is no bigger than what is needed to store the first command-line argument and the NUL position. Hence, the relationship between the buffer size and the NUL position can be expressed as $x_n = x_s - 1$.
- Line 10 assigns the pointer to this memory region to `str`. C allows so-called pointer arithmetic in that the address stored in `str` can be modified as if it were an integer variable. In our example, line 15 increments `str` by one and hence introduces an offset x_o relative to the beginning of the buffer; that is, x_o denotes the difference between the pointers `str` and `argv[1]`.
- From the offset x_o and the null position x_n, we can check if the loop invariant holds. As long as $x_o < x_n$, the value of `*str` is non-zero and the loop is executed. As soon as $x_o = x_n$, the loop body is not entered again and the execution of the loop stops. If we can further infer that $x_o = x_n$

holds every time the loop stops, we have shown that the buffer pointed to by `argv[1]` is never accessed beyond its bound because all offsets $0, \ldots, x_o$ during the execution of `*str` are no larger than x_s since $x_o \leq x_n = x_s - 1$.

- The values of characters read by `*str` are not known, except that they are non-zero with the exception of the last element. However, the value must be within the range of the C **char** type; that is, the index into the `dist` array, x_d, is restricted by CHAR_MIN $\leq x_d \leq$CHAR_MAX. The access to `dist` is within bounds if $0 \leq x_d \leq 255$ holds; that is, if CHAR_MIN$= 0$ and CHAR_MAX$= 255$.
- Finally, the correctness of the access `dist[i]` in line 19 can be ensured if the loop invariant $0 \leq x_i \leq 255$ can be guaranteed, where x_i represents the value of i within the loop body.

Note that the given chain of reasoning mainly relies only on linear inequalities that can be rewritten to $a_1x_1 + \ldots + a_nx_n \leq c$, where $a_1, \ldots, a_n, c \in \mathbb{Z}$, and $x_1, \ldots x_n$ represent variables or properties of variables in the program. In particular, the state of a program can be described by a conjunction of inequalities; that is, a set of inequalities all of which hold at the given program point. Note that in this representation an equality such as $x = y + z$ can be represented as two inequalities, $x - y - z \leq 0 \wedge -x + y + z \leq 0$. Simple toy languages consisting of assignments of linear expressions can easily be abstracted into operations on inequalities [62]. The next section introduces some of the subtleties that arise in the analysis of real-world languages.

1.3 Analysing C

Implementing a static analysis that is faithful to the semantics of a real-world programming language requires that the semantics of the language be well (or even formally) defined. Giving a formal semantics to an evolving language that already has undergone several standardisations is a laborious task [143] and not very practical if C programs do not adhere to any (single) standard. Worse, even the latest C standard [51] leaves certain implementation aspects up to the compiler, such that the answer to the question of whether the program in Fig. 1.2 is correct with respect to memory accesses can only be "maybe": On many platforms, including Linux on IA32 architectures and Mac OS X on PowerPC, the **char** type is signed, and hence $-128 \leq x_d \leq 127$, thereby violating the requirement that the index into `dist` lie within the interval $[0, 255]$. On platforms where **char** is unsigned, such as Linux on PowerPC, the program is correct. Next to implementation-specific semantics, C itself can be quite intricate. The seemingly plausible change of the statement `dist[(int) *str]++;` to `dist[(unsigned int) *str]++;` does not solve the problem: The so-called promotion rules of integers in C will first convert the value of `*str` to **int** (i.e., to a 32-bit value in $[-128, 127]$) and then to an unsigned integer (i.e., to $[2^{32} - 128, 2^{32} - 1] \cup [0, 127]$), leaving the program essentially unchanged.

Designing an analysis that interprets C programs in the same way as a particular mainstream compiler is a major undertaking in itself; see, e.g., [137]. Hence, rather than implementing a C front end for the analysis, we use the open source GNU C compiler as the front end and extract its intermediate representation. We convert this intermediate representation into Core C, a language amenable to our static analysis; Core C, defined in Chap. 2, contains mainly statements (rather than declarations) and attaches type information to operations (rather than to variables), thereby making many implementation-specific details explicit. Its formal semantics forms the basis of a sound abstraction to operations on inequalities, whose principles are explained in the next section.

1.4 Soundness

Given that a program may operate on a plethora of different inputs, it follows that an analysis that automatically proves every possible execution of the program correct must abstract from the actual program states, for instance, by summarising the possible valuations of variables at a given program point. Section 1.2 argued that the property of correct memory management can be expressed with a set of linear inequalities. Indeed, the idea of the analysis is to infer a set of inequalities that describes possible valuations of variables at a certain program point. Furthermore, since we are interested in verification, any such inequality set must be not only sound (correct) but precise enough to infer invariants that show that the program never exhibits a buffer overflow. Hence, the abstraction of sets of inequalities was chosen for its expressiveness. For the sake of this section, however, we will focus on soundness and leave the discussion of the achievable precision to Sect. 1.6.

1.4.1 An Abstraction of C

Simple program statements like `i=2*j+3` are readily translated into linear inequalities: With x_i and x_j representing the values of `i` and `j`, respectively, the assignment can be expressed as $x_i - 2x_j = 3$. However, analysing the full programming language C requires the translation of features such as arrays, pointer arithmetic, unions, etc., into a concise and, in particular, finite representation. To this end, several abstractions are needed. The following list summarises all abstractions applied within this work:

value abstraction: Summarising the possible values of a variable of each run into a finite representation such as an interval is the classic application of abstract interpretation [95]. With respect to the example, we observe that the value of the loop index `i` can be summarised to the interval [32, 127]. Several numeric domains, such as intervals, affine equations [109], and convex polyhedra [62], have been proposed to abstract concrete program values. The analysis presented in this book uses the domain of convex polyhedra in addition to a simple domain of congruences [85]; that is, information on the multiplicity of variable values.

content abstraction: In C, the size of some memory regions is determined by the value of a variable at run-time. At any given program point, all runs of a program (and hence all variable-sized memory regions) must be described by a single abstract state. Since the abstract state is a polyhedron over a fixed, finite number of variables, it is not possible to map each concrete element of a memory region to one variable in the polyhedron. This may seem like a severe limitation, but the example program shows that the content of the `dist` array is irrelevant when proving correct memory management.

l-value abstraction: Each memory region in C has an address that can be inquired and passed around like any other value. These so-called pointers play a crucial role in C and motivated research into so-called points-to analyses [3, 46, 74, 99, 144, 176]. A points-to analysis treats addresses of variables purely symbolically since the actual addresses of variables can, in principle, differ between two program runs. The invariants inferred by a points-to analysis state which (symbolic) addresses may be found in a pointer variable at run-time.

region summary: Due to dynamic memory allocation, C programs can allocate an arbitrary number of distinct memory regions. These must be summarised into a finite set of memory regions to obtain a terminating and efficient analysis.

None of these abstractions are particularly new, although their combination has not been thoroughly explored. We briefly discuss the problems and our improvements of these abstractions, and their combination.

1.4.2 Combining Value and Content Abstraction

A static analysis usually summarises the possible values of variables, while other memory regions are ignored. Compilers, for instance, perform constant propagation and points-to analysis on simple variables – that is, variables that are not arrays or structures. In contrast to simple variables, worst-case values are usually assumed when accessing structures and arrays for variables whose address is taken or that are accessed with incompatible types. Venet and Brat showed how an interval analysis can be defined over so-called fields that are "added" to variables and C **structs** as part of the analysis [182]. The idea is that fields are only added if the access position is unequivocal; that is, if the array index or the pointer offset is constant. Consider the access `dist[i]` in line 19 of our example program. The index variable `i` is always accessed in its entirety and hence at the same offset 0. The initialisation in line 18 therefore adds a field containing the polyhedral variable x_i. In contrast, the variable `dist` is accessed at a variable offset that is calculated from the index `i`.

In this case, the write position is an interval $x_i \in [32, 127]$ (rather than a constant) and therefore no field is added. The approach of adding a new field only if the access offset is constant produces a finite number of fields and hence a finite number of variables in the polyhedron. In Chap. 5, we extend this approach to allow the same part of a memory region to be accessed with different types. These accesses are surprisingly common in C programs. For example, in Fig. 1.2, the call to **`memset`** accesses `dist` as a memory region of **`char`**, whereas line 14 accessed `dist` with its declared type **`int`**. Hence, treating differently typed accesses to the same memory region precisely is important and one novelty in this book.

This approach to finiteness simply ignores the content of memory regions that are accessed at different offsets, thereby resulting in an analysis that is too imprecise for many verification tasks. This problem can be tackled by inferring information about certain properties of a memory region rather than inferring the memory region's actual content. We consider two possibilities:

element summary: Memory regions such as the `dist` array can be summarised by representing all array elements with a single abstract variable. In the case of the example, x_e might represent the values of all elements of `dist`. An analysis might infer that $x_e \in [0, 0]$ after zeroing the array at line 11. During each loop iteration, one element of the array is incremented while the remaining elements stay the same. This operation can be reflected on the abstract variable x_e by incrementing it weakly; that is, by setting x_e to an approximation of the previous value and the previous value plus one [80]. For the example program, $x_e \in [0, x_s]$ could be inferred; that is, each array element has a value between 0 and the size of the string.

meta information: Rather than inferring the values of (elements of) memory regions, it is possible to infer information relating to a certain property of a memory region. For instance, we explicitly state where the NUL character in the `argv[1]` buffer resides. The position of the NUL has been recognised to be the crucial information when analysing C string buffers [189].

In this work, we do not pursue the idea of summarising elements, mainly due to unresolved issues on constructing summary elements, if and how they can be split when overwriting them and hence how to limit the number of summary elements. In contrast, inferring information on the first zero position in a buffer requires a single polyhedral variable for each memory region and hence has no finiteness problems. Tracking NUL positions as part of a polyhedra-based analysis was presented in [71, 167], and the approach is further developed in Chap. 11.

1.4.3 Combining Pointer and Value-Range Analysis

In order to evaluate a read or a write access through a pointer variable, it is necessary to know what memory regions that pointer points to. Several different approaches can be taken to infer this information. During the last decade,

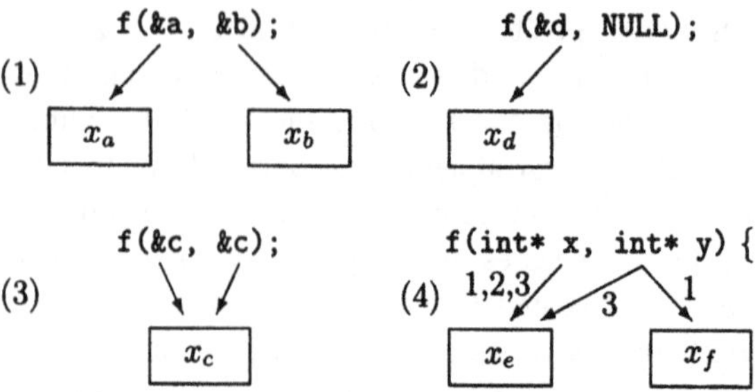

Fig. 1.3. Points-to information from different call sites.

tremendous advances have been made in the field of flow-insensitive points-to analysis [99,176] in which a set of all l-values (addresses of memory regions) is calculated that a given pointer variable may possibly contain during any execution of the program. The precision of points-to analysis can be substantially improved by performing a field-sensitive and/or a context-sensitive analysis.

A field-based analysis treats fields of a C **struct** as independent variables. A sound field-sensitive analysis must cater to pointer arithmetic commonly found in C programs; that is, a pointer might have a non-zero offset added to it before it is dereferenced, thereby accessing a different field from what its original l-value suggests. Chapter 5 shows how pointer arithmetic can be analysed by using the value-range analysis to calculate offsets relative to a base address, thereby giving precise offset information when pointers are dereferenced. In contrast, the most precise field-based points-to analyses distinguish between constant and non-constant offsets [175]. Tracking a points-to set using a points-to domain separately from the numeric offset that is tracking using a numeric domain is not always straightforward, and a formal description of how to combine both analyses is one contribution of this work.

For the sake of scalability, most points-to analyses are context-insensitive; that is, they combine points-to sets from different call sites when analysing a given function. While a context-insensitive approach scales well, it is not suitable for polyhedral analysis. Consider the example in Fig. 1.3, which shows the resulting points-to sets in drawing (4) for a function `f(int* x, int* y)` that was called as (1) `f(&a,&b)`, (2) `f(&d, NULL)`, and (3) `f(&c, &c)`. The pointed-to memory regions are shown as squares that each contain a single field represented by a polyhedral variable x_i that stores the value of the underlying integer. The first invocation seems to imply that the polyhedron at the callee should contain one variable for each parameter, here x_e and x_f in (4), to which the values of x_a and x_b from the caller are assigned. For the second invocation, however, the memory region containing x_f does not exist and x_e should be the only variable in the polyhedron. Hence the variable x_f represents no concrete memory region, which raises some difficult questions as to what a linear relationship between, say, x_f and x_e means. Another problem occurs at the call site (3), where we chose to represent x_c by x_e but x_f would have been equally justifiable.

The problem of different calling contexts also arises in the context of performing a context-sensitive analysis that aims to reuse a previously analysed function. While polyhedra are, in principle, able to express linear relationships between input and output variables of a function that can be substituted at every call site, the C language itself seems to be a major obstacle to a context-sensitive analysis. For instance, Nystrom et al. [139] proposed a two-stage points-to analysis that is fully context-sensitive; that is, their analysis is as precise as inlining each function at all call sites. In a bottom-up pass, their analysis calculates summaries for each function, which are then inserted at each call site before a top-down pass calculates the points-to sets. Each summary describes all side effects that a function has on its local heap. However, for functions that are called with incompatible points-to sets, all statements that are relevant to l-value flow have to be copied to each call site, thereby defying the goal of context-sensitive analysis without inlining function bodies. This observation suggests that a fully context-sensitive analysis of C is likely to be impossible. In this work, we simply expand each function at each call site, which, in principle, incurs an exponential growth in the code size, but has been successfully applied in verification [31]. This choice also prohibits the analysis of recursive functions.

Finally, analysing dynamically allocated memory requires further techniques to ensure finiteness. Allocation sites that are only executed once should simply create a new memory region that can be read and written like declared variables in the program. In contrast, memory regions allocated within a loop must be summarised. We follow the classic approach in that memory regions that are allocated by a `malloc` statement at the same program point are summarised. By transforming the input program such that every function is expanded at its call site, this tactic is automatically refined such that memory regions allocated by a `malloc` statement in a given function are not summarised for different call sites of the function. In the upcoming analysis, functions are only inlined semantically, that is, they are re-analysed for every new call site such that care has to be taken to achieve the same semantics for dynamically allocated memory regions.

This concludes the overview of what we choose to extract from a C program. The details of these abstractions form Part I of this book. We now embark on the question of how to automatically approximate the state space of a C program.

1.5 Efficiency

Any useful program-analysis tool has to be efficient in order to be of practical help to the programmer. Interestingly, an efficient analysis can be implemented on top of semi-decision procedures such as theorem proving by using time-outs [152]. Theorem proving is an attractive approach due to its ability to describe properties over a potentially infinite state space such as the value of a variable or the shape of a heap. However, the ability to create arbitrarily sized descriptions can affect termination of automated proving strategies, hence the use of timeouts. In contrast, classic model checking operates on finite automata (that is, a finite state space) and therefore always terminates [49]. In practice, however, it is difficult to soundly map the state of a program to a finite automaton of acceptable size. Thus, model checking is often impractical in that the size of the finite automaton grows too rapidly with respect to the input program to permit the analysis of larger systems [48]. Rather than using a finite state space, our analysis uses a convex polyhedron to describe a potentially infinite state space, which necessarily implies that some descriptions are approximations to the actual state space. On the positive side, our analysis can be terminating, as the inferred polyhedra are always finite. In this book, we use the framework of abstract interpretation by Cousot and Cousot [56] to describe this approximating analysis. We briefly illustrate the idea of a static analysis based on abstract interpretation before discussing the challenge of implementing such an analysis efficiently.

Consider the **for** loop in lines 18–19 of the running example in Fig. 1.2, whose control-flow graph is depicted in the upper half of Fig. 1.4. The edges of the control-flow graph are decorated with the polyhedra P, Q, R, S, and T, which denote the state at that given program point and which we write as sets of inequalities. In order to illustrate how these polyhedra are incrementally inferred, we write P_j to indicate the jth update of the state P. As before, let x_i denote the value of the program variable `i`. After executing the initialisation statement `i=32`, the initial state of P is given by $P_0 = \{x_i = 32\}$. This state is propagated to $Q_0 = P_0$, where the test `i<128` partitions this state into $S_0 = \{x_i = 32, x_i \leq 127\}$ and $R_0 = \{x_i = 32, x_i \geq 128\}$. Note here that $x_i < 128$ is tightened to $x_i \leq 127$ since all program variables are integral. With respect to the sets of points described by these states, S_0 is equivalent to Q_0 and R_0 is unsatisfiable; that is, the set of points described by R_0 is empty. An unsatisfiable polyhedron implies that the corresponding point in the program is unreachable; here, the state of R_0 implies that the loop will not terminate without iterating at least once. The analysis continues by propagating the satisfiable state S_0. Since the value of x_i in S_0 is 32 and therefore between 0 and 255, the array access `dist[i]` is within bounds. Incrementing the loop counter yields a new state $T_0 = \{x_i = 33\}$, which is propagated back to the beginning of the loop to where the control-flow paths merge. It is at this merge point that the two state spaces P_0 and T_0 are joined to form $Q_1 = P_0 \sqcup T_0 = \{32 \leq x_i \leq 33\}$, where the join operator $\sqcup$ calculates a polyhedron that includes its two arguments. Since the maximum value of x_i is still below 128, another iteration of the loop is calculated, yielding $T_1 = \{33 \leq x_i \leq 34\}$ after the instruction `i++`. This state in turn can be joined to form $Q_2 = P_0 \sqcup T_1 = \{32 \leq x_i \leq 34\}$. Depending on the loop bounds, the

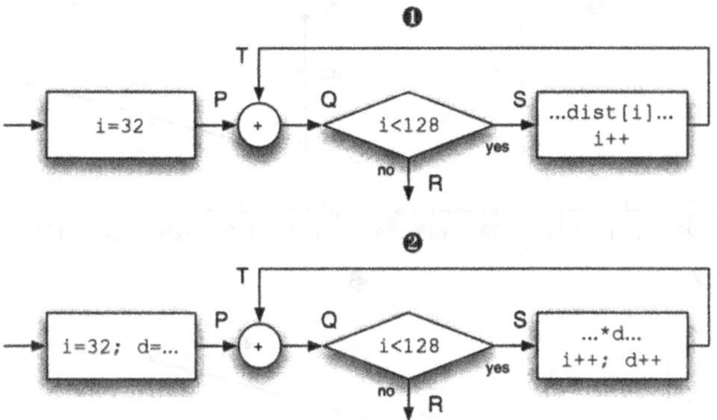

Fig. 1.4. The control-flow graph of the original **for** loop and the modified variant.

analysis may perform an excessive number of iterations, which is unacceptable for an efficient analysis. In order to avoid this, widening can be applied, which accelerates the fixpoint calculation [59]. The principal idea of widening is to compare two state spaces that result from two consecutive iterations and remove those inequalities that are not stable. In the example, we calculate the widened polyhedron $Q_2' = Q_1 \nabla Q_2$; that is, we remove all inequalities from Q_1 that do not exist in Q_2. The result is $Q_2' = \{32 \leq x_i\}$. Enforcing the condition `i<128` yields $S_2 = \{32 \leq x_i \leq 127\}$ for the loop body and $R_2 = \{32 \leq x_i \geq 128\}$, which is equivalent to $\{x_i \geq 128\}$ as state space when the loop exits. Analysing the loop body with S_2 will infer that $x_i \in [32, 127]$ and hence that the index `i` lies within the bounds of the array. Furthermore, after the evaluation of `i++`, the new state space $T_2 = \{33 \leq x_i \leq 128\}$ arises and hence $Q_3 = P_0 \sqcup T_2 = \{32 \leq x_i \leq 128\}$. Intersecting this state with the loop invariant $x_i \leq 127$ yields $S_3 = \{32 \leq x_i \leq 127\}$, which is equivalent to S_2, and hence a fixpoint has been reached. It can be shown that the inferred state includes all possible values that the variable i can take on in the program.

While the calculation above of the loop invariant $x_i \in [32, 127]$ demonstrates the basic technique of inferring a fixpoint of a loop, the real strength of polyhedra lies in the ability to infer relationships between different variables. In order to illustrate this ability, consider the following modified **for** loop that is functionally equivalent to the one in Fig. 1.2:

```
int* d=&dist; d+=32;
for (i=32; i<128; i++, d++)
  printf("'%c'␣:␣%i\n", i, *d);
```

Instead of recalculating the array index, the access position is calculated incrementally by advancing the pointer `d` by one element in each loop iteration. The corresponding control flow is shown in the second graph of Fig. 1.4.

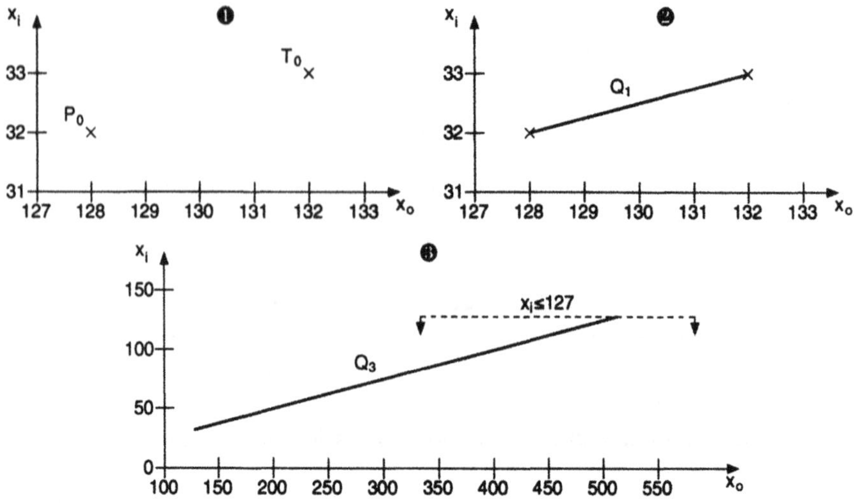

Fig. 1.5. Inferring the state space within the loop using polyhedral analysis.

Let x_o denote the byte offset of pointer `d` relative to the beginning of `dist`. Assuming that each **int** element of the array requires four bytes of storage, the statement `d+=32` increments x_o from 0 to 128 and the abstract state with which the loop is entered is given by $P_0 = \{x_i = 32, x_o = 128\}$. After evaluating the test `i<128`, x_i is incremented by one, while x_o is incremented by the size of one element of `dist`, namely 4 bytes. Thus, the state after executing the loop body once is $T_0 = \{x_i = 33, x_o = 132\}$.

While the join $P_0 \sqcup T_0$ can be described by $\{32 \leq x_i \leq 33, 128 \leq x_o \leq 132\}$, a more precise set of inequalities exists that includes P_0 and T_0. Consider the geometric interpretation of the state space in the first graph of Fig. 1.5. A more precise (but still concise) characterisation of the state space is given by the convex hull of the two points; that is, the smallest closed, convex space that includes P_0 and T_0. In our example, a set of inequalities that includes the convex hull of P_0 and T_0 is $Q_1 = \{32 \leq x_i \leq 33, x_o = 4x_i\}$, as depicted in the second graph of the figure.

As with the example before, we evaluate another loop iteration in which $S_1 = Q_1$ and where $T_1 = \{33 \leq x_i \leq 34, x_o = 4x_i\}$. The join $P_0 \sqcup T_2$ is again the convex hull of the two polyhedra, yielding $Q_2 = \{32 \leq x_i \leq 34, x_o = 4x_i\}$, which differs from Q_1 only in the upper bound on x_i. Applying widening yields the infinite state $Q_2' = Q_1 \nabla Q_2 = \{32 \leq x_i, x_o = 4x_i\}$. The loop invariant ensures that the next iteration yields $Q_3 = Q_2' \cup \{x_i \leq 127\}$, as shown in the third graph of Fig. 1.5.

The polyhedron $Q_3 = \{32 \leq x_i \leq 127, x_o = 4x_i\}$ describes an invariant that is sufficient to show that the pointer `d` has an offset between 32 and 512 and hence lies within the `dist` array, which contains 1024 bytes. Note that this invariant required reasoning about the relationship between x_i and x_o: Without this relational information, widening would have resulted in $\{32 \leq x_i, 128 \leq x_o\}$, which, by adding the loop invariant `i<128`, would give the less precise loop invariant $\{32 \leq x_i \leq 127, 128 \leq x_o\}$, which leaves x_o unrestricted.

We chose to infer relational information, as previous work based solely on intervals yielded results that were too imprecise for the verification of string buffer operations [184]. One drawback of using convex polyhedra as the basis for a static analysis is their inherent complexity. Specifically, calculating the convex hull of two polyhedra is an exponential operation [24]. This book therefore presents a novel sub-class of general polyhedra that is based on the idea of decomposing polyhedra into sets of planar polyhedra. To this end, Chap. 7 introduces efficient algorithms for planar polyhedra; in particular, we present a novel convex hull algorithm for planar polyhedra. By building on these planar algorithms, Chap. 8 presents the Two-Variables-Per-Inequality (TVPI) domain, which provides an efficient way of manipulating polyhedra in which each inequality has at most two variables. The following chapter presents techniques to refine polyhedra around the contained set of integral points, a process that is required to ensure that coefficients of inequalities do not grow indefinitely. Such a guarantee cannot currently be given for general polyhedra. As such, the TVPI domain presents, to our knowledge, the most precise polyhedral domain with a performance guarantee.

Given an abstraction from C and an efficient domain to calculate an over-approximation of its state space, we proceed to detail improvements in the precision of the analysis.

1.6 Completeness

For the sake of staying focussed on relevant aspects of finding buffer overflows, we chose to test and refine our analysis against a program called `qmail-smtp`, which is part of a mail transfer agent (MTA) whose task it is to forward email traffic. As this program parses incoming emails from the network, it is susceptible to buffer-overflow attacks and therefore a prime candidate for inspection. It is also simple enough in that it is single-threaded, does not make use of recursive functions, and uses few library functions.

The verification of real-world programs opens up many challenges, some of which are not clear until the analysis is run the first time on the chosen input program. While the aspects of soundness and efficiency need to be addressed before an analysis is implemented, the precision of an analysis (or the lack thereof) often manifests itself when the analysis is run. When precision is unduly lost, the analyser emits warnings that do not correspond to actual mistakes in the program. These so-called false positives then motivate a refinement of the analysis. Note that the way the C program is abstracted and the choice of the polyhedral domain both significantly affect the ability to infer precise results. However, this section presents three aspects of our analysis that are solely dedicated to improving the precision. These aspects are the ability to argue about NUL positions in string buffers, an improved widening strategy, and a refinement of the points-to analysis using Boolean flags in the polyhedral domain. We discuss each aspect in turn.

1.6.1 Analysing String Buffers

A basic idiom in the context of string buffer operations is to iterate over the contents of a string until the NUL position, which terminates the string, is reached. An example of this operation is given in lines 13–16 in Fig. 1.2. Here, the buffer `argv[1]` denotes the first command-line argument, which, if it exists, contains a NUL -terminated string of arbitrary size. Due to the unknown size, it is not possible to model individual elements of this buffer with polyhedral variables. Instead, the buffer `argv[1]` can be treated as a dynamically allocated memory region whose size is given by the polyhedral variable x_s. Furthermore, it is known that a NUL character exists that terminates the input argument. Without loss of generality, we can assume that this NUL character resides in the last element of the buffer. Suppose that the polyhedral variable x_n denotes this NUL position; then $x_n = x_s - 1$. As a third parameter, let x_o denote the offset of the pointer `str` relative to the address `argv[1]`. To the programmer, it is rather obvious that the pointer `str` is incremented in line 15 until the NUL position has been reached. To the analysis, this information is only indirectly available since the test `*str` in line 13 does not query the NUL position x_s but merely accesses the buffer. Let x_c denote the character returned by `*str`. The idea of a string buffer analysis is to encode the NUL position by refining x_c to $[1, 255]$ if the access position x_o is in front of the NUL position x_n and to refine x_c to 0 if $x_o = x_n$. By using the ability of polyhedra to express linear relationships between variables, we relate x_c, x_o, and x_n such that testing the loop condition `*str` results in refining x_n and x_o. In particular, testing if `*str` is true corresponds to adding $x_c > 0$ to the state, which recovers the information that $x_o < x_n$. Similarly, testing if `*str` is false corresponds to adding $x_c = 0$ to the state, which recovers the information that $x_o = x_n$. In fact, this test partitions the state space (since x_c is positive) and thereby infers that $x_o = x_n$ on loop exit and that $x_o < x_s$ within the loop, which proves that `argv[1]` is never accessed out-of-bounds. The details of this analysis are given in Chap. 11, which presents the string buffer analysis as a refinement of the basic analysis of C that is described in the first part of the book.

1.6.2 Widening with Landmarks

The key to an efficient polyhedral analysis is to accelerate the fixpoint calculation to overcome slowly growing coefficients in inequalities. This process is known as widening [59] and was already applied in Sect. 1.5. The principal idea of widening is to remove inequalities that have changed between two consecutive iterations. The full removal of inequalities, however, incurs a substantial precision loss, as witnessed by the R_i states from the last section that describe the state space at the end of the **for**-loop in lines 18–19. While the actual value of the loop index `i` on exit of the loop is 128, applying widening can only infer that $x_i \geq 128$. While an operation called narrowing [59] can be applied to refine this state again, we pursue a different strategy in which widening is modified in that changing inequalities are not removed but merely relaxed. The amount by which inequalities are relaxed is inferred by observing conditionals (so-called landmarks) in the analysis. For instance, the test $x_i \leq 127$, which stems from the loop condition `i<128` in line 18, conveys the information that the upper bound of x_i in $Q_1 = \{32 \leq x_i \leq 33\}$ and $Q_2 = \{32 \leq x_i \leq 34\}$ should be increased by another 94 units to yield $Q_2' = \{32 \leq x_i \leq 128\}$. Using this state instead of the fully widened state $Q_2' = \{32 \leq x_i\}$ enables the analysis to infer that $R_2 = \{x_i = 128\}$ rather than $R_2 = \{x_i \geq 128\}$. Widening with landmarks is presented in Chap. 12, where it is shown to be crucial for analysing string buffers in a precise way.

1.6.3 Refining Points-to Analysis

Using a standard points-to analysis is often too imprecise when it comes to the verification of programs. Next to field sensitivity and context sensitivity, a points-to analysis can be categorised with respect to its flow sensitivity. A flow-insensitive analysis infers a single points-to set for each program variable that is valid for the whole program. In contrast, a flow-sensitive points-to analysis infers a points-to set at each program location. Only the latter analysis can therefore determine that a statement such as `if (p!=NULL) *p=42;` does not dereference a `NULL` pointer. While the analysis as presented in the first part performs a flow-sensitive analysis, Chap. 13 details a refinement of this flow-sensitive analysis where the content of points-to sets can be related with the numeric values of program variables, thereby substantially improving the precision of standard flow-sensitive points-to analysis.

1.6.4 Further Refinements

The refinements presented allow our analysis to verify non-trivial examples. Unfortunately, even with the techniques described so far, the program in Fig. 1.2 still evades verification. One problem is the argument `argv` to `main`, which constitutes an array of pointers. In order to express that this array contains an arbitrary number of pointers to NUL -terminated strings, it requires the ability to state that all pointers in the `argv` array have an offset of zero, which is beyond the abstraction techniques of our analysis. However, it is possible to analyse the memory management of the example precisely if a string constant is assigned to `str` in line 10. Interestingly, the verification of the example evades the current implementation of our analysis when `argv` is fixed to contain a single pointer to a NUL -terminated buffer of *arbitrary* size as described in Sect. 1.6.1. Chapter 14 details this and other shortcomings and suggests efficiency and precision improvements in order to make the analysis applicable to real-world C programs.

We conclude the introduction with an overview of related tools and a summary of our contributions to the field of soundly analysing C programs.

1.7 Related Tools

In this section, we present other tools to analyse C or C++ programs, which can generally be partitioned into sound analyses and unsound analyses. While both approaches create false positives, the unsound analyses may also miss errors and thus cannot prove program correctness. We focus mostly on sound analyses, as their techniques are more relevant to the analysis presented in this book.

1.7.1 The Astrée Analyser

The Astrée analyser [30,31,60] is a value-range analysis in the sense of Sect. 1.2 with the aim of proving the absence of run-time errors; that is, range overflows, division by zero, out-of-bounds array accesses, and other memory management errors. The analysis is sound and precise even for floating-point calculations. However, it is restricted to embedded systems in that dynamic memory allocation is only allowed at start-up (no heap-allocated data structures) and recursion is only allowed if it is bounded, as all functions are semantically inlined. The target of the analysis is flight-control software for Airbus A340 and A380 aeroplanes, which features, for instance, second-order digital filters that are repeatedly evaluated. The analysis is able to prove the absence of run-time errors on programs as large as 400,000 lines of code in less than 12 hours using 2.2 GB of memory. In order to estimate the valuations of floating-point variables that arise in the digital filter code, special domains such as the Ellipsoid domain were defined [31]. Linear relationships are inferred using the Octagon domain [130], which can express relationships of the form $\pm x \pm y \leq c$, where c can either be an integer or a floating-point number. Since the largest programs analysed contain about 80,000 global variables, the variables on which relational information is required are divided into packs – that is, sets that potentially overlap. Relational information is only inferred within each pack. This is important for scalability since the Octagon domain, like our TVPI domain, stores information for each pair of variables and thus has a memory footprint that is necessarily quadratic in the number of variables. In order to improve upon the precision of the relatively weak Octagon domain, symbolic propagation of variable values is used [131]. The ability to infer that no floating-point overflow occurs is based on the assumption that the number of iterations of the outermost control loop is bounded by a constant, namely $6 \cdot 10^6$. In order to achieve the required precision and scalability, the analyser is able to partition the set of traces at a particular program point; that is, it is possible to track separate states for different values of a variable and to keep states separate where control-flow edges join [123]. The latter can be used to unroll loops, which is crucial for the kind of embedded code the analysis targets, where loops often initialise variables in the first loop iteration. Arrays that are of static size are either fully unfolded (each element is represented in the analyser) or smashed (all elements are represented by one abstract element). The effect of integer calculations that exceed the range of the target variable is either reported as erroneous or the wrapping that occurs in the concrete program is made explicit, depending on the specification of the user. Finding fixpoints of loops is guided by widening thresholds, which are values that indicate likely bounds on variables. All parameters regarding trace partitioning, array handling, and widening thresholds can be communicated to the analyser by using annotations in the source code. From the experience of finding these parameters, heuristics have been devised that work well for programs that have similar structures, thereby reducing the burden on the user. The implementation of the Astrée analyser uses a configurable hierarchy of modules that implement trace partitioning, track memory layout, etc., and the various numeric domains [61]. This design facilitates the addition of new domains and thereby the adaptation to new classes of programs.

1.7.2 SLAM and ESPX

The unsound PREfix tool [38] was one of the first tools deployed by Microsoft in order to uncover faults in device drivers. The tool was eventually replaced by SLAM [18], which was created at Microsoft Research and is now integrated into the development process of Microsoft Windows. It is commercially available as part of Microsoft Developer Studio, where it serves to reveal programming errors related to locking, handling of files, memory allocation, and other temporal properties [20]. Using a simple specification language called SLIC to describe the effects of certain function calls, a tool called C2BP translates the input C program into a Boolean program. This Boolean program is then checked using the BEBOP model checker [19]. Unless the model checker can verify the properties that were described using the SLIC specification, another tool, called NEWTON, is run in order to refine the Boolean program using the counterexample from BEBOP. The idea is that repeated refinement of the abstraction will eventually create a Boolean program that is precise enough to prove the program correct. The abstraction done in C2BP is meant to be sound but incorrectly handles memory accesses through pointers when they denote overlapping memory regions. Furthermore, it is incorrectly assumed that integer variables cannot wrap, although this issue is being addressed, as pointed out in the talk of [52].

With respect to buffer overflows, Microsoft uses an approach similar to SLAM in that a new specification language, SAL, was defined, with which the programmer is able to annotate C programs. In order to aid the programmer in annotating source code, an unsound tool called SALinfer can provide likely invariants, which the programmer can adapt. An inter-procedural checker called ESPX checks the annotations and is sound if all buffer operations are correctly annotated. To date, Microsoft developers have inserted 400,000 annotations into the next version of Windows, of which 150,000 were inferred automatically [90]. This labour-intensive approach led to the detection of 3,000 potential buffer overflows, which means that approximately 100 annotations are needed to detect a single buffer overflow.

1.7.3 CCured

CCured is a pragmatic approach that combines static analysis with run-time checks. The input C program is parsed using a front end written in O'Caml that exhibits either the semantics of the GNU C compiler or that of Microsoft's Visual C compiler and translates to the so-called C Intermediate Language (CIL) [137]. The intermediate representation is then analysed using dependent types with the intent of proving most memory operations correct [136]. Any pointer whose properties cannot be statically guaranteed is converted into a fat pointer that includes the beginning and end of the buffer that the pointer points to. The code is then transformed by adding run-time checks to all memory accesses that the static analysis cannot guarantee to be safe. In order to avoid dangling pointers caused by freeing dynamically allocated memory regions too early, all calls to `free` are removed and a garbage collector [32] is used. The resulting program is then translated back to C and compiled using a normal C compiler. Recent work has addressed the task of verifying the binary output with the invariants generated during the static analysis [94].

1.7.4 Other Approaches

A vast number of tools have been proposed that use heuristics to highlight locations in the C source code that are likely to be erroneous. By using heuristics, these tools are simpler than sound analysers but may miss faults. An important aspect of all unsound approaches is that their precisions are difficult to compare. For sound approaches, it is sufficient to compare the number of false positives. Unsound approaches, however, can to a certain degree trade off the number of false positives with the number of missed bugs. Since the number of missed bugs is not known, the comparison of the number of false positives is mostly meaningless.

LClint [75] and its variants [115, 116] use lightweight annotations that are added to the C programs to find buffer overflows and other faults. In contrast, Wagner proposed a fully automatic buffer-overflow analysis based on intervals [184], which, however, is not very precise. Dor et al. were the first to analyse pointer accesses to string buffers using polyhedra [71]. However, their work turned out to be unsound [167], which triggered their work on soundly analysing C string functions aided by user annotations [72]. Ghosh et al.

use fault injection to find buffer overflows; that is, a tool repeatedly creates strings with the aim of overflowing a specified buffer in the stack. The process is guided by inspecting the dynamic run-time behaviour of the program [77]. Haugh and Bishop claim that automatically verifying the absence of buffer overflows is impossible. Their STOBO tool instruments a program to observe program behaviour when run on normal input data. The inferred data are then used to characterise possible overflow conditions [98]. Archer is a static analysis for detecting memory access errors using a custom constraint solver that can express linear relations between two variables [189]. The tool uses heuristics on function names to infer relationships between function arguments and return values. The authors observe that a major deficit of their analysis is the inability to track NUL positions. Eau Claire is another checker for buffer overflows that uses theorem provers [45]. Elgaard et al. show how to find all NULL pointer dereferences using a model checker [73]. Unfortunately, their technique is only sound and complete on straight-line code. User annotations are necessary to handle loops, in which case completeness is lost. Jones and Kelly [108] augment programs with run-time information on buffer bounds. Further afield is the analysis of format string vulnerabilities [162], where an input string is passed to `printf`. The observation is that any percent character in the first argument to `printf` determines how many arguments are read from the stack. Thus, a program may be vulnerable if the user can pass an arbitrary string as the first argument to `printf`. In practice, these vulnerabilities can easily be found and removed syntactically by ensuring that the first argument to `printf` is a constant.

There exists great interest in removing memory management faults, both in academia and industry. Hence, this overview of related tools only includes the most predominant tools that we became aware of during our research.

2

A Semantics for C

The C programming language has evolved substantially since its inception and repeatedly undergoes standardisation, the latest of these efforts resulting in the C99 standard [51]. Defining the formal semantics of an evolving language such as C is difficult and has given rise to a Ph.D. thesis in its own right [143]. Hence, for the sake of conciseness, the interpretation of the C standard is left to the GNU C compiler, which translates the input program into a small intermediate language with well-defined semantics. While this approach ties the validity of our analysis to a single compiler and platform, it not only allows us to make some simplifying assumptions but also provides us with a way to soundly analyse programs – like the one considered in our experimental section – that do not strictly comply with any standard and whose exact semantics are defined by the compiler. This is deemed to be no disadvantage, as GNU C is available on most platforms and because a fully automatic analysis can easily be rerun on all platforms of interest.

For conciseness, the analysis is not coupled directly to the intermediate representation of the chosen compiler. Instead, this chapter details how it is translated into a language called Core C that merely contains the basic operations of the C language. After describing the Core C language and the execution environment commonly found in current platforms such as Linux or Microsoft Windows, we define the notion of correct memory management. The chapter concludes by stating the concrete semantics of Core C from which a collecting semantics is derived.

2.1 Core C

The static analyser presented in this book operates on a simplified version of C called Core C. In fact, the pragmatics of Core C are closer to assembler than to any high-level language in that variables are declared simply by stating their size, whereas each expression in the program is explicitly typed. Furthermore, the control flow is specified by a sequence of conditionals and jump targets at the end of each basic block. This simplistic structure allows a complete yet concise definition of the semantics of Core C, thereby providing a foundation for a sound static analysis.

Besides normative descriptions, the C standard [51] marks certain aspects as implementation-defined, unspecified, or undefined. Most undefined behaviour is treated as erroneous in our analysis. Other aspects (such as sizes of integers) are fixed by the compiler during the translation into Core C; some carry over into the semantics of Core C. In summary, the following parameters must be addressed:

integer and pointer sizes: The integer and pointer types of C are allowed to have different sizes, depending on the underlying architecture. The C compiler makes the actual sizes explicit in its intermediate representation.

signed arithmetic: There exist few architectures today that implement negative integers using an explicit sign bit in addition to an unsigned integer (so-called sign magnitude) or an explicit sign bit in addition to a negated unsigned integer (so-called one's complement). Both approaches have two representations for zero, the result of a rather cumbersome semantics. The de facto standard for implementing signed integers is the two's complement representation, which allows the reuse of unsigned arithmetic operations (if overflows are of no concern). However, two's-complement arithmetic requires a second set of comparison functions for signed integers, which was not available on earlier architectures. Since the GNU C compiler can only be ported to architectures that use two's-complement arithmetic, the semantics of Core C assumes two's-complement arithmetic, too. We believe that this restriction is no obstacle in practice, and it seems imprudent to develop an analysis for obsolete architectures.

signed right shifts: Closely related to the question of how signed integers are implemented is the availability of a right-shift operation that preserves the sign of a value. The intermediate representation of the GNU C compiler makes the distinction between signed and unsigned right shifts explicit and hence determines the semantics of the right-shift C operation.

rounding of integer division: The C standard caters to two different kinds of rounding for integer division: rounding towards zero and rounding towards minus infinity (truncating the result). Although division operations on most architectures round towards zero, it is again the C compiler that determines the semantics of C integer division.

invalid address 0: The memory address 0 (or `NULL` in C syntax, not to be confused with the character NUL) is reserved as a special tag in C. No implementation may use this address for storing data. In practice, operating systems disallow any access to the first virtual memory page (which includes address 0), thereby causing a segmentation fault each time the memory address 0 is accessed. Our analysis is parameterised by the page size of the target architecture and assumes that all valid memory lies above the first page.

alignment of data: Some architectures impose restrictions on accessing data types that are larger than a single byte. Specifically, memory accesses are usually required to be aligned; that is, the address of a datum must be a multiple of the byte size of the datum. For example, 4-byte integers are often required to be stored at addresses that are multiples of 4. While some architectures (such as Intel platforms) are able to perform unaligned accesses, they require several accesses to memory incurs a performance penalty. Other platforms, such as the PowerPC platform, raise an exception on unaligned accesses of data, sometimes with the ability to emulate the access in software. We assume that the programs to be analysed run even on architectures, such as Sparc, where all memory accesses have to be aligned. While this assumption implies that our analysis flags unaligned memory accesses as erroneous (even on Intel platforms), it greatly simplifies the underlying memory model.

copying padded structures: The C compiler will automatically pad structures with unused bytes to enforce certain alignment restrictions. According to the C standard, an assignment from one structure to another does not have to copy the padded areas between fields. However, for efficiency, most C compilers copy the whole structure, and we require this behaviour for our analysis.

endianness: Endianness denotes the way the bytes that comprise a larger integer value are stored. Architectures are split between storing integers in little-endian format (less significant bytes are stored at lower addresses) or in big endian (more significant bytes are stored at lower addresses). A C program is allowed to infer and exploit the endianness that the underlying architecture implements. The endianness is a parameter of our analysis, and a C program has to be reanalysed for a different endianness.

While the C compiler fixes all of the above implementation-specific parameters above, the semantics of our analysis varies with the pointer size and the endianness of the architecture. To simplify the exposition, we assume that the architecture has 32-bit pointers and uses big-endian byte ordering – a valid assumption for the Intel IA-32 architecture, for example. The analysis can easily be adapted to a different pointer size and endianness, such as little-endian PowerPC and Sparc architectures.

By shifting the burden of defining the semantics of a C program to the GNU C compiler, the intermediate language Core C of our analysis becomes relatively simple. In the following grammar, we use $\langle \text{S} \rangle$ to denote a non-terminal symbol S, [S] to denote an optional symbol, and $(\text{S})^*$ to denote an arbitrary sequence of S. In order to simplify the grammar, we use the syntactic categories in Fig. 2.1 as terminal symbols.

For convenience, we define *Label* to be a finite set of tags and require $l \in \mathit{Label}$. Given these definitions, Fig. 2.2 presents the grammar that defines the Core C language $\mathcal{L}(\langle \text{Core C} \rangle)$. The individual productions are explained below.

category name	meaning
v	variable name
f	function name
l	program label
string	string literal
n	integral number literal

Fig. 2.1. Syntactic categories used in the Core C grammar.

```
⟨CoreC⟩ :: [⟨Vars⟩] (⟨Stmt⟩ ;)* ⟨FDecl⟩ (; ⟨FDecl⟩)* ;
⟨Vars⟩  :: var v : n (, v : n )* ;
⟨FDecl⟩ :: f ( [ v (, v )*] ) [⟨Vars⟩] ⟨Block⟩*
⟨Block⟩ :: l: (⟨Stmt⟩ ;)* ⟨Next⟩
         | l: f ( [ v . n (, v . n )*] ) ; ⟨Next⟩
         | l: *v  ( [ v . n (, v . n )*] ) ; ⟨Next⟩
⟨Next⟩  :: return ;
         | jump l;
         | if ⟨Type⟩ v . n ⟨Op⟩ ⟨Expr⟩ then jump l; ⟨Next⟩
⟨Op⟩    :: < | ≤ | = | ≠ | ≥ | >
⟨Expr⟩  :: n | n * v . n + ⟨Expr⟩
⟨Stmt⟩  :: ⟨Size⟩ v . n = ⟨Expr⟩
         | ⟨Size⟩ v → n = ⟨Expr⟩
         | ⟨Size⟩ v . n = v → n
         | ⟨Size⟩ v . n = ⟨Type⟩ v . n
         | structured n v . n = v . n
         | structured n v → n = v . n
         | structured n v . n = v → n
         | v . n = & f
         | v . n = & v . n
         | v = "stringId"
         | l: v = malloc ( v . n )
         | free ( v . n )
         | [ ⟨Type⟩ ⟨Expr⟩ =]
           "string"( [ ⟨Type⟩ ⟨Expr⟩ (, ⟨Type⟩ ⟨Expr⟩)*] )
⟨Type⟩  :: (uint | int) ⟨Size⟩
⟨Size⟩  :: 1 | 2 | 4 | 8
```

Fig. 2.2. The syntax of Core C.

A Core C program commences with an optional list of global variable declarations and a sequence of statements to initialise them. In contrast to C, variables are introduced by merely stating their size in bytes rather than by specifying a type. A non-empty sequence of function declarations defines the program. Each function declaration may have a list of variable declarations, which has to include all function arguments. The body of a function is a sequence of blocks where each block is uniquely identified by a label $l \in Label$. Each function call, either direct or via a pointer, constitutes a basic block in itself. Each basic block is followed by a sequence of control-flow instructions: that is, either a **return**, an unconditional jump, or a conditional jump followed by yet another control-flow instruction. Functions that return a value in the original C program are modified when translated to Core C by adding a pointer parameter that points to the corresponding call-site variable, thereby alleviating the need for return values.

The first three productions of ⟨Stmt⟩ correspond to the three basic assignments that are necessary in a language with pointers. More complicated pointer operations are broken down during translation. The notation $v\ .\ n$ denotes the nth byte of the memory region occupied by the variable v. This notation corresponds to a typical access to a C **struct**. Furthermore, the notation $v \rightarrow n$ accesses the memory region pointed to by v at the given byte offset. Note that $v \rightarrow 0$ is equivalent to $*v$ in C. The productions of the grammar are carefully chosen to match the more common cases of C programs; a simpler production such as $*v$ would not allow an access to a single field of a structure in $*v$ without introducing intermediate variables. To completely specify the semantics of an access, assignment statements are preceded by a size or type unless it is clear that the variable is a pointer – that is, a **uint4**. For readability, we write sized types as one token; that is, the type of a pointer is written as **uint4**. All variables in an assignment access the same number of bytes given by ⟨Size⟩ except for the fourth production, which allows coercion from one type to another. Here, we follow the C convention that the type of the expression's right-hand side determines whether or not sign extension is used. The next three productions allow the assignment of whole structures; that is, they copy the given number of bytes from one variable to another without interpretation. The following production takes the address of a function or (a field of) a variable. The next production stores a constant string with a terminating zero in the given array v. Note that this statement corresponds to `char v[] = "string"` in C. The more common C string literals, which are expressions in their own right, are translated into global, byte-sized arrays that are initialised through this statement. At each occurrence of the string literal, a variable containing the address of the global array is substituted. The parameters of the memory allocation functions **malloc** and **free** are implicitly of type **uint4**. The last statement calls a primitive, required for non-linear operators, for example. A translation of the introductory example to Core C is shown in the appendix.

Core C programs are executed, just like their C counterparts, in a virtual memory environment that provides an array of byte-wise accessible memory locations. After introducing some basic notation in the next section, Sect. 2.3 formally defines this environment, which is then used to define the semantics of Core C.

2.2 Preliminaries

This section introduces basic notation for Boolean vectors and arithmetic. Let $\mathbb{B} = \{0,1\}$ denote the set of Booleans. A vector $\mathbf{b} = \langle b_{w-1}, \dots b_0 \rangle \in \mathbb{B}^w$ is interpreted as an unsigned integer by $val^{w,\mathbf{uint}}(\mathbf{b}) = \sum_{i=0}^{w-1} b_i 2^i$ and as integer by $val^{w,\mathbf{int}}(\mathbf{b}) = (\sum_{i=0}^{w-2} b_i 2^i) - b_{w-1}2^{w-1}$. Conversely, $bin^w : \mathbb{Z} \to \mathbb{B}^w$ converts an integer to the lower w bits of its Boolean representation. Formally, $bin^w(v) = \mathbf{b}$ iff there exists $\mathbf{b}' \in \mathbb{B}^q$ such that $val^{(q+w),\mathbf{int}}(\mathbf{b}' \| \mathbf{b}) = v$, where $\|$ denotes the concatenation of bit vectors. For instance, $bin^3(15) = \langle 1,1,1 \rangle$ since $val^{5,\mathbf{int}}(\langle 0,1,1,1,1 \rangle) = 15$. We write v instead of $bin^w(v)$ if w is obvious from the context. In order to distinguish calculations on Boolean vectors from standard arithmetic, let $+^w, *^w : \mathbb{B}^w \times \mathbb{B}^w \to \mathbb{B}^w$ denote addition and multiplication that truncate the result to the lower w bits, for instance $\langle 1,1,1,1 \rangle +^4 \langle 0,0,0,1 \rangle = \langle 0,0,0,0 \rangle$. Note that the signedness of the arguments of $+^w$ and $*^w$ does not affect the results of these operations.

Core C programs operate in a virtual memory environment that we formalise as a sequence of bytes. Let $\mathcal{B} = \mathbb{B}^8$ denote the set of bytes and $\Sigma = \mathcal{B}^{2^{32}}$ all states of 4 GB that a program on a 32-bit architecture can take on. Let $\sigma \in \Sigma$ denote a given memory state of a program, and define $\sigma^s : [0, 2^{32} - 1] \to \mathcal{B}^s$ to denote a read access at the given byte offset, where $s \in \{1,2,4,8\}$ is the number of bytes to be read. All accesses $\sigma^s(a)$ must be aligned; that is, the address a must be divisible by s. We further assume a little-endian architecture, which requires that $\sigma^{2s}(a) = \sigma^s(a+s) \| \sigma^s(a)$. A write operation is formalised as a substitution $\sigma[a \overset{s}{\mapsto} v]$, where $a \in [0, 2^{32}-1]$ and $v \in \mathbb{B}^{8s}$. The resulting store $\sigma' = \sigma[a \overset{s}{\mapsto} v]$ satisfies $\sigma'^s(a) = v$, and furthermore $\sigma'^1(b) = \sigma^1(b)$ for all addresses $b \notin \{a, \dots a+s-1\}$. The little-endian architecture imposes a similar invariant on the write operation. The notation $\sigma[a+i \overset{1}{\mapsto} \sigma^1(b+i)]_{i=0}^{n-1}$ copies n bytes from address b to a.

2.3 The Environment

For the sake of defining the concrete semantics of a program $P \in \mathcal{L}(CoreC)$, let V^g denote the global variables of P and let S^g its initialisation statements. Moreover, let $f_1 \dots f_n$ be the functions that constitute P and define $lookupFunc(f_i) = \langle P^{f_i}, V^{f_i}, l^{f_i} \rangle$ for each function. Here, P^{f_i} denotes the formal parameters of f_i, V^{f_i}, the locally declared variables, and l^{f_i} the label of the first basic block of f_i. We use *lookupBlock* and *lookupNext* to map a block label $l \in Label$ to the statement sequence or function call of a basic block and to the control-flow instructions, respectively. For example, given a basic block of the form $l: s_1; \dots s_n;$ **return**, the functions return $lookupBlock(l) = l : s_1; \dots s_n;$ and $lookupNext(l) =$**return**.

Every parameter of a function also has to be declared as a local variable of that function; that is, $P^{f_i} \subseteq V^{f_i}$. Hence, the set of all declared variables of P is $\mathcal{M} := V^g \cup \bigcup_{i=1}^{n} V^{f_i}$. Let $size : \mathcal{M} \rightarrow \mathbb{N}$ represent the declared size of each variable in bytes. Note that the C compiler accepts (and generates) Boolean variables, which are translated into **uint8**. They contain 0 to denote false and one to denote true. The association between a variable $v \in V^{f_i}$ of the current function f_i and its address a in memory is given through the pair $\langle v, a\rangle \in \mathcal{M} \times [0, 2^{32} - 1]$. A set of these pairs $A \in \mathcal{P}(\mathcal{M} \times [0, 2^{32} - 1])$ assigns addresses to all local variables of a function. To describe the concrete semantics of a function, this set is paired with a label $l \in Label$ to form a stack frame $\Theta = Label \times \mathcal{P}(\mathcal{M} \times [0, 2^{32} - 1])$. The label l in a given stack frame $c = \langle l, A\rangle \in \Theta$ denotes the basic block of the call site, where execution continues on return. The call stack at a specific point in execution is given by $\theta \in \Theta^*$, where $\Theta^* = \{c_0 \cdots c_n \mid n \in \mathbb{N}, c_i \in \Theta\}$ is the set of all sequences of stack frames. Given a call stack $\theta = c_0 \cdot c_1 \cdots c_s$, the frame c_s corresponds to the current function, while the address map A in the first frame $c_0 = \langle l, A\rangle$ assigns addresses to all globally declared variables. Note that a variable can only occur once in each stack frame but several times (with different addresses) in a given stack, namely during a recursive function call. Thus, at any given point in the execution of a program, the only accessible entities are functions, global variables (in c_0), and variables of the current function (in c_s). This is reflected in the symbol table, which is a (partial) function $addr_\theta : (\{f_1, \ldots f_n\} \cup \mathcal{M}) \rightarrow [0, 2^{32} - 1]$ that assigns a fixed address to each function f_i and an address to all variables of the contexts c_s and c_0 of θ. Note that it is possible to access variables in the inner stack frames by passing their addresses as parameters to callees.

During execution of a C program, further memory can be allocated on the heap. These dynamically allocated memory regions are characterised by $\Delta = \mathcal{P}(Label \times [0, 2^{32} - 1] \times \mathbb{N})$, specifically, let $\delta \subseteq \Delta$ denote the set of currently allocated memory regions. Then each region $\langle l, a, s\rangle \in \delta$ that has been allocated at position l in the program occupies $s \in \mathbb{N}$ bytes of memory starting at address $a \in [0, 2^{32} - 1]$. The way that memory on the heap is allocated guarantees that $a + s \in [0, 2^{32} - 1]$; that is, a dynamically allocated memory region never exceeds the virtual state space. Note that several memory regions may be allocated at the same program position l. In fact, the allocation site l is only tracked to facilitate the static analysis of dynamic memory.

In order to allocate a new region of memory, it is necessary to define what regions are already in use. Given a call stack θ and the set of dynamically allocated memory regions δ, we define the set of used addresses as $used(\theta, \delta)$, which is the union of the following sets of addresses:

- $\{a, \ldots a + (size(v) - 1)\}$ for all $\langle l, A\rangle \in \theta$ such that $\langle v, a\rangle \in A$
- $\{a, \ldots a + (s - 1)\}$ for all $\langle l, a, s\rangle \in \Delta$

As mentioned before, $size(v)$ denotes the size of the declared variable $v \in \mathcal{M}$. Besides program-specific address ranges, a few memory regions are also reserved by the execution environment; that is, they may not be accessed by the program itself. Let $reserved_\theta \subseteq [0, 2^{32} - 1]$ denote these reserved addresses, which are comprised of the following regions:

- A program may not access the regions that store the program itself. Hence, for each function f_i with $a = addr_\emptyset(f_i)$ and that occupies s bytes, the range $\{a, \ldots a + (s - 1)\}$ is reserved.
- The operating system disallows accesses to the first memory page, which for our chosen architecture resides at addresses $\{0, \ldots 4095\}$. This ensures that the processor automatically stops programs that dereference `NULL` pointers.
- At the high end, one gigabyte of memory is reserved for the operating system, and no pointer to this region is ever passed to the application program. (This is true for current versions of all major 32-bit operating systems.) Hence, the address range $[(2^{31} + 2^{30}), 2^{32} - 1]$ is not available to the program.
- Any address in the current sequence of stack frames $\theta \in \Theta^*$ that does not serve to store program variables may not be accessed from the program. Examples of such locations include the return addresses of functions. These addresses are in $reserved_\theta$ which is therefore parameterised by the stack θ.
- The locations and sizes of each dynamically allocated memory region $\delta \in \Delta$ are stored in a region in $reserved_\theta$ of fixed size. Access to this region is only permitted by the primitives **malloc** and **free** and not by the program itself.

Given the information on memory regions that are reserved and those that are used for calculations by the program, it is now possible to define how new memory is allocated. For the sake of being independent of a particular machine, no actual algorithm for finding an unused memory region is given. We merely require the existence of a function $fresh_\theta^\delta : \mathbb{N} \to [0, 2^{32}-1]$ that retrieves free memory to fulfill the request for dynamically allocated memory or a new stack frame. Specifically, $a = fresh_\theta^\delta(s)$ allocates s bytes at memory address a such that $\{a, \ldots a + (s - 1)\} \cap (used(\theta, \delta) \cup reserved_\theta) = \emptyset$. If no such address exists, $fresh_\theta^\delta$ returns 0. Note that the generality of the allocation functions reflects the freedom of the compiler and the run-time system to choose how to lay out variables in memory. This reflects one of the few obligations on the programmer in that no assumption may be made on the absolute or relative positioning of either stack- or heap-allocated variables. Given the functions *used* and $fresh_\theta^\delta$, it is now possible to state when all operations of a Core C program are correct with respect to memory management.

Definition 1. *A program $P \in \mathcal{L}(\langle CoreC \rangle)$ exhibits correct memory management in the state $\langle \sigma, \theta, \delta \rangle$ if in any read access $\sigma^s(a)$ and in any write access $\sigma[a \overset{s}{\mapsto} n]$ the set inclusion $\{a, \ldots a + (s - 1)\} \subseteq used(\theta, \delta)$ holds for any implementation of $fresh_\theta^\delta$ that adheres to the specification above.*

Conversely, a Core C program exhibits incorrect memory management whenever it accesses an address that is currently unused or in $reserved_\theta$.

While most accesses to free memory regions can be intercepted by the virtual memory system by disallowing access to the corresponding pages, most regions in $reserved_\theta$ are accessed by the C run-time during each function call and return, which makes protection difficult. Furthermore, these regions are too small to be protected at the hardware level. In order to illustrate this, consider the return address in the stack frame shown in Fig. 1.1, which is embedded between function arguments and local variables. A hardware protection against buffer-overflow vulnerabilities would require that a write access to the few bytes in $reserved_\theta$ that constitute the return address be disallowed during the execution of the function. Furthermore, the location of the return address must be writable before the function is entered.

A recent reaction to the severity of buffer-overflow vulnerabilities resulted in protecting the whole stack from being executable; that is, disallowing the processor from fetching code from the pages that constitute the stack. However, certain run-time environments, such as the Java Virtual Machine, require an executable stack. For this reason, operating systems must provide a way to circumvent this protection. For instance, in Windows Vista it is possible to make the stack executable again by calling an undocumented function in `ntdll.dll`, an essential dynamic library that is loaded with every program. An attacker who was previously able to overwrite the return address of a function with the address of the stack-allocated buffer now merely overwrites the return address with the function to disable stack execution. The address of the code in the buffer is then positioned such that upon return the code in the buffer is executed [174]. Thus, preventing the execution of code in the stack segment is not sufficient in practice to prevent buffer-overflow attacks.

At the software level, Vista makes this attack harder by using "address space layout randomization"; that is, it loads the `ntdll.dll` library at one of 256 different locations in memory, thereby decreasing the chances that the attacker picks the correct address of the function that enables the stack execution. On the downside, loading dynamic libraries at different addresses can carry a performance penalty. For instance, the Mac OS X operating system minimises application start-up time by a process called pre-binding, which fixes the address of each dynamic library to an address that does not overlap with any other dynamic library used by any single application.[1] As a consequence, using dynamic libraries at run-time requires no relocation, and temporarily unused pages containing library code can simply be discarded rather than being swapped out to disk, thereby increasing the responsiveness of the overall system. Loading libraries at randomised addresses would make this performance trick impossible.

Other compile-time program transformations have been proposed to make access from the C program to these reserved regions less likely [22,63,108,183]. These transformations rearrange the order of local variables and insert array bound checks in cases where it is clear that a function-local buffer is accessed.

[1] This is called "optimising system performance" when installing new software.

These techniques generally incur a run-time overhead and only reduce the likelihood of a successful attack, which renders them far from ideal. For instance, Windows Vista uses stack cookies, which are inserted by the compiler whenever a function contains a stack-allocated array. A stack cookie is an integer placed between the array and the return address that is set to a random number on function entry. Before the function exists, it is checked that this integer contains the same number, which makes it very likely that the array was not written beyond its bounds. One downside of all of the techniques presented is that a prevented attack still aborts a running dæmon or service of a server, which opens up the possibility of a denial-of-service attack.

Attacks that overwrite administrative information on dynamically allocated memory regions are more difficult and therefore less widespread. Their strategy often relies on the wrapping of integer variables that feed into allocation functions such as `malloc`. By allocating a buffer that is smaller than what the program assumes, it might be possible to overwrite administrative information that is stored beyond the end of the allocated memory region. Again, protecting administrative information in the memory region that surrounds dynamically allocated memory from accesses by the C program is difficult.

Hence, a more laudable goal is to remove all buffer-overflow vulnerabilities from a program by ensuring that it exhibits correct memory management. The definition of correct memory management is expressed in terms of accesses to the store σ. The next section introduces the semantics of Core C programs in order to define how these memory accesses are carried out.

2.4 Concrete Semantics

Given the memory model from the last section, we now embark on describing how a Core C program transforms the concrete state $\langle\sigma, \theta, \delta\rangle \in \Sigma \times \Theta^* \times \Delta$. Specifically, Figs. 2.3a and 2.3b define a set of concrete transfer functions of the form $[\![\cdot]\!]^\natural_{\mathrm{N}}$ for different productions N, namely:

$[\![\cdot]\!]^\natural_{\mathrm{Block}} : (\Sigma \times \Theta^* \times \Delta) \to (\Sigma \times \Theta^* \times \Delta)$ specifies how a basic block is executed. A basic block contains either a sequence of statements or a direct or indirect function call.

$[\![\cdot]\!]^\natural_{\mathrm{Next}} : (\Sigma \times \Theta^* \times \Delta) \to (\Sigma \times \Theta^* \times \Delta)$ evaluates conditionals and executes the indicated basic block if the condition holds.

$[\![\cdot]\!]^\natural_{\mathrm{Stmt}} : (\Sigma \times \Theta^* \times \Delta) \to (\Sigma \times \Theta^* \times \Delta)$ defines the semantics of a Core C statement.

$[\![\cdot]\!]^{\natural,s}_{\mathrm{Expr}} : (\Sigma \times \Theta^* \times \Delta) \to \mathbb{B}^s$ evaluates the value of a linear expression, yielding a bit vector of s bits. All occurring variables are accessed as variables of width s bits.

Here, the $\natural$ superscript indicates that these functions define the concrete (or "natural") semantics. Before describing these functions in detail, we define how execution of a program starts. Let $P = V^g s_0; \ldots s_n; F$ be the program that is to be run. Here, V^g denotes the declaration of global variables, $s_0; \ldots s_n;$ the initialisation statements and F a sequence of function definitions including the function **main** where execution starts. The semantics of this Core C program is defined by

$$[\![\, \mathbf{main}() \,]\!]^{\natural}_{\mathsf{Block}}([\![\, s_n \,]\!]^{\natural}_{\mathsf{Stmt}}(\cdots [\![\, s_0 \,]\!]^{\natural}_{\mathsf{Stmt}}\langle \sigma^{init}, \langle l, A^g \rangle, \emptyset \rangle \ldots))$$

where σ^{init} is an arbitrary store and $\langle l, A^g \rangle$ is the first stack frame. Here, $l \in \mathit{Label}$ is a label that does not occur in the program and A^g associates global variables with memory regions as follows. Given the declaration of global variables $V^g = v_1 : s_1; \ldots v_n : s_n;$, let $A^g = A_n$, $A_0 = \emptyset$, and $A_i = A_{i-1} \cup \{\langle v_i, \mathit{fresh}^{\emptyset}_{\langle l, A_{i-1} \rangle}(s_i) \rangle\}$. Execution stops after the **return** statement is evaluated, and $\langle l, A^g \rangle$ is the current context. While the **main** function takes no argument in the invocation above, it is possible to add these for any particular set of program arguments by adding statements to the initialisation sequence. A return value can be added in a similar way.

Figure 2.3a defines the semantics for basic blocks, control flow, function calls, and simple assignments. The first rule specifies how a sequence of statements $s_1; \ldots s_n;$ transforms the initial state $\langle \sigma, \theta, \delta \rangle$, which is then propagated to the $[\![\cdot]\!]^{\natural}_{\mathsf{Next}}$ function, where the next basic block is executed. In contrast, a function call will execute the first basic block l_t of the called function f_i without ever evaluating the following control-flow instructions. The parameters $p_1, \ldots p_n$ are assigned as if they were structures; the underlying memory regions are simply copied. This step requires that all locally declared variables (which include parameters) be allocated in memory, which is performed by the recursive definition of the list A. The last variant of a basic block is the indirect function call. Note that this definition is partial, as there might not be a function with the given address $\mathit{val}^{32,\mathbf{uint}}(\sigma^4(\mathit{addr}_\theta(v)))$. The analysis presented later will flag an error every time the concrete semantics is unspecified.

Another partial definition is that of the **return** keyword in that it requires at least one frame on the stack. Note that the allocation map A is removed when a function returns, thereby effectively freeing all local variables in that frame since the lookup function for variables addr_θ is parameterised by the current stack.

Similarly to the jump instruction, which simply executes the denoted basic block (which has to be part of the current function), the conditional will branch similarly to the jump instruction if the condition is met; otherwise, it will evaluate the next control-flow instruction. The evaluation of the condition draws upon several new constructs. Every access to a variable may have a byte offset to facilitate accesses to structures. Hence, the access $v.o$ will read the memory at $\mathit{addr}_\theta(v)+o$. The resulting binary value that is read from memory, namely $\sigma^s(\mathit{addr}_\theta(v)+o)$, is interpreted as an integral value by $\mathit{val}^{8s,t}(\cdot)$, and the normal relational operators can be applied. Note that the signedness of

Basic Blocks.

$$[\![\, l : s_1; \ldots s_n; \,]\!]^{\natural}_{\mathsf{Block}}\langle \sigma, \theta, \delta \rangle =$$
$$\quad [\![\, \mathit{lookupNext}(l) \,]\!]^{\natural}_{\mathsf{Next}}([\![\, s_n \,]\!]^{\natural}_{\mathsf{Stmt}}(\cdots([\![\, s_1 \,]\!]^{\natural}_{\mathsf{Stmt}}\langle \sigma, \theta, \delta \rangle)\ldots))$$

$$[\![\, l : f_i(a_1, \ldots a_n); \,]\!]^{\natural}_{\mathsf{Block}}\langle \sigma, \theta, \delta \rangle = [\![\, \mathit{lookupBlock}(l_t) \,]\!]^{\natural}_{\mathsf{Block}}\langle \sigma', \theta', \Delta' \rangle$$
$$\begin{aligned}
\text{where } & \langle\, \langle p_1, \ldots p_n \rangle, \langle v_1, \ldots v_m \rangle, l_t \rangle = \mathit{lookupFunc}(f_i) \\
& s_1 = \mathit{size}(p_1), \ldots s_n = \mathit{size}(p_n) \\
& A = A_m, A_0 = \emptyset, A_i = \{\langle v_i, \mathit{fresh}^{\delta}_{\theta.\langle l, A_{i-1}\rangle}(s_i)\rangle\} \cup A_{i-1} \\
& \text{if } \exists \langle v, 0 \rangle \in A_m \text{ then stop due to stack overflow} \\
& \langle \sigma', \theta', \Delta' \rangle = [\![\, \textbf{structure}\ s_n\ p_n.0 = a_n; \,]\!]^{\natural}_{\mathsf{Stmt}}(\cdots(\\
& \qquad [\![\, \textbf{structure}\ s_1\ p_1.0 = a_1 \,]\!]^{\natural}_{\mathsf{Stmt}}\langle \sigma, \theta \cdot \langle l, A \rangle, \Delta \rangle \\
& \qquad)\ldots)
\end{aligned}$$

$$[\![\, l : *v(a_1, \ldots a_n); \,]\!]^{\natural}_{\mathsf{Block}}\langle \sigma, \theta, \delta \rangle =$$
$$\quad [\![\, l : f_i(a_1, \ldots a_n); \,]\!]^{\natural}_{\mathsf{Block}}\langle \sigma, \theta, \delta \rangle \text{ where } \mathit{addr}_\theta(f_i) = \mathit{val}^{32,\mathbf{uint}}(\sigma^4(\mathit{addr}_\theta(v)))$$

Control Flow.

$$[\![\, \textbf{return} \,]\!]^{\natural}_{\mathsf{Next}}\langle \sigma, \theta \cdot \langle l, A \rangle, \Delta \rangle = [\![\, \mathit{lookupNext}(l) \,]\!]^{\natural}_{\mathsf{Next}}\langle \sigma, \theta, \delta \rangle$$

$$[\![\, \textbf{jump}\ l \,]\!]^{\natural}_{\mathsf{Next}}\langle \sigma, \theta, \delta \rangle = [\![\, \mathit{lookupBlock}(l) \,]\!]^{\natural}_{\mathsf{Block}}\langle \sigma, \theta, \delta \rangle$$

$$[\![\, \textbf{if}\ v.o\ t\ s\ \mathit{op}\ \mathit{exp}\ \textbf{then jump}\ l\ ;\ n \,]\!]^{\natural}_{\mathsf{Next}}\langle \sigma, \theta, \delta \rangle =$$
$$\begin{cases} [\![\, \mathit{lookupBlock}(l) \,]\!]^{\natural}_{\mathsf{Block}}\langle \sigma, \theta, \delta \rangle \\ \qquad\qquad \text{if } \mathit{val}^{8s,t}(\sigma^s(\mathit{addr}_\theta(v) + o))\ \mathit{op}\ \mathit{val}^{8s,t}([\![\, \mathit{exp} \,]\!]^{\natural,s}_{\mathsf{Expr}}\langle \sigma, \theta, \delta \rangle) \\ [\![\, \mathit{nxt} \,]\!]^{\natural}_{\mathsf{Next}}\langle \sigma, \theta, \delta \rangle \quad \text{otherwise} \end{cases}$$

Expressions.

$$[\![\, n \,]\!]^{\natural,s}_{\mathsf{Expr}}\langle \sigma, \theta, \delta \rangle = \mathit{bin}^{8s}(n)$$

$$[\![\, n * v.o + e \,]\!]^{\natural,s}_{\mathsf{Expr}}\langle \sigma, \theta, \delta \rangle = \left(\mathit{bin}^{8s}(n) *^{8s} \sigma^s(\mathit{addr}_\theta(v) + o)\right) +^{8s} [\![\, e \,]\!]^{\natural,s}_{\mathsf{Expr}}$$

Assignment.

$$[\![\, s\ v.o = \mathit{exp} \,]\!]^{\natural}_{\mathsf{Stmt}}\langle \sigma, \theta, \delta \rangle = \langle \sigma', \theta, \delta \rangle$$
$$\quad \text{where } \sigma' = \sigma[\mathit{addr}_\theta(v) + o \overset{s}{\mapsto} [\![\, \mathit{exp} \,]\!]^{\natural,s}_{\mathsf{Expr}}\langle \sigma, \theta, \delta \rangle]$$

$$[\![\, s\ v \to o = \mathit{exp} \,]\!]^{\natural}_{\mathsf{Stmt}}\langle \sigma, \theta, \delta \rangle = \langle \sigma', \theta, \delta \rangle$$
$$\quad \text{where } \sigma' = \sigma[\mathit{val}^{32,\mathbf{uint}}(\sigma^4(\mathit{addr}_\theta(v))) + o \overset{s}{\mapsto} [\![\, \mathit{exp} \,]\!]^{\natural,s}_{\mathsf{Expr}}\langle \sigma, \theta, \delta \rangle]$$

$$[\![\, s\ v_1.o_1 = v_2 \to o_2 \,]\!]^{\natural}_{\mathsf{Stmt}}\langle \sigma, \theta, \delta \rangle = \langle \sigma', \theta, \delta \rangle$$
$$\quad \text{where } \sigma' = \sigma[\mathit{addr}_\theta(v) + o \overset{s}{\mapsto} \sigma^s(\mathit{val}^{32,\mathbf{uint}}(\sigma^4(\mathit{addr}_\theta(v))) + o)]$$

Fig. 2.3a. Concrete semantics of Core C: basic blocks, control flow, and assignments.

Assignment of Structures.

$$[\![\ \textbf{structure}\ \ s\ v_1.o_1 = v_2.o_2\ ]\!]^{\natural}_{\textsf{Stmt}}\langle\sigma,\theta,\delta\rangle = \langle\sigma',\theta,\delta\rangle$$
$$\text{where } \sigma' = \sigma[addr_\theta(v_1) + o_1 + i \overset{1}{\mapsto} \sigma^1(addr_\theta(v_2) + o_2 + i)]_{i=0}^{s-1}$$

$$[\![\ \textbf{structure}\ \ s\ v_1 \rightarrow o_1 = v_2.o_2\ ]\!]^{\natural}_{\textsf{Stmt}}\langle\sigma,\theta,\delta\rangle = \langle\sigma',\theta,\delta\rangle$$
$$\text{where } \sigma' = \sigma[val^{32,\textsf{uint}}(\sigma^4(addr_\theta(v_1))) + o_1 + i \overset{1}{\mapsto} \sigma^1(addr_\theta(v_1) + o_2 + i)]_{i=0}^{s-1}$$

$$[\![\ \textbf{structure}\ \ s\ v_1.o_1 = v_2 \rightarrow o_2\ ]\!]^{\natural}_{\textsf{Stmt}}\langle\sigma,\theta,\delta\rangle = \langle\sigma',\theta,\delta\rangle$$
$$\text{where } \sigma' = \sigma[addr_\theta(v) + o_1 + i \overset{1}{\mapsto} \sigma^1(val^{32,\textsf{uint}}(\sigma^4(addr_\theta(v))) + o_2 + i)]_{i=0}^{s-1}$$

Type Casts.

$$[\![\ s_1\ v_1.o_1 = t\ s_2\ v_2.o_2\ ]\!]^{\natural}_{\textsf{Stmt}}\langle\sigma,\theta,\delta\rangle = \langle\sigma',\theta,\delta\rangle$$
$$\text{where } \sigma' = \sigma[addr_\theta(v_1) + o_1 \overset{s_1}{\mapsto} bin^{8s_1}(val^{8s_2,t}(\sigma^{s_2}(addr_\theta(v_2) + o_2)))]$$

Address-Of Operators.

$$[\![\ v_1.o_1 = \&v_2.o_2\ ]\!]^{\natural}_{\textsf{Stmt}}\langle\sigma,\theta,\delta\rangle = \langle\sigma',\theta,\delta\rangle$$
$$\text{where } \sigma' = \sigma[addr_\theta(v_1) + o_1 \overset{4}{\mapsto} bin^{32}(addr_\theta(v_2) + o_2)]$$

$$[\![\ v.o = \&f\ ]\!]^{\natural}_{\textsf{Stmt}}\langle\sigma,\theta,\delta\rangle = \langle\sigma',\theta,\delta\rangle$$
$$\text{where } \sigma' = \sigma[addr_\theta(v) + o \overset{4}{\mapsto} bin^{32}(addr_\theta(f))]$$

String Constants.

$$[\![\ v = \texttt{"}c_0c_1\ldots c_{k-1}\texttt{"}\ ]\!]^{\natural}_{\textsf{Stmt}}\langle\sigma,\theta,\delta\rangle = \langle\sigma',\theta,\delta\rangle$$
$$\text{where } \sigma' = (\sigma[addr_\theta(v) + i \overset{1}{\mapsto} c_i\ ]_{i=0}^{k-1})[addr_\theta(v) + k \overset{1}{\mapsto} 0]$$

Dynamically Allocated Memory.

$$[\![\ l:\ v_1 = \textbf{malloc}(v_2)\ ]\!]^{\natural}_{\textsf{Stmt}}\langle\sigma,\theta,\delta\rangle = \langle\sigma',\theta,\delta'\rangle$$
$$\begin{aligned}\text{where } s &= val^{32,\textsf{uint}}(\sigma^4(addr_\theta(v_2)))\\ a &= fresh^{\delta}_{\theta}(s)\\ \delta' &= \text{ if } a = 0 \text{ then } \delta \text{ else } \delta \cup \{\langle l, a, s\rangle\}\\ \sigma' &= \sigma[addr_\theta(v_1) \overset{4}{\mapsto} bin^{32}(a)]\end{aligned}$$

$$[\![\ \textbf{free}(v)\ ]\!]^{\natural}_{\textsf{Stmt}}\langle\sigma,\theta,\delta\rangle = \langle\sigma,\theta,\delta'\rangle$$
$$\begin{aligned}\text{where } a &= val^{32,\textsf{uint}}(\sigma^4(addr_\theta(v)))\\ \delta' &= \begin{cases}\delta & \text{if } a = 0\\ \delta \setminus \{\langle l, a, s\rangle\} & \text{if } \exists s \in \mathbb{N}, l \in \textit{Label} \text{ such that } \langle l, a, s\rangle \in \delta\end{cases}\end{aligned}$$

Fig. 2.3b. Concrete semantics of Core C: statements and primitives for dynamically allocated memory. The notation $s\ v$ specifies the size s of the variable v; $t\ s\ v$ specifies type t and size s of the variable v.

a type can influence the outcome of a comparison and that converting the binary vectors to integer values makes this difference explicit.

The conditional involves the evaluation of a linear expression. This semantic action is defined next and is parameterised by the width s of the result in bytes. Interestingly, the sign of the resulting type when accessing the underlying memory is irrelevant since the results modulo 2^{8s} are the same.

Linear expressions are also allowed in assignment statements. In the first case, the value of an expression exp is evaluated in the current state and the result is written to the address of v with the displacement o added. Assigning to a pointed-to value is similar, except the result is written to the address contained in the variable rather than the address of the variable itself. Note that the fact that pointers are 32 bits is made explicit here in the access $\sigma^4(addr_\theta(v))$. This 32-bit-wide vector is converted to a value in $[0, 2^{32} - 1]$ by $val^{32,\mathbf{uint}}(\cdot)$. The third assignment transfers a value pointed to by v_2 to the left-hand side. Note that C statements of the form `*x = *y` have to be broken down into two statements, using an intermediate variable to store the contents of `*y`.

Figure 2.3b defines the evaluation of more complex Core C statements. An interesting observation about the semantics of C is that the language implementation must be able to copy whole regions of memory, which is not an operation that is directly available to the programmer. For example, it is not possible to assign one fixed-sized array to another; however, it is possible when these arrays are wrapped in a C **struct**. Due to this oddity, the three principal ways of assignment have to be reimplemented for memory regions, with the difference that the right-hand side can only be a simple variable (rather than an expression).

A consequence of the ad hoc overloading of arithmetic operations is that the conversion between different-sized integers is a core operation of C. Note that the rule for type casts cannot be reformulated by accessing a memory region with a different size or signedness. The reason for this is the necessity of doing sign extension on signed integers and zero padding for unsigned integers. These adjustments happen implicitly when the read value of type t_2 and size s_2 is converted to an integer before the conversion back to a Boolean vector of size s_1 is applied using $bin^{s_1}(\cdot)$.

Taking the address of a variable or a function merely requires the conversion of the address to a 32-bit Boolean vector and thus introduces no novel notation. Assigning a string to a memory region is a simple byte-copying loop as in structure assignment, except for the final zero byte, which is implicit in the string constant. Thus, the size of the variable that is assigned to has to be of size $k + 1$ for a string of k letters.

The last two functions are the primitives for allocating and freeing memory on the heap. The **malloc** function stores the newly allocated region in δ' (if sufficient memory is available). The **free** function is partial in that it will not proceed if its argument does not correspond to the beginning of a previously allocated memory region. The analysis will flag such a situation as erroneous.

<table>
<tr><td><,<=,==,!=,>=,></td><td>relational expressions</td></tr>
<tr><td>&,|,^,~</td><td>bit-wise expressions</td></tr>
<tr><td>&&,||,!</td><td>logical expressions</td></tr>
<tr><td><<,>></td><td>bit-shift expressions</td></tr>
<tr><td>%,/</td><td>integer division and modulus expressions</td></tr>
</table>

Fig. 2.4. Primitives of C whose implementation is omitted.

This concludes the discussion of the Core C language constructs. While a full analysis requires other C operations and primitive functions, we omit their presentation since their implementations are merely technical and thus would not add to the Core C semantics already presented . For the sake of completeness, Fig. 2.4 lists the primitives of C that are necessary to cover the full C language without libraries.

The concrete semantics presented so far serves as a reference for specifying the abstract semantics that defines the actual analysis. In order to convey the underlying ideas and to motivate the abstract semantics, it is useful to summarise several runs of a concrete program, which is the topic of the next section.

2.5 Collecting Semantics

The single-step semantics presented in the last section describes how a single state is modified by executing a Core C statement. By operating on a single state at a time, primitives that read data from the operating system can only return a single value, reflecting one specific run of the program. Inferring a property that holds for all possible runs of a program therefore requires a way to define a primitive that can return all possible inputs. To this end, we lift the single-step semantics from transforming a single memory state to transforming sets of memory states. This enables input primitives to map a single memory state to many memory states, one for each possible input. Allowing input primitives to return all possible values can be seen as a first abstraction in that it disregards all input data. Indeed, it enables the definition of the collecting semantics of a program – that is, the set of all states that are possible at any given program point. Given the collecting semantics, the question of whether a program exhibits correct memory management on all inputs reduces to inspecting the inferred states. While this is not a practical approach in general, it serves to illustrate the idea of a static analysis that infers an abstract state for each program point. In particular, each abstract state summarises a set of concrete states, and hence calculating the set of all possible concrete states is similar to calculating a single abstract state.

In order to lift the semantic equations of Core C statements to sets of states, define $\overline{\Sigma} = \mathcal{P}(\Sigma \times \Theta^* \times \Delta)$ to be the set of all sets of concrete states.

Note that the cardinality of $\overline{\Sigma}$ is finite since Σ is finite and any $\theta \in \Theta^*$ and $\delta \in \Delta$ can be encoded in Σ at the locations $reserved_\theta$ that were set aside for this purpose. Thus, using explicit triples $\langle\sigma, \theta, \delta\rangle$ as opposed to embedding θ and δ into the current state $\sigma \in \Sigma$ is a mere convenience. The actual lifting from a state $\langle\sigma, \theta, \delta\rangle$ to a set of states $\overline{\sigma} \in \overline{\Sigma}$ is defined for each $N = \langle\text{Block}\rangle$, $\langle\text{Next}\rangle$, $\langle\text{Stmt}\rangle$, $\langle\text{Expr}\rangle$ as follows:

$$[\![\, s \,]\!]^{\natural}_{\mathrm{N}}\overline{\sigma} = \{[\![\, s \,]\!]^{\natural}_{\mathrm{N}}\langle\sigma, \theta, \delta\rangle \mid \langle\sigma, \theta, \delta\rangle \in \overline{\sigma} \\ [\![\, s \,]\!]^{\natural}_{\mathrm{N}}\langle\sigma, \theta, \delta\rangle \text{ is defined}\}$$

Here, we reuse the notation of a transfer function on a single state $[\![\, s \,]\!]^{\natural}_{\mathrm{N}}$ for the transfer function on a set of states. Note that on some states the results of the transfer functions are undefined. Undefinedness in the concrete semantics indicates an error condition and is similar to the C notion of "undefined behaviour" [51]. In the context of the concrete semantics presented so far, the following erroneous conditions lead to undefined behaviour:

illegal memory accesses: Any read access $\sigma^s(a)$ and any write access $\sigma[a \overset{s}{\mapsto} n]$ where $a \notin [0, 2^{32} - 1]$ constitutes an illegal memory access. Note that an access for which the inclusion $\{a, \ldots a + (s - 1)\} \subseteq used(\theta, \delta)$ does not hold is actually well defined; however, any such access will be flagged as incorrect by the analysis.

calling an invalid function: Whenever a pointer **p** is dereferenced in order to call a function, the address stored in the pointer must be the result of an address-of function operation such as `p=&f;` where `f` is a function. No offset may be added to **p**.

jumping out of a function: If a **jump** instruction specifies a label outside the current function, the execution continues with a mismatched stack. In particular, the $addr_\theta$ function may be undefined for certain program variables. Note that C allows cross-function jumps through the `setjmp` and `longjmp` functions. For presentational purposes, the semantics of the **jump** instruction does not cater to these since a formal specification is more technical than insightful.

freeing memory that is not allocated: Calling **free** with a pointer that is neither `NULL` nor denotes the beginning of a dynamically allocated memory block causes undefined behaviour in C. In particular, this case occurs if the same pointer is freed twice, which is a mistake in C.

overflowing the stack: Allocating memory on the stack using $fresh^\delta_\theta$ may find that no more memory is available. In this case, the program terminates.

Undefined behaviour is excluded from the resulting set of states because we assume that execution stops as soon as an erroneous condition occurs. The alternative to the assumption that the program aborts is to explicitly state how the state space develops when a C program exhibits undefined behaviour. Such an approach, however, is not only very implementation dependent but also has little merit in that such a trace is irrelevant to a static analysis that

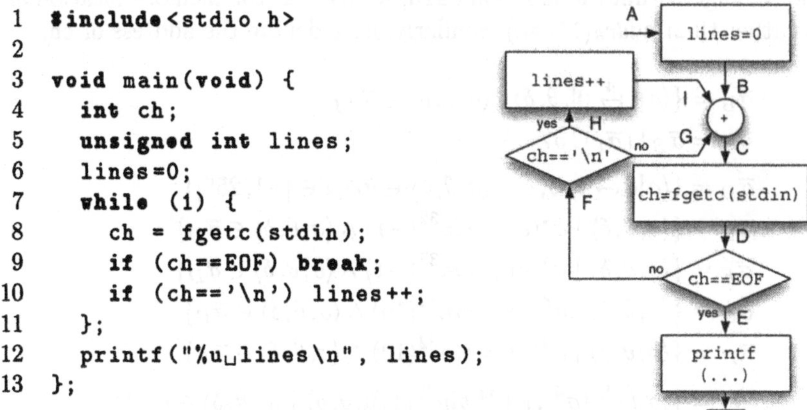

Fig. 2.5. Example C program that counts the number of newlines encountered on `stdin`. The corresponding control flow graph is decorated with nine edges $A, \ldots I$, for which the collecting semantics describes the possible states $\overline{\sigma}_A, \ldots \overline{\sigma}_I$.

is designed to prove that a program does not exhibit undefined behaviour at run-time. The drawback of expressing run-time errors as undefined behaviour is that a proof of correctness has to ensure that each application of a concrete function is well defined in order not to inadvertently ignore a run-time error.

Given the set of transfer functions on sets of states, it is now easy to define primitives that retrieve input from the operating system. For instance, consider the function `fgetc(FILE *stream)` that retrieves a single character from the given file `stream` and returns a C `int` containing either `EOF` (which is a macro defined as -1) or a value $0 \ldots 255$, depending on the character read. The corresponding Core C primitive takes a pointer parameter *res* through which the result is written and the `stream` parameter *stream*, which is ignored:

$$\llbracket \text{"fgetc"}(\mathbf{uint4}\ res, \mathbf{uint4}\ stream) \rrbracket^{\natural}_{\mathrm{Stmt}} \overline{\sigma} = \{\langle \sigma[val^{32,\mathbf{uint}}(\sigma^4(addr_\theta(res))) \overset{4}{\mapsto} i], \theta, \delta\rangle \mid \langle \sigma, \theta, \delta\rangle \in \overline{\sigma}, i \in [-1, 255]\}$$

Here, $val^{32,\mathbf{uint}}(\sigma^4(addr_\theta(res)))$ denotes the content of `res`; that is, the address to which the result is written. For each input state $\langle \sigma, \theta, \delta\rangle \in \overline{\sigma}$, the primitive creates 257 new states, namely one state for each possible return value $i \in [-1, 255]$, where -1 denotes the `EOF` value and $0, \ldots 255$ denotes the value of the character retrieved. Given this function, consider calculating the collecting semantics of the line counting program on the left of Fig. 2.5.

The corresponding flow graph on the right possesses nine edges $A, \ldots I$. Each edge has one possible set of states, namely $\overline{\sigma}_A, \ldots \overline{\sigma}_I$, whose definitions can be derived by translating the C program into Core C and applying the concrete semantics for each statement. For brevity, we only show the result in the form of the semantic equations lifted to sets of states. Since the stack θ

does not change within the function `main`, we use l as the memory location of `lines` rather than $addr_\theta$(`lines`); similarly, let c denote the address of `ch`.

$$\begin{aligned}
\overline{\sigma}_B &= \{\langle \sigma[l \overset{4}{\mapsto} 0], \theta, \delta\rangle \mid \langle \sigma, \theta, \delta\rangle \in \overline{\sigma}_A\} \\
\overline{\sigma}_C &= \overline{\sigma}_B \cup \overline{\sigma}_I \cup \overline{\sigma}_G \\
\overline{\sigma}_D &= \{\langle \sigma[c \overset{4}{\mapsto} i], \theta, \delta\rangle \mid \langle \sigma, \theta, \delta\rangle \in \overline{\sigma}_C, i \in [-1, 255]\} \\
\overline{\sigma}_E &= \{\langle \sigma, \theta, \delta\rangle \mid \sigma^4(c) = bin^{32}(-1) \wedge \langle \sigma, \theta, \delta\rangle \in \overline{\sigma}_D\} \\
\overline{\sigma}_F &= \{\langle \sigma, \theta, \delta\rangle \mid \sigma^4(c) \neq bin^{32}(-1) \wedge \langle \sigma, \theta, \delta\rangle \in \overline{\sigma}_D\} \\
\overline{\sigma}_G &= \{\langle \sigma, \theta, \delta\rangle \mid \sigma^4(c) \neq bin^{32}(10) \wedge \langle \sigma, \theta, \delta\rangle \in \overline{\sigma}_F\} \\
\overline{\sigma}_H &= \{\langle \sigma, \theta, \delta\rangle \mid \sigma^4(c) = bin^{32}(10) \wedge \langle \sigma, \theta, \delta\rangle \in \overline{\sigma}_F\} \\
\overline{\sigma}_I &= \{\langle \sigma[l \overset{4}{\mapsto} (\sigma^4(l) +^{32} bin^{32}(1))], \theta, \delta\rangle \mid \langle \sigma, \theta, \delta\rangle \in \overline{\sigma}_H\}
\end{aligned}$$

The solution to the equation system, which characterises all possible memory states of the program, depends on the initial states in $\overline{\sigma}_A$. In general, these states are defined by executing the initialisation statements of a program, starting on the set of all possible memory states Σ and the empty stack and an empty set of dynamically allocated memory regions. Thus define

$$\overline{\sigma}_A = [\![s_n]\!]^{\natural}_{\text{Stmt}}(\ldots [\![s_0]\!]^{\natural}_{\text{Stmt}}\{\langle \sigma, \langle l, A^g\rangle, \emptyset\rangle \mid \sigma \in \Sigma\} \ldots)$$

where $s_0, \ldots s_n$ are the initialisation statements of the Core C program and A^g maps globally defined variables to addresses in memory. The definition above resembles the one described in Sect. 2.4 except that it is cast in terms of sets rather than a single memory state. In particular, we choose all possible states $\sigma \in \Sigma$ as the starting state, which ensures that any invocation of the program, which starts on some random memory state, is included in the collecting semantics.

A constructive way to infer a solution to the equation system above is a fixpoint calculation. That is, we start off with $\overline{\sigma}_B = \cdots = \overline{\sigma}_I = \emptyset$ and apply any of the equations above and augment the set of states $\overline{\sigma}_A, \ldots \overline{\sigma}_I$ with the results until no more changes arise. For example, given a new set $\overline{\sigma}_A$, a new value can be calculated for $\overline{\sigma}_B$. Since the actual set of $\overline{\sigma}_B$ is too large to explicitly write out, we will merely give a description of the set. Specifically, we only present the possible contents of the memory locations $\sigma^4(l) = x_l$ and $\sigma^4(c) = x_c$ in the form of tuples $\langle x_l, x_c\rangle$. For instance, the result of evaluating the first equation $\overline{\sigma}_B = \{\langle \sigma[l \overset{4}{\mapsto} 0], \theta, \delta\rangle \mid \langle \sigma, \theta, \delta\rangle \in \overline{\sigma}_A\}$ is written as $\langle 0, -2^{31}\rangle, \ldots \langle 0, 2^{31} - 1\rangle$. New state sets for an edge are derived by evaluating its equation with the most recent state sets of other edges, a strategy also known as Gauss-Seidel iteration [55, 58]. State sets that have stabilised are marked with a star on the left.

$$\begin{aligned}
* \quad \overline{\sigma}_B &= \langle 0, -2^{31}\rangle, \ldots \langle 0, 2^{31} - 1\rangle \\
\overline{\sigma}_C &= \langle 0, -2^{31}\rangle, \ldots \langle 0, 2^{31} - 1\rangle
\end{aligned}$$

$$
\begin{aligned}
\overline{\sigma}_D &= \langle 0,-1\rangle, \ldots \langle 0,255\rangle \\
\overline{\sigma}_E &= \langle 0,-1\rangle \\
\overline{\sigma}_F &= \langle 0,0\rangle, \ldots \langle 0,255\rangle \\
\overline{\sigma}_G &= \langle 0,0\rangle, \ldots \langle 0,9\rangle, \langle 0,11\rangle, \ldots \langle 0,255\rangle \\
\overline{\sigma}_H &= \langle 0,10\rangle \\
\overline{\sigma}_I &= \langle 1,10\rangle \\
\overline{\sigma}_C &= \langle 0,-2^{31}\rangle, \ldots \langle 0,2^{31}-1\rangle, \langle 1,10\rangle \\
\overline{\sigma}_D &= \langle 0,-1\rangle, \ldots \langle 0,255\rangle, \langle 1,-1\rangle, \ldots \langle 1,255\rangle
\end{aligned}
$$

At this point, the states $\overline{\sigma}_E, \ldots \overline{\sigma}_H$ are updated in sequence, resulting in states identical to those above except that all states are duplicated for the case $\sigma^4(l) = 1$. This leads to the new state at edges I and C:

$$
\begin{aligned}
\overline{\sigma}_I &= \langle 1,10\rangle, \langle 2,10\rangle \\
\overline{\sigma}_C &= \langle 0,-2^{31}\rangle, \ldots \langle 0,2^{31}-1\rangle, \langle 1,-2^{31}\rangle, \ldots \langle 1,2^{31}-1\rangle, \langle 2,10\rangle
\end{aligned}
$$

After another loop iteration, a new range of tuples $\langle 2,-2^{31}\rangle, \ldots \langle 2,2^{31}-1\rangle$ is added. We characterise the states C more concisely as follows:

$$\overline{\sigma}_C = \langle 0,-2^{31}\rangle, \ldots \langle 2,2^{31}-1\rangle, \langle 3,10\rangle$$

More valuations for $\sigma^4(l)$ are added until all possible values for the variable `lines` are exhausted. The calculation finishes as follows:

$$
\begin{array}{lll}
 & \overline{\sigma}_C = \langle 0,-2^{31}\rangle, \ldots \langle 2^{32}-2,2^{31}-1\rangle, \langle 2^{32}-1,10\rangle & \\
* & \overline{\sigma}_D = \quad \langle 0,-1\rangle, \ldots \langle 0,255\rangle, & \\
 & \qquad \langle 1,-1\rangle, \ldots \langle 1,255\rangle, & \\
 & \qquad \vdots & \\
 & \langle 2^{32}-1,-1\rangle, \ldots \langle 2^{32}-1,255\rangle & \\
* & \overline{\sigma}_E = \langle 0,-1\rangle, \ldots \langle 2^{32}-1,-1\rangle & \\
* & \overline{\sigma}_F = \quad \langle 0,0\rangle, \ldots \langle 0,255\rangle, & \\
 & \qquad \langle 1,0\rangle, \ldots \langle 1,255\rangle, & \\
 & \qquad \vdots & \\
 & \langle 2^{32}-1,0\rangle, \ldots \langle 2^{32}-1,255\rangle & \\
 & \overline{\sigma}_G = \quad \langle 0,0\rangle, \ldots \langle 0,9\rangle, & \langle 0,11\rangle, \ldots \langle 0,255\rangle, \\
 & \qquad \langle 1,0\rangle, \ldots \langle 1,9\rangle, & \langle 1,11\rangle, \ldots \langle 1,255\rangle, \\
 & \qquad \vdots & \vdots \\
 & \langle 2^{32}-1,0\rangle, \ldots \langle 2^{32}-1,9\rangle, & \langle 2^{32}-1,11\rangle, \ldots \langle 2^{32}-1,255\rangle \\
* & \overline{\sigma}_H = \langle 0,10\rangle, \ldots \langle 2^{32}-1,10\rangle & \\
* & \overline{\sigma}_I = \langle 0,10\rangle, \ldots \langle 2^{32}-1,10\rangle & \\
* & \overline{\sigma}_C = \langle 0,-2^{31}\rangle, \ldots \langle 2^{32}-1,2^{31}-1\rangle &
\end{array}
$$

In the final update, $\overline{\sigma}_C$ is updated with the states where `lines` is $2^{32}-1$ and `ch` is not equal to 10, thereby reaching a fixpoint to the equation system. The fact that the equalities with a star indeed define a solution for the initial equation system can be seen by observing that calculating $\overline{\sigma}_D$ from $\overline{\sigma}_C$ results in the same state space $\overline{\sigma}_D$ that is marked with a star above. Hence, all dependent equations starting with $\overline{\sigma}_D$ will retain the same solution set, and a fixpoint is reached.

Note that $\langle\overline{\Sigma}, \subseteq\rangle$ forms a complete partial order (cpo) since $\overline{\Sigma}$ is finite. Augmenting the sets $\overline{\sigma}_B, \ldots \overline{\sigma}_I$ by evaluating the equalities always converges onto the minimal fixpoint of the equation system, as shown by Kleene; see [59].

The above procedure for calculating the collecting semantics is in fact a decision procedure for the task of determining whether an out-of-bounds buffer access may happen at run-time. Undecidability is not an issue since the domain $\overline{\Sigma}$ is finite, albeit very large. Thus, due to the size of the sets, the amount of time that it would take to calculate the collecting semantics makes this procedure impractical. The following chapters introduce an abstract view of the collecting semantics – that is, a symbolic representation of the states that can be calculated more readily.

2.6 Related Work

Designing a program analysis on any real-world language is a major undertaking due to the sheer complexity of implementing all syntactic constructs and examining the usually ill-defined semantics of each language construct. One approach to circumvent this burden is to restrict an analysis to a toy language in order to show the merit of a new idea. A more compelling technique is to translate a real-world programming language into a simpler intermediate language that only exposes the properties of interest. Prime examples of the latter approach are most works on points-to analysis, where the intermediate language consists of as little as four flow equations [99,176]. Alas, a translation into such a small language is usually an approximation in itself and thus subject to correctness concerns. For instance, the points-to analysis of Heintze and Tardieu [99] is unsound if the C program contains pointer arithmetic. Steensgaard showed how to conservatively incorporate pointers with offsets into a points-to analysis [175]. The analysis is sound if all pointers have offsets that lie within the bounds of the underlying memory region. This assumption allows for a simpler intermediate language that distinguishes only between assignments between two variables and assignments of expressions that may contain non-zero offsets. Thus, this simplification is based on partial correctness of the C program in that it may not exhibit certain undefined behaviour such as out-of-bounds accesses. Since the goal of our analysis is to show the absence of undefined behaviour, no simplification with respect to values of expressions is possible.

Since the values of expressions cannot be dismissed in an analysis that infers maximum bounds on variables, there seems to be little room for simplifying the source language, except for removing syntactic sugar. The advantage of using an intermediate language with the same expressiveness as C is that the translation from C to Core C is purely syntactic without any approximation that requires a correctness argument. In fact, the only concern regarding the preservation of the source code semantics is the interpretation of the C source program, a task that is delegated to the GNU C compiler.

The downside of Core C being as expressive as C is that the collecting semantics is formulated in terms of machine concepts, namely in terms of bits and bytes in memory. Due to the low-level character of the collecting semantics, the task of relating it to abstract entities such as polyhedra becomes a challenge in itself.

Despite the low-level nature of Core C, its collecting semantics only expresses an approximation of the program behaviour [59]. Firstly, the collecting semantics assumes that the input to the program can be arbitrary. In the context of verifying embedded systems, assuming that input is arbitrary is often too strong and can lead to imprecise results. A sensor, for instance, might always return a floating-point number within a given range and, in particular, it might never return infinity or "not a number" [30]. Secondly, calculating the union of memory configurations removes any information about causality between memory states. For example, given an erroneous state, the collecting semantics does not provide enough information to read off a program input that leads to this state since the actual execution trace that led to the erroneous state is part of a set of all possible execution traces. Another consequence is that the collecting semantics cannot be used to prove termination. Even in the light of these approximations, calculating the collecting semantics is infeasible except for very small systems [155]. The next chapter introduces abstract domains that can summarise states in the collecting semantics succinctly.

Part I

Abstracting Soundly

3

Abstract State Space

A typical C program, like the one presented in the introduction, can be run on arbitrary inputs. Thus, the collecting semantics cannot be calculated since the input of the program may fill up all of the 4-GB state space. Even in cases in which the program only operates on a fixed amount of memory, the number of memory configurations is usually too large to be enumerated exhaustively. Hence, an automatic prover for a certain property needs to summarise the possible memory configurations into a symbolic state. However, a summary that finitely describes the exact concrete states of a program solves the Halting Problem, which itself is undecidable [181]. Thus, we circumvent this problem by mapping a concrete state (that is, the collecting semantics at a particular program point) to an abstract state, which in turn maps back to a set of concrete states that includes the original set but may include other, spurious concrete states. Hence, an abstract state over-approximates the set of possible concrete states, a concession that makes a proof of undecidable program properties possible in many cases. The challenge in devising a static analysis is therefore to find an abstract representation of program states that is precise in order to verify as many programs as possible. In other words, a precise analysis requires an abstract representation that includes as few spurious concrete states as possible to ensure that questions like "Does the pointer access *p lie within the bounds of the pointed-to buffer?" can be answered by the abstract domain with "yes" whenever this is true in the actual C program.

This section presents two abstract domains that summarise the concrete state into a finite and tractable representation while being precise enough to verify non-trivial programs. The two abstract domains are the following:

points-to domain *Pts*: The points-to domain tracks which memory regions a given pointer possibly points to. For the question above, this domain may state that the pointer p may point to the memory regions v or w.

numeric domain *Num*: The numeric domain expresses bounds on variables and relations between variables using linear inequalities. For the question above, this domain can show that the offset (that might have been added

```
1 char* strings[] = {"Three", "string", "constants." };
2 char* s = NULL;
3 int i;
4 for (i=0; i<sizeof(strings)/sizeof(strings[0]); i++) {
5   s = strings[i];
6   printf("%s\n", s);
7 };
```

Fig. 3.1. A loop that requires arguing about numeric properties and l-value flow.

to the pointer p) is greater than or equal to zero and less than the size of v and less than the size of w, thereby providing the required "yes" answer.

Neither of these domains is new as such. The information regarding whether two pointers may reference the same program location is a prerequisite for many compiler optimisations [1]. This so-called alias analysis was later generalised to points-to analysis [74]. The second domain deployed in our analysis is the numeric domain of convex polyhedra [62]. While this domain is well known, too, it is not widely used due to its scalability problems. While scalability is addressed in Part II of this book, the focus of this part lies in how the domains interface with the value-range analysis. In particular, this chapter presents the operations on the points-to and the numeric domains.

Before presenting the points-to domain of the analysis in Sect. 3.2 and the numeric domain in Sect. 3.3, we give an introductory example in order to illustrate how an abstract domain is put to use in a value-range analysis.

3.1 An Introductory Example

The collecting semantics presented in the last chapter demonstrated how the set of all possible states of a given program can be defined. This section applies the same techniques to abstract states in order to demonstrate how an over-approximation of the collecting semantics can be inferred. To this end, consider the C fragment in Fig. 3.1. The `strings` array contains three pointers to three different strings that are to be printed in line 6. The loop initialises the index `i` to zero such that the first element of `strings` is accessed in the first loop iteration. The variable is advanced until the last entry of the array is reached, at which point the loop terminates.

In order to show how an abstract state is calculated for this program, consider its simplified flow graph in Fig. 3.2. The aim of our analysis is to characterise the values that can arise along the edges of the control-flow graph. To this end, each edge is decorated with an abstract state P, Q, R, S,

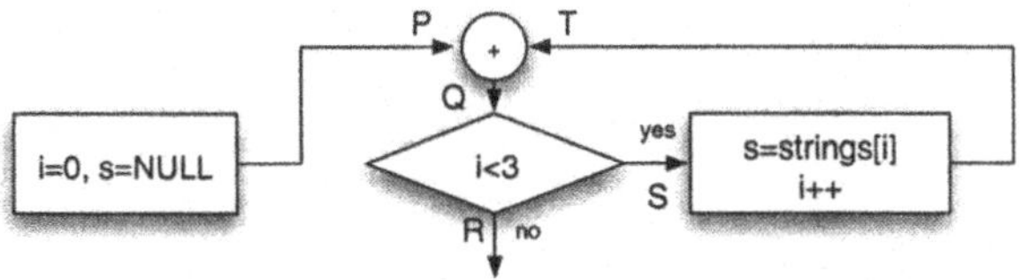

Fig. 3.2. Flow graph of the program in Fig. 3.1.

and T. For the sake of analysing the program, let P^i, Q^i, R^i, S^i, and T^i denote the values of `i` and let P^s, Q^s, R^s, S^s, and T^s represent the memory regions that the pointer `s` may point to. We shall merely describe the value of `i` as an interval and the value of `s` as a set of abstract addresses $\mathcal{A}$. In particular, we omit how to represent the array but assume that the statement **`strings[i]`** returns the address of the ith string, which shall be stored at the abstract addresses $s_i \in \mathcal{A}$ for $i \in \{0,1,2\}$. The possible values are characterised by the following equations. Here, the operator $\curlyvee$ is defined as $[l_1, u_1] \curlyvee [l_2, u_2] = [\min(l_1, l_2), \max(u_1, u_2)]$, that is, it calculates the smallest interval that includes the two given intervals. Furthermore, the special tag $\text{NULL} \in \mathcal{A}$ denotes that a pointer contains **`NULL`**.

$$\begin{aligned} P^i &= [0,0] & P^s &= \{\text{NULL}\} \\ Q^i &= P^i \curlyvee T^i & Q^s &= P^s \cup T^s \\ R^i &= Q^i \cap [3, 2^{31}-1] & R^s &= Q^s \\ S^i &= Q^i \cap [-2^{31}, 2] & S^s &= Q^s \\ T^i &= \{v+1 \mid v \in S^i\} & T^s &= \{s_1 \mid 0 \in S^i\} \cup \\ & & & \{s_2 \mid 1 \in S^i\} \cup \\ & & & \{s_3 \mid 2 \in S^i\} \end{aligned}$$

A solution for these equations can be calculated by an upward fixpoint calculation using Jacobi iteration [58]. In the example, this iteration strategy is applied by defining an initial iterate $\langle P_0, Q_0, R_0, S_0, T_0 \rangle$ and calculating $\langle P_{j+1}, Q_{j+1}, R_{j+1}, S_{j+1}, T_{j+1} \rangle$ from the equations above with P_j, Q_j, R_j, S_j, T_j substituting the corresponding variables on the right side. Here, each tuple element X_j is itself composed of the corresponding interval X_j^i and the points-to set X_j^s. Figure 3.3 shows the actual calculation of the fixpoint. The initial state is $P_0^i = Q_0^i = R_0^i = S_0^i = T_0^i = \bot$, where $\bot$ denotes the empty interval and $P_0^s = Q_0^s = R_0^s = S_0^s = T_0^s = \emptyset$. Stability of the loop is attained in iteration 13 since $S_{13}^i \subseteq S_{12}^i$ and $S_{13}^s \subseteq S_{12}^s$. Given this simple example, we can infer the domain operations that are required for a static analysis. The control flow of the program is reflected in the semantic equations, in particular, a join of two control-flow paths is represented by a join operation in the equation. These join operations depend on the underlying

j	P^i_j ,	P^s_j	Q^i_j ,	Q^s_j	R^i_j ,	R^s_j	S^i_j ,	S^s_j	T^i_j ,	T^s_j
0	$\bot$,	$\emptyset$	$\bot$,	$\emptyset$	$\bot$,	$\emptyset$	$\bot$,	$\emptyset$	$\bot$,	$\emptyset$
1	$\underline{[0,0]}$,	$\underline{\{\text{NULL}\}}$	$\bot$,	$\emptyset$	$\bot$,	$\emptyset$	$\bot$,	$\emptyset$	$\bot$,	$\emptyset$
2	$[0,0]$,	$\{\text{NULL}\}$	$\underline{[0,0]}$,	$\underline{\{\text{NULL}\}}$	$\bot$,	$\emptyset$	$\bot$,	$\emptyset$	$\bot$,	$\emptyset$
3	$[0,0]$,	$\{\text{NULL}\}$	$[0,0]$,	$\{\text{NULL}\}$	$\bot$,	$\underline{\{\text{NULL}\}}$	$\underline{[0,0]}$,	$\underline{\{\text{NULL}\}}$	$\bot$,	$\emptyset$
4	$[0,0]$,	$\{\text{NULL}\}$	$[0,0]$,	$\{\text{NULL}\}$	$\bot$,	$\{\text{NULL}\}$	$[0,0]$,	$\{\text{NULL}\}$	$\underline{[1,1]}$,	$\underline{\{s_1\}}$
5	$[0,0]$,	$\{\text{NULL}\}$	$\underline{[0,1]}$,	$\underline{\{\text{NULL}, s_1\}}$	$\bot$,	$\{\text{NULL}\}$	$[0,0]$,	$\{\text{NULL}\}$	$[1,1]$,	$\{s_1\}$
6	$[0,0]$,	$\{\text{NULL}\}$	$[0,1]$,	$\{\text{NULL}, s_1\}$	$\bot$,	$\underline{\{\text{NULL}, s_1\}}$	$\underline{[0,1]}$,	$\underline{\{\text{NULL}, s_1\}}$	$[1,1]$,	$\{s_1\}$
7	$[0,0]$,	$\{\text{NULL}\}$	$[0,1]$,	$\{\text{NULL}, s_1\}$	$\bot$,	$\{\text{NULL}, s_1\}$	$[0,1]$,	$\{\text{NULL}, s_1\}$	$\underline{[1,2]}$,	$\underline{\{s_1, s_2\}}$
8	$[0,0]$,	$\{\text{NULL}\}$	$\underline{[0,2]}$,	$\underline{\{\text{NULL}, s_1, s_2\}}$	$\bot$,	$\underline{\{\text{NULL}, s_1, s_2\}}$	$[0,1]$,	$\{\text{NULL}, s_1\}$	$[1,2]$,	$\{s_1, s_2\}$
9	$[0,0]$,	$\{\text{NULL}\}$	$[0,2]$,	$\{\text{NULL}, s_1, s_2\}$	$\bot$,	$\{\text{NULL}, s_1, s_2\}$	$\underline{[0,2]}$,	$\underline{\{\text{NULL}, s_1, s_2\}}$	$[1,2]$,	$\{s_1, s_2\}$
10	$[0,0]$,	$\{\text{NULL}\}$	$[0,2]$,	$\{\text{NULL}, s_1, s_2\}$	$\bot$,	$\{\text{NULL}, s_1, s_2\}$	$[0,2]$,	$\{\text{NULL}, s_1, s_2\}$	$\underline{[1,3]}$,	$\underline{\{s_1, s_2, s_3\}}$
11	$[0,0]$,	$\{\text{NULL}\}$	$\underline{[0,3]}$,	$\underline{\{\text{NULL}, s_1, s_2, s_3\}}$	$\bot$,	$\underline{\{\text{NULL}, s_1, s_2, s_3\}}$	$[0,2]$,	$\{\text{NULL}, s_1, s_2\}$	$[1,3]$,	$\{s_1, s_2, s_3\}$
12	$[0,0]$,	$\{\text{NULL}\}$	$[0,3]$,	$\{\text{NULL}, s_1, s_2, s_3\}$	$\underline{[3,3]}$,	$\underline{\{\text{NULL}, s_1, s_2, s_3\}}$	$[0,2]$,	$\underline{\{\text{NULL}, s_1, s_2, s_3\}}$	$[1,3]$,	$\{s_1, s_2, s_3\}$
13	$[0,0]$,	$\{\text{NULL}\}$	$[0,3]$,	$\{\text{NULL}, s_1, s_2, s_3\}$	$[3,3]$,	$\{\text{NULL}, s_1, s_2, s_3\}$	$[0,2]$,	$\{\text{NULL}, s_1, s_2, s_3\}$	$[1,3]$,	$\{s_1, s_2, s_3\}$

Fig. 3.3. Fixpoint calculation using Jacobi iteration. States that have changed are underlined.

domain: While it is simple set union for the l-values of x_s, it is the more complex operation $\curlyvee$ for the interval domain. Similarly, a split in the control flow by a conditional is reflected with an intersection operation in the semantic equations. This operation is commonly called a meet operation. In the example, the conditional expresses a property from which only an intersection of the numeric domain can be deduced. A third operation, which tests for inclusion, is required in order to detect stability. In general, the domains and their operations form a lattice. In the context of our analysis, these are $\langle Num, \sqsubseteq_N, \sqcup_N, \sqcap_N \rangle$ and $\langle Pts, \sqsubseteq_A, \sqcup_A, \sqcap_A \rangle$, where inclusion $\sqsubseteq$ tests for stability, $\sqcup$ is the join, and $\sqcap$ is the meet operation. In addition, the symbol $\bot$ (bottom) is used to denote the empty set of states, which implies that the corresponding program point is unreachable. Using two domains simultaneously gives rise to domain interaction; that is, the ability to propagate information from one domain to the other. For instance, the points-to sets R_3^s up to R_{11}^s are meaningless since the code is unreachable as the numeric states R_3^i up to R_{11}^i are $\bot$. Thus, the values of R_3^s up to R_{11}^s can be refined to $\emptyset$. A more subtle interaction between domains is presented in Sect. 3.3.4 in the context of the numeric domain *Num*. Before we embark on the latter, we explore the design space of the points-to domain *Pts* and formally define its operations.

3.2 Points-to Analysis

A points-to analysis infers a set of l-values that a given pointer may contain at run-time. In the context of our analysis, the term l-value is used to indicate an abstract (i.e. symbolic) address of a variable or memory region rather than merely an expression that can be used on the left-hand side of an assignment. A points-to analysis has traditionally been used in optimising compilers to infer that a pointer cannot point to a certain variable. This information makes it possible to store the variable in a CPU register while the pointer is accessed. In the context of our analysis, it simply serves to infer which memory regions are accessed when reading or writing through a pointer. An idiosyncrasy of the C programming language is that pointers might have an offset added to them, which is a common way to access arrays. Even without considering pointer offsets, a points-to analysis is necessarily approximate [39]. In order to illustrate the difficulties of determining an exact points-to relation, consider the code fragment in the left column of Fig. 3.4.

In this example, the pointer variable p contains the value zero after line 2 is executed, as the constant `NULL` in C is defined as `(void*) 0`. The **`if`**-statement in the third line will always replace the content of p with the address of either a or b, depending on the return value of the random number function `rand()`. A standard points-to analysis, which ignores the numeric values of program variables, can analyse the above program only approximately; that is, it has to assume that `*p` writes to either a or b. In order to express approximate

```
int a, b, *p;
p=NULL;
if (rand()) p=&a;
       else p=&b;
*p=42;
```

```
1 int a, b, *p;
2 p=NULL;
3 if (rand()) p=&a;
4 if (p) {
5   *p=42;
6 }
```

Fig. 3.4. Tracking NULL values in pointers.

points-to relationships, a points-to domain must be able to infer a set of possibly pointed-to memory regions. These points-to relationships are also called may-aliases, as an element a in the points-to set of p indicates that p may point to a but does not necessarily do so. Worse, may points-to analyses implicitly assume that a variable can be NULL, and hence a singleton points-to set also merely implies a may-alias relation. This is a severe limitation in the context of verification, as illustrated by a variation of the example shown on the right of Fig. 3.4. In this case, the points-to set of p after the execution of line 3 contains merely a, indicating that p may or may not point to a. In particular, if NULL is not explicitly represented in the points-to set of p, then it cannot be shown that *p=42 is legal; i.e., that the statement does not dereference NULL. The reason why a possible NULL value of a pointer is not tracked by scalable points-to analyses is that they are usually flow-insensitive; that is, the inferred points-to set is valid at all points in the program. Yet, when a pointer variable comes into scope, it initially contains no l-values. This fact must be retained by an analysis that infers points-to sets that are valid at all program points. An easy solution is to assume that any pointer can always contain NULL. Note, however, that a NULL pointer may still have an offset, which is a pure value. Neglecting such an offset renders an analysis unsound [142].

However, the idiom of testing a pointer for NULL is a common one in C programs, and proving the absence of NULL-pointer dereferences is a requirement to show correct memory management. Thus, to prove the example above correct, an analysis is needed in which the points-to set of a variable can be replaced (by assignment) and restricted (by conditionals), resulting in an analysis that infers points-to sets on a per-statement granularity rather than as a global property of a pointer variable. Such an analysis is commonly referred to as a flow-sensitive points-to analysis [46, 74]. Given that elements can be removed from a points-to set, we introduce a special address NULL, which can be stored in points-to sets to denote that a variable contains NULL rather than an address. Thus, in the example above, the points-to set of p at the end of line 3 consists of a and the tag NULL. The latter is removed by the **if**-statement in line 4. Using this tag, it is possible to infer not only that p is not NULL, but also that the access through p definitely writes to a, as it is now the only possible value of p.

While the precision of flow-sensitive points-to analyses is superior to that of flow-insensitive analyses, they incur a higher cost, especially in memory usage, since each program point may feature a different points-to set. However, a flow-sensitive analysis requires a fixpoint calculation that is similar to that of the numeric domain presented below, whereas a flow-insensitive analysis requires a closure or unification [100] to calculate a solution to a set of flow equations. To illustrate this, Fig. 3.5 shows a sequence of statements and the inferred points-to sets.

statement	flow-sensitive	flow-insensitive
`x=&u;`	$y \mapsto \{\text{NULL}\}, x \mapsto \{u\}$	$y \mapsto \emptyset, x \mapsto \{u\}$
`y=x;`	$y \mapsto \{u\}, x \mapsto \{u\}$	$y \mapsto \{u\}, x \mapsto \{u\}$
`x=&v;`	$y \mapsto \{u\}, x \mapsto \{v\}$	$y \mapsto \{u, v\}, x \mapsto \{u, v\}$

Fig. 3.5. Flow-sensitive versus flow-insensitive points-to analysis.

Both flow-sensitive and flow-insensitive algorithms initially create a points-to set that indicates that y does not point to any variable but may be NULL. Note that flow-insensitive points-to sets implicitly include NULL. Due to the second statement, the flow-sensitive analysis replaces the points-to set of y with that of x, which contains u due to the first statement. In contrast, the flow-insensitive analysis creates a flow relation $y \leftarrow x$, implying that all addresses that are stored in x can also flow to y. A closure algorithm is required to actually propagate the addresses from x to y. Running such a closure after the second statement yields $y \mapsto \{u\}$; however, rerunning the closure after executing the third statement propagates all possible addresses that x may contain (namely those of u and v) to y, whereas the points-to set of y in the flow-sensitive analysis remains unchanged. The superior efficiency of a closure-based points-to analysis lies in the ability to completely collect flows between variables and to apply the closure on the full set of constraints. This advantage is lost when flow-insensitive points-to analysis is combined with a separate numeric analysis to infer offsets of pointers. To illustrate this, reconsider the code fragment in Fig. 3.1. Assume again that the three strings are stored at addresses denoted by the l-values s_1, s_2, and s_3, respectively. A numeric analysis will infer a range of values for i by iteratively analysing the loop. In particular, during the first iteration, i is zero and a flow $s \leftarrow s_1$ will be created; in the second iteration i is in $[0, 1]$ and the flow $s \leftarrow s_2$ is added; and finally $s \leftarrow s_3$ is added when $i \in [0 \ldots 2]$. In order to infer which strings can be passed to `printf` in line 5, the last iteration of the loop must be analysed with an up-to-date points-to set to ensure that the points-to set of s contains the l-values of all strings. However, it is difficult to anticipate which iteration will lead to a fixpoint, and thus the closure algorithm must be rerun during each loop iteration to ensure that a correct points-to set is used in the last loop iteration. Alas, calculating the closure on the whole points-to domain is expensive to the extent that the cost becomes prohibitive for a practical analysis. Our analysis therefore features a flow-sensitive points-to analysis, which turns out to fit much better with a numeric analysis. We now introduce the notation necessary to manipulate points-to sets for a flow-sensitive analysis.

3.2.1 The Points-to Abstract Domain

In this section, we formally define the points-to domain *Pts*. To this end, let $\mathcal{X}$ denote the finite set of abstract variables that represent the content of program variables and let $\mathcal{A}$ denote the finite set of abstract addresses, also called l-values. Let $Pts = \mathcal{X} \to \mathcal{P}(\mathcal{A})$ denote the set of points-to maps, where each $A \in Pts$ maps an abstract variable $x \in \mathcal{X}$ to a set of symbolic addresses $A(x) \subseteq \mathcal{A}$. Updating $A \in Pts$ to $A' = A[x \mapsto a]$, where $a \subseteq \mathcal{A}$, results in $A'(x) = a$ and $A'(y) = A(y)$ for all $y \neq x$. The following operations on *Pts* are defined: For any $A_1, A_2 \in Pts$, define $A' = A_1 \sqcup_A A_2$ such that $A'(x) = A_1(x) \cup A_2(x)$ for all $x \in \mathcal{X}$. Let $A_1 \sqsubseteq_A A_2$ iff $A_1(x) \subseteq A_2(x)$ for all $x \in \mathcal{X}$. Note that no meet operator is defined on *Pts*, as a refinement of a given domain $A \in Pts$ can be done by using the update notation. Since *Pts* is finite, the structure $\langle Pts, \sqsubseteq_A \rangle$ is a complete partial order; that is, for any family of sets $\mathbb{A} \subseteq Pts$, there exists $A \in Pts$ such that $A = \bigsqcup_A \mathbb{A}$. Thus, the solution of a set of semantic equations exists and can be calculated in a finite number of steps using a standard upward fixpoint calculation (Kleene iteration).

Let $\text{NULL} \in \mathcal{A}$ be a tag that is distinct from all other abstract addresses. If a variable $x \in \mathcal{X}$ can contain a value (in contrast to an address), then $\text{NULL} \in A(x)$. Define the projection operator $\exists_X : Pts \to Pts$, which resets the points-to set of each variable $x \in X$; specifically, if $A' = \exists_X(A)$, then $A'(x) = \{\text{NULL}\}$ and $A'(y) = A(y)$ for all $y \notin X$. Furthermore, let $\mathcal{A}^{\mathcal{F}} \subseteq \mathcal{A}$ denote a set of function addresses such that, for all functions f in the Core C program, there exists exactly one $a_f \in \mathcal{A}^{\mathcal{F}}$. A points-to domain $A \in Pts$ is called unsatisfiable if $A(x) = \emptyset$ for some $x \in \mathcal{X}$. An unsatisfiable domain A at a certain location in the program implies that this location is unreachable.

Implementation of the Points-to Domain

The points-to map $A \in Pts$ described is readily implementable as a tree or hash table. An implementation using balanced binary trees, for example, has the advantage that the join $\sqcup_A$ and entailment $\sqsubseteq_A$ operations can be implemented efficiently in the average case: Rather than recursively joining or comparing two trees, both operations can stop short whenever two nodes reside at the same physical address since their content is then necessarily identical [30, Sect. 6.2]. Our implementation follows a different route, in which the join and inclusion operations of the points-to domain are incorporated into the numeric domain. This approach is described in detail in Chap. 13. For simplicity of presentation, the following chapters are cast in terms of the points-to domain above.

3.2.2 Related Work

Points-to analysis has received a vast amount of attention in the last two decades. Analyses can be classified by three criteria that describe the sensitivity towards certain flows of l-values. These criteria are flow sensitivity (the ability to differentiate between different branches in a program), context sensitivity (the ability to differentiate between different call sites of a function), and field sensitivity (the ability to differentiate between different fields in a structure). One strand in the literature focuses on improving precision by applying flow-sensitive analysis [46, 74] and improved analysis of recursive data structures [42, 43, 70] up to the combination with numeric analysis for improved shape analysis [80], an approach also taken in Chap. 13. Another direction for improving precision is represented by flow- and context-sensitive analyses [43, 118]. A rather different approach is represented by the aim of scaling up points-to analysis to very large programs using flow-insensitive analyses based on fast union-find algorithms [176]. These unification-based analyses cannot distinguish between the two assignments `x=y` and `y=x`, which leads to a loss of precision [100] that can partly be recovered by adding some directionality at an additional cost [65]. Flow-insensitive analyses that fully obey the directionality of assignments were pioneered by Andersen [3]. This approach formulates the l-value flow between variables as subset relationships. Algorithms that solve these constraint systems run in $O(n^3)$, which is unacceptable for large programs. However, intelligent removal of chains and cycles can put subset-based approaches on a par with unification-based algorithms [99]. A major challenge in analysing C is the presence of pointer arithmetic and structures. While some initial analyses were unsound with respect to pointer arithmetic [99,176], correctness can be achieved by combining a standard points-to analysis with a simple value-range analysis [144,175]. The value-range analysis infers that a pointer variable has either a constant offset or any offset. In the first case a flow of l-values is only generated from or to the field at the inferred offset. If the offset is not constant, an l-value flow from or to all fields in the structure or array is created. This approach assumes that the underlying C program never accesses a memory region out-of-bounds. As such, an unknown pointer offset can only access fields within the same structure.

Furthermore, efficient context-sensitive points-to analysis of C is often jeopardised by the effect of disparate aliasing at call sites, which requires the partial inlining of functions [139]. Interestingly, context-sensitive analysis in some circumstances only marginally improves precision [66], whereas treating calls to memory allocation functions such as `malloc` in a context-sensitive way can considerably improve precision [140]. Furthermore, while context-sensitive analysis is considered to be expensive, BDD-based implementations exist that perform fully context-sensitive analyses remarkably efficiently [188], even though BDD-based approaches are very sensitive to variable ordering [26].

3.3 Numeric Domains

Besides inferring possible l-values for each pointer variable, it is important to know what offset a given pointer has whenever it is dereferenced in order to ensure that the access is within bounds of the underlying memory region. Since any computation in a program can feed into the calculation of a pointer offset, it is necessary to infer values for all variables in the program. To this end, this section presents the numeric domain *Num* of our analysis, which is, in fact, the combination of two domains, namely a polyhedral domain, *Poly*, as proposed by Cousot and Halbwachs [62], and a congruence domain, *Mult*, proposed by Granger [85]. While *Poly* allows the analysis to infer that an access to a memory region is within bounds, the *Mult* domain can guarantee that an access to an array is aligned to element boundaries. We introduce each domain in turn and discuss their combination afterwards.

3.3.1 The Domain of Convex Polyhedra

Our analysis expresses numeric constraints over the set of abstract variables $\mathcal{X}$, the same set over which the points-to domain in the last section was defined. For the sake of this section, let $\mathbf{x}$ denote the vector of all variables in $\mathcal{X}$, thereby imposing an order on $\mathcal{X}$. Let $Lin^{\mathbb{R}}$ denote the set of linear expressions of the form $\mathbf{a} \cdot \mathbf{x}$, where $\mathbf{a} \in \mathbb{R}^n$, and let $Ineq^{\mathbb{R}}$ denote the set of linear inequalities $\mathbf{a} \cdot \mathbf{x} \leq c$, where $c \in \mathbb{R}$. For simplicity, let, for example, $6x_3 \leq x_1 + 5$ abbreviate $\langle -1, 0, 6, 0, \ldots 0 \rangle \cdot \mathbf{x} \leq 5$ and $x_2 = 7$ abbreviate the two opposing inequalities $x_2 \leq 7$ and $x_2 \geq 7$, the latter being an abbreviation of $-x_2 \leq -7$. As the analysis only infers integral properties, the notation $e_1 < e_2$ is used to abbreviate $e_1 \leq e_2 - 1$. Each inequality $\mathbf{a} \cdot \mathbf{x} \leq c \in Ineq^{\mathbb{R}}$ induces a half-space $[\![\mathbf{a} \cdot \mathbf{x} \leq c]\!] = \{\mathbf{x} \in \mathbb{R}^{|\mathcal{X}|} \mid \mathbf{a} \cdot \mathbf{x} \leq c\}$. A set of inequalities $I \subseteq Ineq^{\mathbb{R}}$ induces a closed, convex space $[\![I]\!] = \bigcap_{\iota \in I} [\![\iota]\!]$. Let $\mathcal{S} = \{[\![I]\!] \mid I \subseteq Ineq^{\mathbb{R}}\}$ denote the set of all convex spaces and $S = S_1 \curlyvee S_2$ denote the topological closure of the convex hull of $S_1, S_2 \in \mathcal{S}$; that is, the smallest closed, convex space S such that $S_1 \subseteq S$ and $S_2 \subseteq S$. Together with inclusion $\subseteq$ and intersection $\cap$, $\mathcal{S}$ forms a complete lattice $\langle \mathcal{S}, \subseteq, \curlyvee, \cap \rangle$. Thus, the solution of a set of semantic equations such as those presented in Sect. 3.1 exists and can be formulated as a fixpoint. However, such a fixpoint cannot always be calculated in finite time for reasons given in the following two paragraphs.

Unlimited Number of Inequalities

There exist convex sets $S \in \mathcal{S}$ such that $|I| \notin \mathbb{N}$ for all $I \subseteq \mathit{Ineq}$ with $[\![I]\!] = S$. However, any computer implementation that stores inequality sets can only represent convex spaces that are generated by the intersection of a finite number of inequalities, known as polyhedra. However, if S_1 is a triangle, S_2 a hexagon, S_3 a dodecagon, and so forth such that the vertices of S_i are contained within those of S_{i+1}, then $S_1 \subseteq S_2 \subseteq S_3 \ldots$ is an ascending chain that converges onto a disc. While every S_i is a polyhedron, $\bigcup_i S_i$ is not since a disc cannot be represented by a finite set of inequalities. Hence, the lattice of polyhedra is incomplete; that is, a fixpoint calculation can converge onto a convex space that is not a polyhedron.

Unlimited Growth of Coefficients

While $\mathit{Lin}^{\mathbb{R}}$ and $\mathit{Ineq}^{\mathbb{R}}$ are defined over $\mathbb{R}$, computer-representable elements of these sets must be confined to finite elements of $\mathbb{R}$. Consider the sequence $x_i \in \mathbb{Q}$, which is defined such that $x_0 = 1$, $x_{n+1} = (x_n + 2/x_n)/2$. The values $x_0, \ldots x_j$ are included in the convex spaces $S_j = [\![\{1 \leq x \leq x_j\}]\!]$. The sequence $S_0 \subseteq S_1 \subseteq \ldots$ is an ascending chain that converges onto $[\![\{1 \leq x \leq \sqrt{2}\}]\!]$. Thus, a fixpoint computation may create inequalities that can, in general, contain coefficients and constants of infinite size. Restricting coefficients and constants to rational numbers again leads to an incomplete domain.

Curtailing Infinite Growth

The above-mentioned sources of infinite ascending chains can be overcome by using an acceleration technique called widening [56]. This technique ensures that an infinite sequence of convex state spaces $S_0 \subseteq S_1 \subseteq \ldots$ is eventually stable by removing inequalities that are new or whose coefficients have changed with respect to the previous state. Removing new inequalities limits the total number of inequalities, while removing inequalities whose coefficients change prevents unlimited growth of coefficients. When widening is applied to increasing chains of convex states, the resulting state is in the set of convex states that can be finitely described. This refined subset of $\mathcal{S}$ can be characterised as follows: Let *Lin* denote the set of linear expressions whose coefficients are drawn from $\mathbb{Z}$, and let *Ineq* denote the inequalities constructed from *Lin* and constants drawn from $\mathbb{Z}$. Allowing only a finite number of inequalities only admits convex spaces that range over $\mathbb{Q}$ such that each $\mathbf{a}\cdot\mathbf{x} \leq c \in \mathit{Ineq}$ induces a half-space $[\![\mathbf{a} \cdot \mathbf{x} \leq c]\!] = \{\mathbf{x} \in \mathbb{Q}^{|\mathcal{X}|} \mid \mathbf{a} \cdot \mathbf{x} \leq c\}$. By applying widening, the attainable convex spaces are defined by $\mathit{Poly} = \{[\![I]\!] \mid I \in \mathit{Ineq} \wedge |I| \in \mathbb{N}\}$, namely the set of (finitely generated) convex polyhedra. In order to guarantee that a fixpoint calculation only converges onto elements in *Poly* rather than $\mathcal{S} \supset \mathit{Poly}$, a widening operator $\nabla : \mathit{Poly} \times \mathit{Poly} \to \mathit{Poly}$ with the following properties is required [59, 62]:

1. $\forall P, Q \in \mathit{Poly} \,.\, P \subseteq P \nabla Q$
2. $\forall P, Q \in \mathit{Poly} \,.\, Q \subseteq P \nabla Q$
3. For all increasing chains $P_0 \subseteq P_1 \subseteq \ldots$, the increasing chain defined by $R_0 = P_0$ and $R_{i+1} = R_i \nabla P_{i+1}$ is ultimately stable.[1]

In order to reflect the restriction of *Poly* to finitely representable state spaces in the operations on *Poly*, we use $\sqcup_P$ for $\curlyvee$, and furthermore $\sqcap_P$ for $\cap$ and $\sqsubseteq_P$ for $\subseteq$, which make up the lattice of convex polyhedra $\langle \mathit{Poly}, \sqsubseteq_P, \sqcup_P, \sqcap_P \rangle$. This lattice is incomplete, as neither the join nor meet of an arbitrary number of polyhedra is necessarily a polyhedron.

The use of an incomplete lattice together with a widening operator has an effect on the quality of the attainable fixpoints. A stable polyhedron obtained in an upward iteration with widening is in general a post-fixpoint – that is, a polyhedron that is larger than the actual fixpoint. This is obvious from the example above, where $[\![\{1 \leq x \leq \sqrt{2}\}]\!]$ is the actual fixpoint; however $[\![\{1 \leq x \leq \sqrt{2}\}]\!] \notin \mathit{Poly}$. Post-fixpoints for this example are $[\![\{1 \leq x \leq 2\}]\!]$ or even $[\![\{1 \leq x\}]\!]$, or any other polyhedron that entails the actual fixpoint.

Widening ensures that infinitely ascending chains in the domain cannot impede an analysis from reaching a fixpoint. In particular, it guarantees that the unlimited growth that may occur when analysing loops is curtailed. Suppose the induction variable x is initialised to zero and incremented thereafter for each loop iteration. The corresponding sequence of states $S_0 = [\![\{x = 0\}]\!]$, $S_1 = [\![\{0 \leq x \leq 1\}]\!]$, $S_2 = [\![\{0 \leq x \leq 2\}]\!]$, etc., exhibits an unlimited growth of the constant that bounds x from above. Widening will remove the changing bound on x, yielding $S_\infty = [\![\{0 \leq x\}]\!]$ as a post-fixpoint. However, even when infinite ascending chains are over-approximated by widening, the coefficients may still grow beyond a manageable size as a result of applying the $\sqcup_P$ operation. One possible approach for dealing with inequalities with very large coefficients and constants is to merely discard them during the analysis [169], thereby incurring a precision loss that is difficult to understand and anticipate when interpreting the results of an analysis. A more semantic approach is to observe that all variables of interest are in fact integral, which makes it possible to restrict the domain of *Poly* further to the set of convex spaces over $\mathbb{Z}$. These so-called $\mathbb{Z}$-polyhedra can be characterised by the fact that all their vertices (that is, all extreme points of the convex space) have integral coordinates. As a consequence, the inequalities that define $\mathbb{Z}$-polyhedra have coefficients whose size is bounded by the coordinates of the vertices they connect. In particular, by equating polyhedra that contain the same set of integral points, it is possible to define a lattice of $\mathbb{Z}$-polyhedra $\langle \mathit{Poly}_{\equiv\mathbb{Z}}, \sqsubseteq_P^{\mathbb{Z}}, \sqcup_P^{\mathbb{Z}}, \sqcap_P^{\mathbb{Z}} \rangle$. If each equivalence class is represented by its smallest polyhedron, it is possible to set $\sqsubseteq_P^{\mathbb{Z}} = \sqsubseteq_P$ and $\sqcup_P^{\mathbb{Z}} = \sqcup_P$. However, the meet

[1] This definition deviates from the original definition by stating that the chain should be "ultimately stable" rather than "not strictly increasing". We felt that the latter could be misread: Rather than $\exists i \,.\, \bigsqcup_{j \in \mathbb{N}} R_j \subseteq R_i$, it could be interpreted as $\exists i \,.\, R_{i+1} \subseteq R_i$, which would not guarantee convergence.

operation $\sqcap_P$ is not closed for $\mathbb{Z}$-polyhedra. In order to illustrate this, consider Fig. 3.6. The state space $P \in Poly_{\equiv\mathbb{Z}}$ over x_1, x_2 in the first graph is transformed by evaluating the conditional $x_2 \neq 5$, which is implemented by calculating $P' = (P \sqcap_P [\![x \leq 4]\!]) \sqcup_P (P \sqcap_P [\![x \geq 6]\!])$. Observe that the input P as well as the two half-spaces $[\![x \leq 4]\!]$ and $[\![x \geq 6]\!]$ are $\mathbb{Z}$-polyhedra. The second graph shows the two intermediate results $P_1 = P \sqcap_P [\![x \leq 4]\!]$ and $P_2 = P \sqcap_P [\![x \geq 6]\!]$, both of which have two non-integral vertices. As a consequence, the join of P_1 and P_2, shown as the third graph, has non-integral vertices as well and is therefore not a $\mathbb{Z}$-polyhedron. However, if the intermediate results were shrunk around the contained integral point sets (e.g., by applying the techniques presented in Chap. 9) as done in the fourth graph, all vertices of the intermediate results would be integral and the join would be a $\mathbb{Z}$-polyhedron, too. However, for general, n-dimensional polyhedra, the number of inequalities necessary to represent a $\mathbb{Z}$-polyhedron can grow exponentially with respect to a polyhedron over $\mathbb{Q}$ that contains the same integral points [157, Chap. 23]. Thus, no efficient algorithm exists to implement the $\sqcap_P$-operation on $\mathbb{Z}$-polyhedra, and hence, for the remainder of this book, the *Poly* domain over $\mathbb{Q}$ will be used to approximate $\mathbb{Z}$-polyhedra. However, in order to limit the growth of coefficients, Chap. 9 will present efficient techniques to approximate the $\sqcap_P^{\mathbb{Z}}$-operation. In anticipation of this chapter, we define all operations that query the value of a polyhedron to return integral bounds.

Given the basic lattice operations on polyhedra, the next section introduces useful operations that allow for a more concise manipulation of polyhedra, thereby providing the foundations for the abstract semantics presented later.

3.3.2 Operations on Polyhedra

In contrast to the points-to domain, the manipulation of polyhedra can be more intricate. Assigning a value to a variable x is easily accomplished whenever the variable does not occur in the inequality set describing a polyhedron $P \in Poly$. In this case, the polyhedron $P \sqcap_P [\![\{x = 42\}]\!]$ corresponds to P except that the value of x is 42. If x is already constrained within the polyhedron, $P \sqcap_P [\![\{x = 42\}]\!]$ implements the semantics of an **if**-statement that tests if x is 42. In order to update x to a new value, the old value of x has to be discarded first. To this end, define the family of projection operators $\exists_x : Poly \to Poly$ such that $\exists_{x_i}(P) = \{\langle x_1, \ldots, x_{i-1}, x, x_{i+1}, \ldots x_n\rangle \mid \langle x_1, \ldots x_n\rangle \in P, x \in \mathbb{R}\}$. Intuitively, $\exists_x(P)$ removes any information pertaining to x from the polyhedron $P \in Poly$. Thus, an assignment `x=42` can be implemented as $\exists_x(P) \sqcap_P [\![\{x = 42\}]\!]$. Observe that an update of the form `x=x+1` needs special treatment since $P \sqcap_P [\![\{x = x + 1\}]\!]$ represents the unsatisfiable (that is, empty) polyhedron. Thus, in cases where the variable that has to be updated appears in the right-hand side expression, we assign the value to an intermediate variable $t \in \mathcal{X}^T$ and afterwards assign t to x. Here,

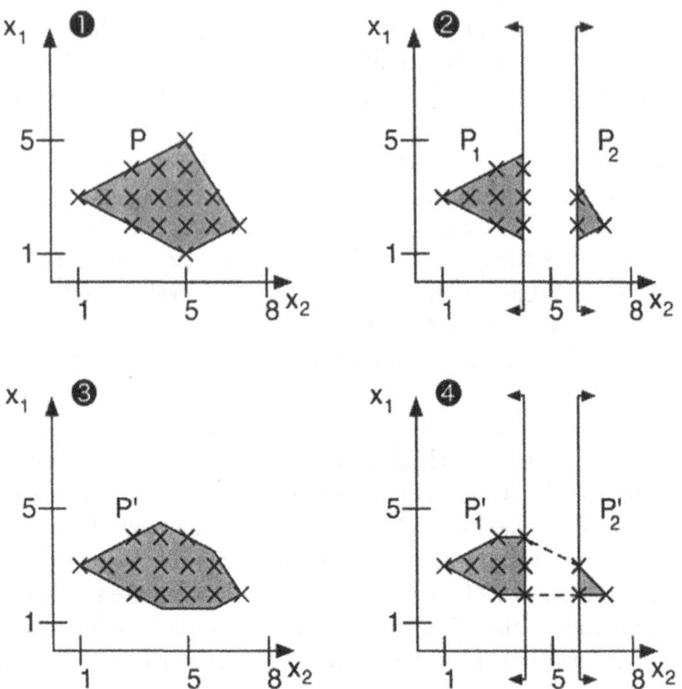

Fig. 3.6. Example to show that $\mathbb{Z}$-polyhedra are not closed under intersection. Integral points are shown as crosses and the half-spaces $[\![x \leq 4]\!]$ and $[\![x \geq 6]\!]$ are indicated by vertical lines with arrows pointing towards the feasible space.

$\mathcal{X}^T \subseteq \mathcal{X}$ is a dedicated set of temporary abstract domain variables that do not correspond to any program variable. Thus, any assignment x=e, where e is a linear expression, can be modelled as $\exists_t([\![\{x = t\}]\!] \sqcap_P \exists_x(P \sqcap_P [\![\{t = e\}]\!]))$ whenever t is unrestricted in P. Note that an assignment x=e where x occurs in e is invertible [62] and can be implemented by an affine transformation of the polyhedron [14]. In practice, however, a uniform implementation of assignment as shown above can be implemented as efficiently; see Chap. 8 and [169]. Since assignment is a reoccurring concept, we write $P \triangleright x := e$ as a convenient abbreviation of the update described above.

While the notation $P \triangleright x := e$ is a concise way to specify most assignment operations, it is not sufficient to express division and right shifts. To remedy this, let $P \triangleright x := y \gg n$, $n \in \mathbb{N}$ denote a right shift of n bits; that is, it updates x such that P contains integral solutions of $x = \lfloor y/2^n \rfloor$. Linear relations that satisfy this equation are $2^n x = y - d$ where $d \in \{0, \ldots 2^n - 1\}$. Hence define $P \triangleright x := y \gg n$ as $\exists_t([\![\{x = t\}]\!] \sqcap_P \exists_x(P \sqcap_P [\![\{y - (2^n - 1) \leq 2^n t \leq y\}]\!]))$.

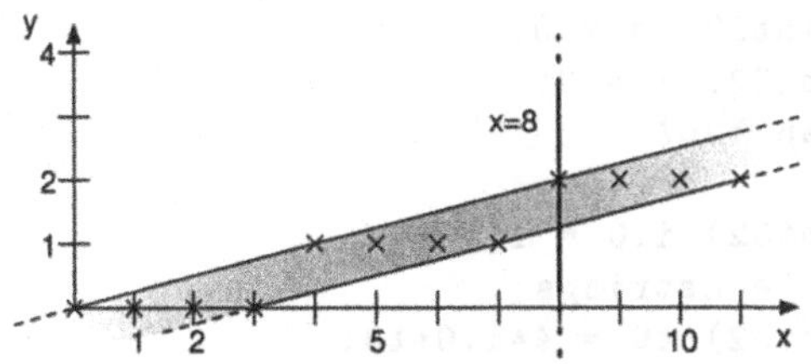

Fig. 3.7. State space for shifting y two bits right.

In order to illustrate this operation, let $P' = P \triangleright y := x \gg 2$. The resulting state space satisfies $P' \sqsubseteq_P [\![\{x - 3 \leq 4y \leq x\}]\!]$. Figure 3.7 depicts P' for the case where x was unrestricted in P. Crosses in the diagram denote integral points in P'. The vertical line depicts the feasible space for $P' \sqcap_N [\![\{x = 8\}]\!]$, which implies $1.25 \leq y \leq 2$, and hence, because x approximates an integral variable, $y = 2 = 8 \gg 2$ follows. Observe that $x \gg n$ does not correspond to $x/2^n$ in C since the latter truncates digits of the absolute value, which corresponds to rounding to zero, whereas the former truncates bits in the two's-complement representation, which corresponds to rounding towards $-\infty$. Note though that integer division by a constant can be defined analogously to the right-shift operator for positive divisors and by relaxing the upper bound to model rounding upwards for negative divisors [165].

In order to find the minimum value of an expression $\mathbf{a} \cdot \mathbf{x} \in Lin$ such that $\mathbf{x} \in P$, we introduce the operation $minExp : Lin \times Poly \to (\mathbb{Z} \cup \{-\infty\})$. To this end, let $C = \{c \in \mathbb{Z} \mid P \sqcap_P [\![\{\mathbf{a} \cdot \mathbf{x} \leq c\}]\!] \neq \emptyset\}$ (that is, C contains all constants c such that the half-space defined by $\mathbf{a} \cdot \mathbf{x} \leq c$ has a non-empty intersection with P) and let

$$minExp(\mathbf{a} \cdot \mathbf{x}, P) = \begin{cases} \min(C) & \text{if } \min(C) \text{ exists} \\ -\infty & \text{otherwise} \end{cases}$$

Observe that $minExp(\mathbf{a} \cdot \mathbf{x}, P)$ can be realised with the simplex algorithm [157]: If there exists $\mathbf{y} \in \mathbb{Q}^n$, $n = |\mathcal{X}|$, that minimises the expression $\mathbf{a} \cdot \mathbf{y}$ over P, then put $minExp(\mathbf{a} \cdot \mathbf{x}, P) = \lceil \mathbf{a} \cdot \mathbf{y} \rceil$; otherwise put $minExp(\mathbf{a} \cdot \mathbf{x}, P) = -\infty$.

Let $P(\mathbf{a} \cdot \mathbf{x})$ denote the set of values that an expression $\mathbf{a} \cdot \mathbf{x} \in Lin$ can take on in a given polyhedron $P \in Poly$. Specifically, define $P(\mathbf{a} \cdot \mathbf{x}) = \emptyset$ if $P = \emptyset$. Otherwise, let $P(\mathbf{a} \cdot \mathbf{x}) = [l, u]$, where $l \in \mathbb{Z} \cup \{-\infty\}$ and $u \in \mathbb{Z} \cup \{\infty\}$ denote the minimum and maximum values that the linear expression can take on in P. The value of u is given by $u = -minExp(-\mathbf{a} \cdot \mathbf{x}, P)$, where $-(-\infty) = \infty$.

This concludes the section on commonly used operations on polyhedra. We now embark on defining a domain that expresses congruence information.

```
1     (uint32) s = 0;
2     (int32) i = 0;
3     jump b207
4   b130
5     (int32) i.0 = i;
6     t9  = &strings;
7     (int32) t9 = 4*i.0+t9;
8     (uint32) s = *t9;
9     t11  = &t10;
10    printf((uint32) t11, (uint32) s)
11    (int32) i = i+1;
12    jump b207
13  b207
14    (uint32) i.1 = i;
15    if (uint32) i.1<=2 then jump b130
16    jump b235
```

Fig. 3.8. The loop of the string printing program translated to Core C. Note that the temporary variables `i.0` and `i.1` are introduced by the GNU C compiler.

3.3.3 Multiplicity Domain

Inferring bounds on the values of variables and pointers is often not enough to prove correct memory management. Consider again the example in Fig. 3.1 in which the array of strings is accessed by the expression `strings[i]`. Here, `i` ranges over the integer values 0, 1, and 2. The corresponding Core C fragment in Fig. 3.8 shows the body of the loop and, in particular, how the array access is translated. Since the array contains pointers that are each four bytes in size, line 7 multiplies the loop counter `i` by four and adds this value to the address of the `strings` array. The result is dereferenced in line 8. Given a pure polyhedral analysis and a points-to analysis, we can only infer that `*t9` will access the array `strings` at the offsets $0, 1, \ldots 8$. This information, however, is not enough to infer that the string pointers in the array are written into `s` since an access to, say, the bytes $3 \ldots 6$ is possible, which would indicate that some bytes of the first and the second pointers can be written to `s`. While this is not the case in the actual program, the analysis does not provide enough information to deduce this. In order to infer that the access always reads whole pointers, it is necessary to infer that `*t9` has an offset that is a multiple of the pointer size.

An analysis that deduces that program variables can only take on the values of the form $c + m\mathbb{Z}$ (that is, values in $\{\ldots -2m+c, -m+c, c, c+m, c+2m, \ldots\}$) was introduced by Granger [85] in the form of a congruence domain. This information is sufficient to infer that an access such as `a[i].f=0` only writes the **f** field of the structures that make up the array a. However, in the context of string buffer analysis, it is sufficient to know that a variable is a multiple of 2^n in order to show that accesses to arrays of pointers are aligned. Hence, we propose a simpler domain than that of Granger, which we call the multiplicity domain. In particular, let $\mathit{Mult} = \mathcal{X} \to \{0, \ldots 64\}$ denote the function space that tracks the number of least-significant bits (LSBs for short) that are zero. A map $M \in \mathit{Mult}$ assigns a value $n = M(x)$ to each variable $x \in \mathcal{X}$ whenever x takes on values that are a multiple of 2^n. We assume that no variable is larger than 64 bits, such that a zero value of a variable x can always be expressed as $M(x) = 64$. Updating M to $M' = M[x \mapsto n']$ results in $M'(x) = n$ and $M'(y) = M(y)$ for all $y \neq x$. The lattice operations on *Mult* are defined as follows. For any $M_1, M_2 \in \mathit{Mult}$, define $M' = M_1 \sqcup_M M_2$ such that $M'(x) = \min(M_1(x), M_2(x))$ for all $x \in \mathcal{X}$. Let $M_1 \sqsubseteq_M M_2$ iff $M_1(x) \geq M_2(x)$ for all $x \in \mathcal{X}$. Note that in this context the mapping $M_\top \in \mathit{Mult}$ with $M_\top(x) = 0$ for all $x \in \mathcal{X}$ is the largest lattice element; that is, it is the least precise in that it states that all variables are divisible by one.

In order to refine a given domain, let $\mathit{Equ} = \mathit{Lin} \times \mathbb{Z}$ denote the set of linear equalities written $e = c$, where $e \in \mathit{Lin}$, $c \in \mathbb{Z}$, and define $\delta : \mathbb{Z} \to \{0, \ldots 64\}$ so that $\delta(x)$ is the number of LSBs that are zero in x or 64, whichever is smaller. We define the meet operator $\sqcap_M : \mathit{Mult} \times \mathit{Equ} \to (\mathit{Mult} \cup \{\bot_M\})$, where $\bot_M$ is a symbolic tag denoting an unsatisfiable domain. The idea is to refine a given domain element $M \in \mathit{Mult}$ to $M' = M \sqcap_M (e = c)$ by adding information expressed by $e = c$. To illustrate how $\sqcap_M$ can be realised, let $e \equiv a_1x_1 + \ldots + a_nx_n$ such that $a_i \neq 0$ for $i = 1, \ldots n$. We refine the multiplicity of a single variable x_j by rewriting $e = c$ to

$$-a_jx_j = a_1x_1 + \ldots a_{j-1}x_{j-1} + a_{j+1}x_{j+1} + \ldots a_nx_n - c$$

Refining the domain using $\sqcap_M$ adds additional information and, hence, the number of least-significant bits that are zero for x_j cannot decrease. On the contrary, the right-hand side of the equation above may imply a larger number of least-significant bits that are zero, which can be used to update $M(x_j)$. Note that the multiplicity of each term a_ix_i is $\delta(a_i) + M(x_i)$, whereas the multiplicity of the constant is simply $\delta(c)$. The multiplicity of the right-hand side must be greater than or equal to that of each individual term; that is, it must be at least

$$\min\left(\delta(c), \min_{i=1,\ldots j-1,j+1,\ldots n}(\delta(a_i) + M(x_i))\right)$$

In the case where $a_j > 1$ or $a_j < -1$, the number given by the expression above has to be reduced by $\delta(a_j)$ in order to infer the number of zero LSBs of x_j. Considering that $M(x_j)$ can only increase, the update of x_j in M can thus be defined as follows:

$$M\left[x_j \mapsto \max\left(M(x_j), \min\left(\delta(c), \min_{i=1,\ldots j-1,j+1,\ldots n}(\delta(a_i) + M(x_i))\right) - \delta(a_j)\right)\right]$$

	x +	y +	$2z$	$= 0$
action	$M(x)$	$M(y)$	$\delta(2)+M(z)$	$\delta(0)$
initial	3	1	1+1	64
update x	3	1	1+1	64
update y	3	2	1+1	64
update z	3	2	1+1	64

Fig. 3.9. Updating multiplicity information on adding an equality constraint.

Since this operation updates the multiplicity information of only one variable, it has to be applied n times to update the multiplicity of all variables in $e = c$.

In order to illustrate the application of this formula, consider updating the multiplicity information of M in Fig. 3.9 when the equation $x + y + 2z = 0$ is added to the domain. The initial multiplicity values in M are such that x is a multiple of 8, whereas y and z are merely even. Updating x cannot improve this information since $\min(\delta(1)+M(y), \delta(2)+M(z), \delta(0)) = \min(1, 2, 64) = 1$. Strengthening the multiplicity information for y is possible, however, since $\delta(2) + M(z) = 2$ is the smallest term. The final update of z cannot improve the bound since $\min(\delta(1)+M(x), \delta(1)+M(y), \delta(c)) = 2$ and, after subtracting $\delta(2) = 1$, is no larger than $M(z) = 1$.

Consider the same example when the constant is set to 1. In this case, none of the updates above can improve any of the multiplicity values. Moreover, the left-hand side of the equation $x + y + 2z = 1$ is known to be at least a multiple of four, which implies that no values exist for x, y, z such that the equality above holds, so calculating $M \sqcap_M (x + y + 2z = 1)$ must result in an unsatisfiable domain. Hence, in general, the meet operation returns $\bot_M$ whenever the following holds:

$$\min_{i=1,\ldots n} (\delta(a_i) + M(x_i)) > \delta(c)$$

To summarise, the algorithm for calculating $M' = M \sqcap_M (e = c)$ returns $\bot_M$ whenever the multiplicity of c is smaller than that of the left-hand side. Otherwise, the multiplicity of each variable is updated in turn. An update of a single variable needs to consider all n variables, and thus updating all variables is quadratic. This is not considered to be a problem in practice since n is never larger than 2 or 3. In fact, usually only a single variable needs updating, namely when intersecting with an equation that stems from an assignment operation: Define $M' = M \triangleright x := e$ analogously to the definition for the polyhedral domain. In this case, the variable x and the temporary variable t are projected out before the intersection; hence they have no zero LSBs and $M(x) = M(t) = 0$. Therefore, updating the terms in e cannot lead to any additional information, and only $M(x)$ and $M(t)$ need updating. Analogously, the assignment $M \triangleright x := y \gg n$ merely implies that $M(x)$ is at least $(M(y) - n)$, and no new information can be inferred on $M(y)$.

In order to test an alignment assumption during verification, it is useful to restrict $M \in \mathit{Mult}$ such that $M(x) \geq n$. This fact can be expressed by calculating $M \sqcap_M (x = 2^n)$. For instance, let x denote the offset of pointer `t#9` in Fig. 3.1. Then $M' = M \sqcap_M (x = 4)$ ensures that $M'(x) \geq 4$ or $M' = \bot_M$. If $M \sqsubseteq_M M'$, then M did not change and the access is aligned. A warning must be emitted otherwise.

With respect to the calculation of fixpoints, observe that *Mult* is finite and that the lattice $\langle \mathit{Mult}, \sqsubseteq_M, \sqcup_M, \sqcap_M \rangle$ is complete. Standard Kleene iteration is therefore sufficient to calculate a fixpoint. In accordance with the points-to domain and the polyhedral domain, we define a family of projection operators $\exists_x : \mathit{Mult} \to \mathit{Mult}$ such that if $M' = \exists_x(M)$ then $M'(x) = 0$ and $M'(y) = M(y)$ for all $y \neq x$.

The next section combines the multiplicity and polyhedral domains into a single *Num* domain that expresses all numeric properties in the analysis.

3.3.4 Combining the Polyhedral and Multiplicity Domains

In this section, we combine the polyhedral and multiplicity domains into a single numeric domain *Num*. Let $\mathit{Num} = (\mathit{Poly} \times \mathit{Mult}) \cup \{\bot_N\}$, where $\bot_N$ denotes the empty state that corresponds to an unreachable point in the program. The two domains can be combined by lifting their operations pointwise. Thus, define:

- $\langle P, M \rangle \sqsubseteq_N \langle P', M' \rangle$ iff $P \sqsubseteq_P P'$ and $M \sqsubseteq_M M'$
- $\langle P', M' \rangle = \langle P_1, M_1 \rangle \sqcup_N \langle P_2, M_2 \rangle$ iff $P' = P_1 \sqcup_P P_2$ and $M' = M_1 \sqcup_M M_2$
- $\langle P', M' \rangle = \langle P, M \rangle \rhd x := e$ iff $P' = P \rhd x := e$ and $M' = M \rhd x := e$
- $\langle P', M' \rangle = \langle P, M \rangle \rhd x := e \gg n$ iff $P' = P \rhd x := e \gg n$ and $M' = M \rhd x := e \gg n$
- $\langle P', M' \rangle = \exists_x(\langle P, M \rangle)$ iff $P' = \exists_x(P)$ and $M' = \exists_x(M)$

Defining the meet operator for intersection requires a case distinction depending on the outcome of the operations on the individual domains. Thus, define

$$\langle P, M \rangle \sqcap_N \{e = c\} = \begin{cases} \bot_N & \text{if } P' = \emptyset \vee M' = \bot_M \\ \langle P', M' \rangle & \text{otherwise} \end{cases}$$

where $P' = P \sqcap_P [\![\{e = c\}]\!]$ and $M' = M \sqcap_M \{e = c\}$. The meet operation $\langle P, M \rangle \sqcap_N \{e \leq c\}$ is defined analogously except that in this case $M' = M$. Both variants of the meet operator can be lifted to sets of inequalities by interpreting $N \sqcap_N \{e_1 \leq c_1, \ldots e_n \leq c_n\}$ as $N \sqcap_N \{e_1 \leq c_1\} \sqcap_N \ldots \sqcap_N \{e_n \leq c_n\}$, where $N \in \mathit{Num}$. The meet operator can return the special bottom value $\bot_N \in \mathit{Num}$, which is returned verbatim by all operations above, except for $\bot_N \sqcup_N N = N \sqcup_N \bot_N = N$. Furthermore, $\bot_N \sqsubseteq_N N$ and $N \sqsubseteq_N \bot_N$ only if $N = \bot_N$. Define $\equiv$ to refine M such that $\langle P, M \rangle \sqcap_N \{x \equiv s\} = \bot_N$ if $M' = M \sqcap_M \{x = s\} = \bot_M$ and $\langle P, M' \rangle$ otherwise. For brevity, let $\exists_X(N)$ abbreviate $\exists_{x_1}(\ldots \exists_{x_n}(N) \ldots)$, where $X = \{x_1, \ldots x_n\} \subseteq \mathcal{X}$. The next section observes how returning $\bot_N$ instead of $\langle \emptyset, M \rangle$ is a way to improve precision.

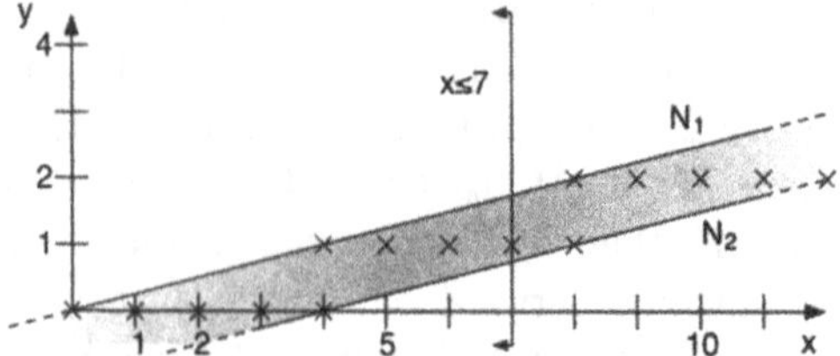

Fig. 3.10. Propagating information from the *Poly* to the *Mult* domain.

The Reduced Product between *Poly* and *Mult*

The meet operator $\sqcap_P$ is interesting because it reduces the two domains *Poly* and *Mult* in the sense that a state of the form $\langle \emptyset, M\rangle$ or $\langle P, \bot_M\rangle$ is equated with $\bot_N = \langle \emptyset, \bot_M\rangle$, which is a smaller state in the lattice *Num*. This reduction avoids the propagation of unsatisfiable domains, as witnessed in Fig. 3.3. Reducing two domains is usually a trade-off between the computational effort of propagating information from one domain to the other and the gain in efficiency and precision of the analysis. In some cases, it is possible to create a so-called reduced product – that is, a combination of domains that are implemented as one and thereby only provide states where no further reduction is possible [50]. Such a reduction is possible between the polyhedral domain *Poly* and the multiplicity domain *Mult*.

In order to show how information from one domain can refine the existing information in another, consider the statements `x=4*y; if (rand()) y--;`. Let N denote the initial state in which x is unbounded. The first statement defines $N_1 = N \triangleright x := 4y$, whereas the statement `y--`, which is guarded by the **if**-statement, transforms this state to $N_2 = N_1 \triangleright y := y - 1$. These two states are shown as black lines in Fig. 3.10, while the grey area depicts the polyhedron of $N_{12} = N_1 \sqcup_N N_2$, which corresponds to the state space after the **if**-statement. Note that the join introduced three new integral points for $y = 1$ in addition to the points $\langle 4, 1\rangle$ and $\langle 8, 1\rangle$ that stem from N_1 and N_2, respectively. As a consequence, the intersection with an inequality such as $x \leq 7$ restricts the maximum value x in $N_{12} \sqcap_N \{x \leq 7\}$ to 7. However, according to the multiplicity domain M_1 in N_1, the value of x is a multiple of 4; i.e., $M_1(x) = 2$. Similarly, a linear translation by 4 implies that x is still a multiple of 4 in N_2 and therefore in the join N_{12}, too. Thus, the maximum value of x after intersecting with $x \leq 7$ is 4. Hence, a straightforward propagation of information from the multiplicity domain to the numeric domain is to refine inequalities according to the multiplicity of the contained variables. For instance, the calculation $N_{12} \sqcap_N \{x \leq 7\}$ above can be refined to $N_{12} \sqcap_N \{x \leq 4\}$ since x is a multiple of 4. However, suppose the C program executes `z=x+1; if (z<=8) {}`, which leads to the state $N_3 = N_{12} \triangleright z := x + 1$ and $N_4 = N_3 \sqcap_N \{z \leq 8\}$. Semantically, this is equivalent to testing for $x \leq 7$.

However, none of the lower bits of z are known to be zero, and the evaluation of $N_3 \sqcap_N \{z \leq 8\}$ cannot be refined without examining all linear relationships that z has with other variables in N_3.

An automatic way to incorporate the information of *Mult* into the polyhedral domain is to scale each variable x in the polyhedron by $1/2^{M(x)}$. Thus, an intersection $\langle P, M\rangle \sqcap_N \{\mathbf{a} \cdot \mathbf{x} \leq c\}$ is executed by calculating $P' = P \sqcap_P [\![\{\langle 2^{M(x_1)}a_1, \dots 2^{M(x_n)}a_n\rangle \cdot \mathbf{x} \leq c\}]\!]$, where $\langle a_1, \dots a_n\rangle = \mathbf{a}$. The entailment check $\sqsubseteq_N$ and the join $\sqcup_N$ also face the challenge of operating on polyhedra that correspond to different multiplicity domains M and M'. In this case, the axes of the polyhedron that correspond to the variable x with $M(x) > M'(x)$ have to be scaled by $2^{M(x)-M'(x)}$. The benefit of this representation shows when a tightening around the $\mathbb{Z}$-grid of the polyhedron is applied. For the example above, $P_3 \sqsubseteq_P [\![2^{M_3(z)}z = 2^{M_3(x)}x + 1]\!] = [\![z = 4x + 1]\!]$, where $N_3 = \langle P_3, M_3\rangle$. Enforcing $N_3 \sqcap_P \{z \leq 8\}$ results in $P_3 \sqcap_P \{z \leq 8\}$ (since $2^{M_3(z)} = 1$), which in turn will update the bound of x in P_3 to $x \leq 1\frac{3}{4}$, which is tightened to $x \leq 1$. Not only is the value of x in N_3 now $x \leq 1 \cdot 2^{M(x)} = 4$, but through tightening around the $\mathbb{Z}$-grid in *Poly*, the bound of z in P_3 is automatically refined to $z \leq 5$. Note that incorporating the multiplicity information into the polyhedron reduces the magnitude of coefficients drastically. This is interesting, as Fourier-Motzkin variable elimination, which is used to implement the operator $\exists_x$, is exact on $\mathbb{Z}$-polyhedra if the coefficient a of the removed variable satisfies $|a| \leq 1$ [148]. Hence, while the reduction complicates the join and entailment check, it can increase the precision of other domain operations.

A propagation from the polyhedral domain to the multiplicity domain is also possible. For instance, if $P \sqsubseteq_P [\![\{x = 0\}]\!]$, then the corresponding $M \in \mathit{Mult}$ can be updated such that $M(x) = 64$. In fact, in the context of the reduced domain that stores $1/2^{M(x)}$ of x in *Poly*, this kind of reduction is the only possible way of propagating information from *Mult* to *Poly*.

In order to make the upcoming analysis independent of the implementation of *Num*, define $N(\mathbf{a} \cdot \mathbf{x} + c) = [l, u]_{\equiv d}$ as the set of values $\{l, l + d, \dots u\} \subseteq \mathbb{Z}$ that the expression $\mathbf{a} \cdot \mathbf{x} + c$ can take on in N. Furthermore, define $[\![N]\!] \subseteq \mathbb{Z}^{|\mathcal{X}|}$ to be the set of all points that are feasible in $N \in \mathit{Num}$.

We omit the definitions of the domain operations of *Num* that use *Poly* and *Mult* as a reduced product, as their presentation is merely technical. However, we observe that a reduction between the *Poly* and the *Mult* domains is also necessary when *Num* is not implemented as a reduced product. To this end, consider the definition of the two query operations $N(\mathbf{a} \cdot \mathbf{x} + c)$ and $[\![N]\!]$, and let $\mathit{Num} = (\mathit{Poly} \times \mathit{Mult}) \cup \{\bot_N\}$ be the standard product. In this case, $[\![N]\!] = \{\langle v_1, \dots v_n\rangle \mid \langle v_1, \dots v_n\rangle \in (P \cap \mathbb{Z}^n) \wedge \forall i \in [1, n] \,.\, v_i \bmod 2^{M(x_i)} = 0\}$, where $n = |\mathcal{X}|$. In other words, $[\![N]\!]$ contains all points $\mathbf{v} \in \mathbb{Z}^n$ that are in P and obey the multiplicity information in M. Furthermore, let $N(\mathbf{a} \cdot \mathbf{x} + c) = [l, u]_{\equiv d}$ iff $S = \{\mathbf{a} \cdot \mathbf{v} + c \mid \mathbf{v} \in [\![N]\!]\}$ and $l = \min(S)$, $u = \max(S)$ (if they exist), and $d = \max\{d \in \mathbb{N} \setminus \{0\} \mid \forall v \in S \,.\, v \bmod d = 0\}$. In an actual implementation, $N(\mathbf{a} \cdot \mathbf{x} + c) = [l, u]_{\equiv d}$ would be calculated by evaluating $P(\mathbf{a} \cdot \mathbf{x}) = [l', u']_{\equiv d}$

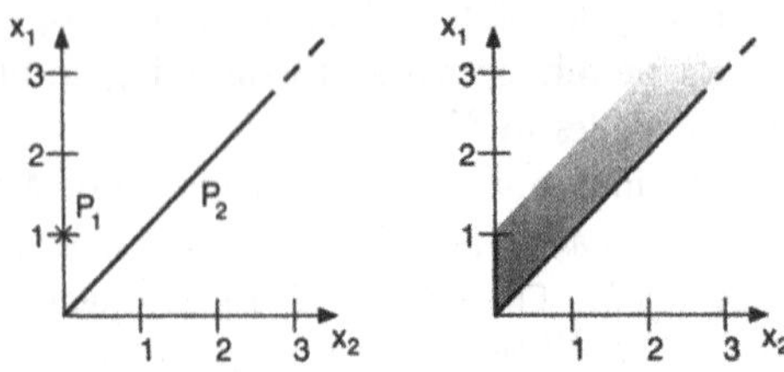

Fig. 3.11. Topological closure is required when calculating $P_1 \sqcup_P P_2$.

and setting $l = d\lceil(l' - c)/d\rceil + c$ and $u = d\lfloor(u' - c)/d\rfloor + c$. In contrast, for a reduced product, $[\![N]\!] = \{\langle 2^{M(x_1)}v_1, \ldots 2^{M(x_n)}v_n\rangle \mid \langle v_1, \ldots v_n\rangle \in (P \cap \mathbb{Z}^n)\}$. In particular, the interval of a linear expression is simply $N(\mathbf{a} \cdot \mathbf{x} + c) = [2^{M(x)}u', 2^{M(x)}l']_{\equiv d}$ (if u' and l' are finite), where d is defined as above. We conclude with an overview of work related to the *Poly* and *Mult* domains.

3.3.5 Related Work

The abstract domain of convex polyhedra was introduced by Cousot and Halbwachs [62] shortly after Karr described an analysis that infers affine relationships between variables [109]. Even though the domain of affine relationships is finite, its implementation is non-trivial, and the best implementation has operators that are cubic in $|\mathcal{X}|$ [133]. However, recent interest in interprocedural analysis has revived interest in this domain when used to construct a power set domain – that is, a set of affine domains [134]. The convex hull operation on polyhedra is exponential, which led to the investigation into sub-classes of polyhedra that provide a performance guarantee [47, 128, 172]. However, the main reason for exponential growth is the calculation of the convex hull via an intermediate representation based on vertices, lines, and rays, which is predominant in most polyhedra libraries [14, 27, 93, 119]. Avoiding this intermediate representation makes for an efficient alternative [169].

While making polyhedral operations more efficient is a prerequisite for large-scale program analysis [30, 31], a different strand of research tries to make polyhedra more expressive. For instance, Bagnara et al. [14] show how to support strict inequalities such as $x < 7$, which is relevant when analysing timed automata that are used to model real-time systems in which clocks can take on continuous values. In Sect. 3.3.1, the join of two polyhedra was defined to be the smallest closed convex space of the inputs rather than the smallest convex space. However, extending polyhedra with strict inequalities in order to include topologically open half-spaces does not remedy the need for closure. Consider the join $P_1 = \{\langle 0, 1\rangle\}$ and the polyhedron $P_2 = [\![\{x = y, x \geq 0\}]\!]$, both depicted on the left of Fig. 3.11. Their convex hull $hull(P_1, P_2)$ on the right is shown with a black border for closed sides of the space. In order to represent the open upper boundary of the space, an inequality $y - x < 1$ is necessary. However, $P_1 \not\subseteq [\![y - x < 1]\!]$, and hence $y - x < 1$ cannot be part of the convex hull. Thus, a topological closure step is always required to ensure that the convex hull is representable by a set of inequalities, which are therefore always non-strict.

By using widening to achieve tractability, polyhedra were successfully applied in areas such as argument-size analysis of Prolog programs [111,120,125], value-range analysis of Pascal programs [33], or the analysis of real-time systems [23, 28, 92]. A way to improve precision is to use sets of polyhedra [11,16,89]. This approach is useful in the context of weaker sub-classes of polyhedra [128]. In particular, the state space can be divided by the value of a variable and each polyhedron in the set is associated with one valuation of that variable [61]. For binary variables and general polyhedra, it is often sufficient and potentially cheaper to use a polyhedral variable as a Boolean flag that distinguishes between the two states, thereby embedding two polyhedra in the same polyhedron, an alternative that is sometimes overlooked [89]. Yet, Chap. 13 shows how joining two polyhedra P_1 and P_2 using a Boolean flag, namely by calculating $P = (P_1 \sqcap_P [\![f = 0]\!]) \sqcup_P (P_2 \sqcap_P [\![f = 1]\!])$, may cause a loss of precision unless both P_1 and P_2 are bounded and integer tightening methods are in place. Alternatively, an analysis using sets of polyhedra can be done without distinguishing individual polyhedra. While a widening on general sets of polyhedra exists [11], we are unaware of lattice operators that limit the size of polyhedral sets in a principled way. Such operators would enable the use of polyhedra in backwards analysis – that is, an analysis where under-approximated states are propagated against the control flow. Such an analysis can infer counterexamples, or input data that lead to erroneous behaviour (in contrast to possible counterexamples [151]). Note that a single polyhedron does not support backwards analysis: Polyhedra are not meet distributive; that is, $P \sqcap_P (P_1 \sqcup_P P_2) \neq (P \sqcap_P P_1) \sqcup_P (P \sqcap_P P_2)$ due to the fact that the convex hull operation may introduce points on the left-hand side that are not present on the right-hand side (see Sect. 4.5.1 for an example). Meet distributivity is a prerequisite for the domain to be pseudo-complemented [29, Chap. IX, Theorem 15]. The existence of a pseudo-complement, in turn, is required to calculate a weakest precondition, which forms the basis of inferring a counterexample [110, Sec. 3.3]. In fact, the weakest preconditions in the context of linear relations can only be inferred in some very restricted sub-classes of linear inequality systems [121].

In the context of analysing congruences, most work is due to Granger, who developed several classes of increasing expressiveness, ranging from simple arithmetic congruences of the form $x \equiv a (mod\ b)$, where $x \in \mathcal{X}$, $a, b \in \mathbb{Z}$ [85], over linear congruence analysis [86] of the form $\sum_{i=0}^{n} a_i x_i \equiv b (mod\ c)$, where $a, b, c \in \mathbb{Z}$, to congruence properties on rationals [87]; that is, $\forall x, y, z \in \mathbb{Q} \,.\, x \equiv y (mod\ z)$ iff $(\exists \lambda \in \mathbb{Z} \,.\, x = y + \lambda z)$. With respect to efficient implementations, Miné [129] observed that congruences of the form $x - y \equiv a (mod\ b)$, where $a, b \in \mathbb{N}$, can be analysed similarly to weakly relational domains such as Octagon [130]. In a similar vein, Bagnara et al. [8] have recast the operations on Granger's congruence domain in terms of operations on generators of polyhedra, thereby improving efficiency.

We are not aware of any work that combines a congruence domain and a polyhedral domain into a reduced product. This reduction has some interesting implications: Given that the congruence domain we employ can only express congruences of the form $x \equiv 0(mod\ 2^m)$, where $m \in [0, 64]$, linear relationships with other variables such as $z = x + 1$ can be inferred that implicitly express that $z \equiv 1(mod\ 2^m)$. Thus, reduced products between congruence and polyhedral domains might yield the same congruence information that the more complex congruence domains [86, 87] can express on their own.
The next chapter details how these domains are used to analyse C programs.

4

Taming Casting and Wrapping

In this chapter, we take a first stab at defining a semantics using the abstract states that were presented in the last chapter. These abstract transfer functions mirror the effect of the concrete semantics except that they modify points-to sets and polyhedra rather than sets of 4-GB memory configurations. The abstract semantics therefore completes the triple of the concrete domain $\Sigma \times \Theta \times \Delta$, the abstract domain *Num* × *Poly*, and the concrete semantics $[\![s]\!]^{\natural}_{\text{Stmt}}$ as shown in the following diagram:

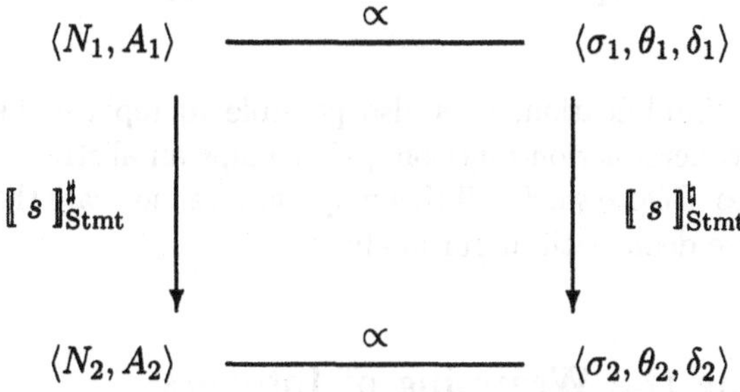

The diagram above introduces a relation $\propto$ (pronounced "approximates") between abstract and concrete states. Using this relation, the analysis of each statement s in Core C can be proved correct by the following observation. Assume that the concrete state $\langle \sigma_1, \theta_1, \delta_1 \rangle$ is included in the collecting semantics before executing statement s. Suppose that the evaluation of the concrete semantics of s yields a new concrete state $\langle \sigma_2, \theta_2, \delta_2 \rangle$. Let the abstract state $\langle N_1, A_1 \rangle$ be in $\propto$ relation to the initial concrete state $\langle \sigma_1, \theta_1, \delta_1 \rangle$, where $\propto$ specifies how the contents of the abstract variables $\mathcal{X}$ and addresses $\mathcal{A}$ in N_1 and A_1 define the values of the variables that reside in the stack frames θ_1 and the dynamic memory regions in δ_1. Given the abstract transfer function $[\![s]\!]^{\sharp}_{\text{Stmt}}$ that transforms $\langle N_1, A_1 \rangle$ into $\langle N_2, A_2 \rangle$, correctness of the analysis is guaranteed if the latter state is in $\propto$ relation with the corresponding concrete state $\langle \sigma_2, \theta_2, \delta_2 \rangle$.

Recall that the concrete semantics is undefined whenever a program error occurs, in which case no new state $\langle \sigma_2, \theta_2, \delta_2 \rangle$ exists. In this case, the evaluation of the abstract semantics $[\![\ s\ ]\!]^{\sharp}_{\mathsf{Stmt}} \langle N_1, A_1 \rangle$ must flag an error. Thus, even in the presence of a partial concrete semantics, a provably correct analysis ensures that the analysed program cannot enter an erroneous state if the abstract analysis flags no warnings.

Rather than providing a full correctness proof of all aspects of the analysis, we demonstrate the principle in this chapter on casting and wrapping by defining Sub C, a strict subset of Core C whose statements only include assignments of linear expressions and casts between variables. The absence of functions, pointers, and structures makes it possible to simplify the diagram above to include only memory configurations σ_i and the numeric domain N_i:

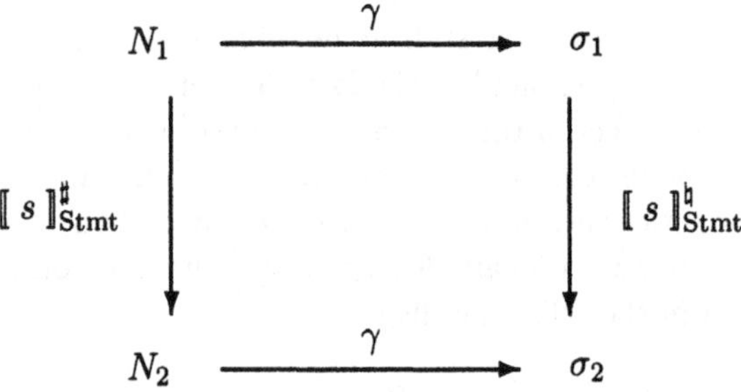

Due to this simplification, it is also possible to replace the abstraction relation with a concretisation function γ that maps an abstract state to a set of concrete states. Before we detail this mapping, we motivate the way casting and wrapping are dealt with in our analysis.

4.1 Modelling the Wrapping of Integers

In contrast to formal methods that verify properties of a high-level specification, a static analysis is complicated by low-level details of source code. For instance, while a specification expresses properties over arbitrary integers, variables in a program are usually confined to finite integer types that are deemed to be large enough to hold all values occurring at run-time. On the one hand, the use of 32-bit and 64-bit variables makes accidental overflows rare, and to add checks to each transfer function of the analysis seems excessive considering the infrequency of variable overflows. On the other hand, programmers often inadvertently introduce wrapping when converting between signed and unsigned variables and deliberately exploit the wrapping effects of two's-complement arithmetic. Thus, wrapping itself should not be considered harmful, particularly when the objective of an analysis is the verification of a different property, such as the absence of out-of-bounds memory accesses.

```
1 while (*str) {
2   dist[*str]++;
3   str++;
4 };
```

Fig. 4.1a. The initial code.

```
1 while (*str) {
2   dist[(int) *str]++;
3   str++;
4 };
```

Fig. 4.1b. Removing the compiler warning.

```
1 while (*str) {
2   dist[(unsigned int) *str]++;
3   str++;
4 };
```

Fig. 4.1c. Observing that **char** may be signed.

In fact, there is a danger in an analysis that flags all wrapping as erroneous, since any intentional use of wrapping generates a warning message, which the developer immediately dismisses as a false positive. This problem is illustrated in the code fragments in Figs. 4.1a to 4.1c that show the loop of the introductory example on p. 5 at different development stages.

Recall that the purpose of the loop is to count the occurrences of each character in `*str`. To this end, the nth element of `dist` is incremented each time a character with the value n is encountered in `str`. Suppose that Fig. 4.1a depicts the initial loop. The compiler emits a warning at line 2 pointing out that the C standard requires that the index type of an array be of type **int** rather than type **char**. The programmer therefore inserts a cast, thereby deriving the loop in Fig. 4.1b. At this point, an observant peer might remark that the **char** type is signed on many platforms, which would enable `dist` to be accessed with negative indices. The program is therefore altered by replacing the cast `(int)` with `(unsigned int)`, with the intent of wrapping character values of the range $[-128, -1]$ to $[128, 255]$.

The resulting program is certainly correct on platforms where **char** is unsigned, such as Linux on PowerPC. However, for Linux on x86 and MacOS X on PowerPC, **char** is signed. Although the programmer intended to convert the value of `*str` to an unsigned value before the extension to a 4-byte quantity took place, the C standard [51] dictates that the value of `*str` first be promoted to **int** before the conversion to an unsigned type is performed. Hence, `dist` can be accessed at indices $[2^{32} - 128, 2^{32} - 1] \cup [0, 127]$, of which the first range is out-of-bounds. A static analysis that considers all wrapping to be erroneous would flag this statement as possibly faulty. However, since the programmer expects that wrapping does occur (namely when converting from a **char** to an unsigned quantity), the warning about wrapping at the **unsigned int**-level may be dismissed as a false positive. Hence, the analysis of the code above should flag the out-of-bounds array access but assume that wrapping itself is intentional.

In order to model wrapping in the analysis, we propose a reinterpretation of classic polyhedral analysis [62]. In particular, rather than checking for wrapping in every transfer function, we refine the approximation relation such that wrapping is mostly implicit; that is, the need for extra polyhedral operations is largely finessed. For the few cases in which wrapping has to be addressed in the transfer functions, we illustrate how to wrap values within the polyhedral domain and propose an algorithm for doing so.

We commence with the definition of a subset of Core C and its concrete semantics. Sections 4.4 and 4.5 explain how wrapping is supported. Section 4.6 presents a wrapping-aware polyhedral analysis. Section 4.7 discusses our approach and relates it to design choices made in other analysers.

4.2 A Language Featuring Finite Integer Arithmetic

The loop shown in Figs. 4.1a–4.1c demonstrates that it is important to clarify where wrapping can arise in a program. In order to define an analysis of wrapping, we introduce the language Sub C, which is a subset of the intermediate language Core C that was presented on p. 26.

4.2.1 The Syntax of Sub C

The Sub C subset only features linear expressions and casts between integers. As before, $(T_1 \ldots T_n)^*$ denotes the repetition of the symbols $T_1 \ldots T_n$.

$$
\begin{array}{lll}
\langle\text{Sub C}\rangle & :: & (\langle\text{Block}\rangle)^* \\
\langle\text{Block}\rangle & :: & l\text{:}\ ((\langle\text{Stmt}\rangle\ ;)^*\ \langle\text{Next}\rangle \\
\langle\text{Next}\rangle & :: & \textbf{jump}\ l\,; \\
 & | & \textbf{if}\ \langle\text{Type}\rangle\ v\ \langle\text{Op}\rangle\ \langle\text{Expr}\rangle\ \textbf{then jump}\ l\,;\ \langle\text{Next}\rangle \\
\langle\text{Op}\rangle & :: & < \mid \leq \mid = \mid \neq \mid \geq \mid > \\
\langle\text{Expr}\rangle & :: & n \mid n\ \text{*}\ v + \langle\text{Expr}\rangle \\
\langle\text{Stmt}\rangle & :: & \langle\text{Size}\rangle\ v = \langle\text{Expr}\rangle \\
 & | & \langle\text{Size}\rangle\ v = \langle\text{Type}\rangle\ v \\
\langle\text{Type}\rangle & :: & (\textbf{uint} \mid \textbf{int})\ \langle\text{Size}\rangle \\
\langle\text{Size}\rangle & :: & \textbf{1} \mid \textbf{2} \mid \textbf{4} \mid \textbf{8}
\end{array}
$$

A Sub C program consists of a sequence of basic blocks, with execution commencing with the first block. Each basic block consists of a sequence of statements and a list of control-flow instructions. The ⟨Stmt⟩ production is restricted to the two statements of interest, namely the assignment of linear

Basic Blocks.
$[\![\, l : s_1; \ldots s_n; \,]\!]^{\natural}_{\mathsf{Block}}\sigma = [\![\, \mathit{lookupNext}(l) \,]\!]^{\natural}_{\mathsf{Next}}([\![\, s_n \,]\!]^{\natural}_{\mathsf{Stmt}}(\ldots([\![\, s_1 \,]\!]^{\natural}_{\mathsf{Stmt}}\sigma)\ldots))$

Control Flow.
$[\![\, \mathbf{jump}\; l \,]\!]^{\natural}_{\mathsf{Next}}\sigma = [\![\, \mathit{lookupBlock}(l) \,]\!]^{\natural}_{\mathsf{Block}}\sigma$

$[\![\, \mathbf{if}\; t\; s\; v\; op\; exp\; \mathbf{then\; jump}\; l\; ;\; nxt \,]\!]^{\natural}_{\mathsf{Next}}\sigma =$

$$\begin{cases} [\![\, \mathit{lookupBlock}(l) \,]\!]^{\natural}_{\mathsf{Block}}\sigma & \text{if } \mathit{val}^{8s,t}(\sigma^{s}(\mathit{addr}(v)))\; op\; \mathit{val}^{8s,t}([\![\, exp \,]\!]^{\natural,s}_{\mathsf{Expr}}\sigma) \\ [\![\, nxt \,]\!]^{\natural}_{\mathsf{Next}}\sigma & \text{otherwise} \end{cases}$$

Expressions.
$[\![\, n \,]\!]^{\natural,s}_{\mathsf{Expr}}\sigma = \mathit{bin}^{8s}(n)$

$[\![\, n * v + exp \,]\!]^{\natural,s}_{\mathsf{Expr}}\sigma = \mathit{bin}^{8s}(n) *^{8s} \sigma^{s}(\mathit{addr}(v)) +^{8s} [\![\, exp \,]\!]^{\natural,s}_{\mathsf{Expr}}\sigma$

Assignment.
$[\![\, s\; v = exp \,]\!]^{\natural}_{\mathsf{Stmt}}\sigma = \sigma[\mathit{addr}(v) \overset{s}{\mapsto} [\![\, exp \,]\!]^{\natural,s}_{\mathsf{Expr}}\sigma]$

Type Casts.
$[\![\, s_1\; v_1 = t\; s_2\; v_2 \,]\!]^{\natural}_{\mathsf{Stmt}}\sigma = \sigma[\mathit{addr}(v_1) \overset{s_1}{\mapsto} \mathit{bin}^{8s_1}(\mathit{val}^{8s_2,t}(\sigma^{s_2}(\mathit{addr}(v_2))))]$

Fig. 4.2. Concrete semantics of Sub C.

expressions to a variable and a type cast. Variables do not need to be declared, but each variable may only be used with one size, which is specified in bytes. In particular, the assignment statement and the conditional require that all variables occurring be of the same size. As before, variables may be used as a **uint** (unsigned integer) in one statement and as an **int** (signed integer) in another.

4.2.2 The Semantics of Sub C

Figure 4.2 presents the concrete semantics of the Sub C language. The absence of functions simplifies the concrete semantics as the mapping from program variable v to its address $\mathit{addr}_{\theta}(\mathrm{v})$ in the call stack θ is replaced by a simpler address map $\mathit{addr} : \mathcal{X} \to [0, 2^{32} - 1]$. This map defines a one-to-one relationship between a polyhedral variable v and a program variable. Like addr_{θ}, the map addr maps different variables to non-overlapping memory regions, an assumption that makes Sub C independent of the endianness of an architecture.

The concrete semantics manipulates the store mainly by operations on bit vectors; only in the conditional and the cast are bit vectors interpreted as numbers. In these cases, the signedness of the variables can actually influence the result. In particular, the type t of the cast determines if the source bit vector is sign-extended (if $t = \mathbf{int}$) or zero-extended (if $t = \mathbf{uint}$) when $s_1 \geq s_2$. We now proceed by abstracting this semantics to an abstract semantics that operates on polyhedra.

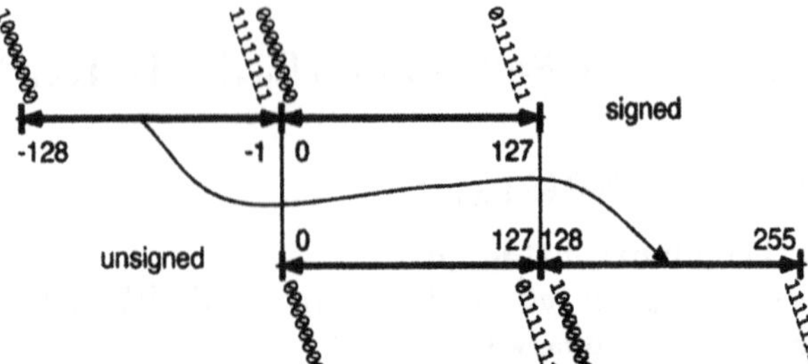

Fig. 4.3. The difference between a signed and an unsigned access can be interpreted as a wrap of negative values to the upper range in an unsigned access.

4.3 Polyhedral Analysis of Finite Integers

Two's-complement arithmetic exploits the wrapping behaviour of integer variables that are confined to a fixed number of bits. For instance, subtracting 1 from an integer is equivalent to adding the largest representable integer value. In fact, the binary representation of the signed integer -1 is identical to that of the largest unsigned integer of the same size. In the context of verification, this dichotomy in interpretation cannot be dismissed since variables in C may be used as both signed and unsigned quantities.

Accessing the same bit sequence as either a signed or unsigned integer corresponds to a wrapping behaviour in that the negative range of the signed integer wraps to the upper range of an unsigned integer, as illustrated in Fig. 4.3. This wrapping behaviour of finite integers creates a mismatch against the infinite range of polyhedral variables. We present our solution to this mismatch in two parts: Section 4.4 presents a concretisation map between the polyhedral domain and the bit-level representation of variables. This map wraps values of abstract variables implicitly to finite sequences of bits, thereby alleviating the need to check for wrapped values each time a variable is read or written. In contrast, Sect. 4.5 details an algorithm that makes the wrapping of program variables explicit in the abstract domain, which is important for casts, and in the conditional statement, whose semantics depend on the size and signedness of the operands.

Revisiting the Domain of Convex Polyhedra

For simplicity, the correctness argument of the analysis of Sub C is formulated in terms of the polyhedral domain *Poly* rather than in the more complex *Num* domain, which is a product of *Poly* and the multiplicity domain *Mult*. In particular, it allows for the following lemma, which states that an update operation on a polyhedron corresponds to the evaluation of a linear expression.

Lemma 1. *Let* $P \in \mathit{Poly}$ *and* $P' = P \triangleright x_i := \mathbf{c} \cdot \mathbf{x} + d$ *with* $\mathbf{c} \in \mathbb{Z}^n$ *and* $d \in \mathbb{Z}$. *Then* $P' = \{\langle v_1, \dots v_{i-1}, v_i', v_{i+1}, \dots v_n\rangle \mid \mathbf{v} = \langle v_1, \dots v_n\rangle \in P \wedge v_i' = \mathbf{c} \cdot \mathbf{v} + d\}$.

4.4 Implicit Wrapping of Polyhedral Variables

This section formalises the relationship between polyhedral variables and bit sequences that constitute the program state. For simplicity, we assume a one-to-one correspondance between the variable names in the program and the polyhedral variables that represent their values. The values of a program variable are merely bit sequences that are prescribed by the possible values of the polyhedral variable. In order to illustrate this, suppose that x is of type **char** and $P(x) = [-1, 2]$. The bit represented patterns are 11111111, 00000000, 00000001, and 00000010, no matter whether x is signed or unsigned. These bit patterns are given by $bin^{8s}(v)$, which turns a value $v \in [-1, 2]$ into a bit sequence of s bytes. Going further, the function $bits_a^s : \mathbb{Z} \to \mathcal{P}(\Sigma)$ produces all concrete stores in which $8s$ bits at address $a = addr(\mathrm{x})$ are set to the value corresponding to $v \in P(x)$ as follows:

$$bits_a^s(v) = \{\langle r_{8*2^{32}} \ldots r_{8(a+s)}\rangle \| \ bin^{8s}(v) \ \| \ \langle r_{8a-1} \ldots r_0\rangle \mid r_i \in \mathbb{B}\}$$

Note that this definition only considers the lower $8s$ bits of the value v. For instance, $bits_a^1(0) = bits_a^1(256)$ since the lower eight bits of 0 and 256 are equal. The mapping $bits_a^s$ can be lifted from the value v of a single variable to the values $\langle v_1, \ldots v_n\rangle \in \mathbb{Z}^n$ of a vector of variables $\langle x_1, \ldots x_n\rangle$, resulting in the stores $\bigcap_{i\in[1,n]} bits_{a_i}^{s_i}(v_i)$. Here $a_i \in [0, 2^{32} - 1]$ denotes the address of the variable x_i in the concrete store and $s_i \in \mathbb{N}$ denotes its size in bytes. Observe that if variables were allowed to overlap, the intersection above might incorrectly collapse to the empty set for certain vectors $\langle v_1, \ldots v_n\rangle \in \mathbb{Z}^n$. Using this lifting, a polyhedron is now related to a set of stores by $\gamma_{\mathbf{a}}^{\mathbf{s}} : Poly \to \mathcal{P}(\Sigma)$, which is defined as

$$\gamma_{\mathbf{a}}^{\mathbf{s}}(P) = \bigcup_{\mathbf{v}\in P\cap\mathbb{Z}^n} \left(\bigcap_{i\in[1,n]} bits_{a_i}^{s_i}(v_i) \right)$$

where $\mathbf{s} = \langle s_1, \ldots s_n\rangle$, $\mathbf{a} = \langle a_1, \ldots a_n\rangle$ and $\mathbf{v} = \langle v_1, \ldots v_n\rangle$.

The definition of $\gamma_{\mathbf{a}}^{\mathbf{s}}$ provides a criterion for judging the correctness of an abstract semantics. In addition, $\gamma_{\mathbf{a}}^{\mathbf{s}}$ permits linear expressions to be evaluated in the abstract semantics without the need to address overflows since $\gamma_{\mathbf{a}}^{\mathbf{s}}$ maps the result of calculations in the polyhedral domain to the correctly wrapped result in the actual program. This property is formalised below.

Proposition 1. *Let $e \in \mathcal{L}(Expr)$ and $e \equiv \mathbf{c}\cdot\mathbf{x}+d$, that is, e is a reformulation of $\mathbf{c}\cdot\mathbf{x}+d$. If $\sigma \in \gamma_{\mathbf{a}}^{\mathbf{s}}(P)$ then $\sigma[a_i \overset{s_i}{\mapsto} [\![e]\!]_{\mathrm{Expr}}^{\natural, s_i}] \in \gamma_{\mathbf{a}}^{\mathbf{s}}(P \rhd x_i := \mathbf{c}\cdot\mathbf{x}+d)$.*

Proof. Define $\pi_i(\langle x_1, \ldots x_{i-1}, x_i, x_{i+1}, \ldots x_n\rangle) = x_i$. Since $\sigma \in \gamma_{\mathbf{a}}^{\mathbf{s}}(P)$ there exists $\mathbf{v} \in P \cap \mathbb{Z}^n$ such that $\sigma = \bigcap_{i\in[1,n]} bits_{a_i}^{s_i}(\pi_i(\mathbf{v}))$. Let $P' = P \rhd x_i :=$ $\mathbf{c}\cdot\mathbf{x}+d$ for some $\mathbf{c} \in \mathbb{Z}^n$ and $d \in \mathbb{Z}$. By Lemma 1, there exists $\mathbf{v}' \in P'$ with $\pi_j(\mathbf{v}') = \pi_j(\mathbf{v})$ for all $j \neq i$. Since $\{a_i, \ldots a_i+s_i-1\} \cap \{a_j, \ldots a_j+s_j-1\} = \emptyset$

for all $j \neq i$, there exists $\sigma' \in \gamma_{\mathbf{a}}^{\mathbf{s}}(P')$ such that $\sigma'^1(a) = \sigma^1(a)$ for $a \in [0, 2^{32} - 1] \setminus \{a_i, \ldots a_i + s_i - 1\}$. Furthermore, the lemma states that $\pi_i(\mathbf{v}') = \mathbf{c} \cdot \mathbf{v} + d$ and, by the definition of $\gamma_{\mathbf{a}}^{\mathbf{s}}$, it follows that $\sigma'^{s_i}(a_i) = bin^{8s_i}(\mathbf{c} \cdot \mathbf{v} + d)$. To show that $\sigma'^{s_i}(a_i) = [\![e]\!]_{\text{Expr}}^{\natural, s_i}\sigma$, we find $\mathbf{a} \in \mathbb{Z}^n$, $d \in \mathbb{Z}$ such that $e \equiv \mathbf{c} \cdot \mathbf{x} + d$ and $[\![e]\!]_{\text{Expr}}^{\natural, s_i}\sigma = bin^{8s_i}(\mathbf{c} \cdot \mathbf{v} + d)$ by induction over e:

- Let $e = n$. By the definition of $[\![\cdot]\!]_{\text{Expr}}^{\natural, s}$, $[\![n]\!]_{\text{Expr}}^{\natural, s_i}\sigma = bin^{8s_i}(n) = bin^{8s_i}$ $(\mathbf{c} \cdot \mathbf{v} + d)$, where $d = n$ and $\mathbf{c} = \langle 0, \ldots 0 \rangle$. Hence $e \equiv \mathbf{c} \cdot \mathbf{x} + d$.
- Let $e = n * x_j + e'$. Suppose that $[\![e']\!]_{\text{Expr}}^{\natural, s_i}\sigma = bin^{8s_i}(\mathbf{c}' \cdot \mathbf{v} + d')$, where $e' \equiv \mathbf{c}' \cdot \mathbf{x} + d'$. By the definition of $[\![\cdot]\!]_{\text{Expr}}^{\natural, s}$, $[\![n * x_j + e]\!]_{\text{Expr}}^{\natural, s_i}\sigma = bin^{8s}(n) *^{8s_i} \sigma^{s_i}(a_j) +^{8s_i} [\![e']\!]_{\text{Expr}}^{\natural, s_i}\sigma$ where $\sigma^{s_i}(a_j) = bin^{8s_i}(v_j)$. By the definition of bin^{8s}, $bin^{8s}(n) *^{8s_i} bin^{8s_i}(v_j) = ((n \mod 2^{8s_i}) * (v_j \mod 2^{8s_i})) \mod 2^{8s_i} = (n * v_j) \mod 2^{8s_i}$; see [67, p. 42]. Similarly, $(n * v_j) \mod 2^{8s_i} +^{8s_i} [\![e']\!]_{\text{Expr}}^{\natural, s_i}\sigma = (n * v_j) \mod 2^{8s_i} +^{8s_i} bin^{8s_i}(\mathbf{c}' \cdot \mathbf{v} + d') = (n * v_j + \mathbf{c}' \cdot \mathbf{v} + d') \mod 2^{8s_i}$. Thus, set $d = d'$ and $\langle c_1, \ldots c_n \rangle = \langle c'_1, \ldots c'_{i-1}, c'_i + n, c'_{i+1}, \ldots c'_n \rangle$, where $\mathbf{c} = \langle c_1, \ldots c_n \rangle$ and $\mathbf{c}' = \langle c'_1, \ldots, c'_n \rangle$. Hence $e \equiv \mathbf{c} \cdot \mathbf{x} + d$.

The force of the result above is that a linear expression ⟨Expr⟩ over finite integer variables can be interpreted as an expression over polyhedral variables without regard for overflows or evaluation order. A prerequisite for this convenience is that all variables occurring in an expression have the same size s. In contrast, assignments between different-sized variables have to revert to a cast statement. In this case, and in the case of conditionals, wrapping has to be made explicit, which is the topic of the next section.

4.5 Explicit Wrapping of Polyhedral Variables

A consequence of the wrapping behaviour of $\gamma_{\mathbf{a}}^{\mathbf{s}}$ is that the effect of a guard such as `x<=y` cannot be modelled as a transformation from a polyhedron P to $P \sqcap_P [\![x \leq y]\!]$. This section explains this problem, discusses possible solutions, and proposes an efficient wrapping algorithm called *wrap*.

4.5.1 Wrapping Variables with a Finite Range

In order to illustrate the requirements on the wrapping algorithm *wrap*, consider Fig. 4.4. The thick line in the upper graph denotes $P = [\![x + 1024 = 8y, -64 \leq x \leq 448]\!]$, which we suppose feeds into the guard `x<=y`, where `x` and `y` both represent variables of type **uint8**. In order to illustrate a peculiarity of modelling the guard, consider the point $\langle x, y \rangle = \langle 384, 176 \rangle \in P$ and let $\sigma \in \gamma_{\mathbf{a}}^{\mathbf{s}}(\{\langle 384, 176 \rangle\})$. Due to implicit wrapping in $\gamma_{\mathbf{a}}^{\mathbf{s}}$, the state σ stipulates that $val^{8,\text{uint}}(\sigma^1(addr(x))) = 128$ and $val^{8,\text{uint}}(\sigma^1(addr(y))) = 176$. Thus, although `x<=y` is true when interpreting `x` and `y` as **uint8** in σ, the polyhedron

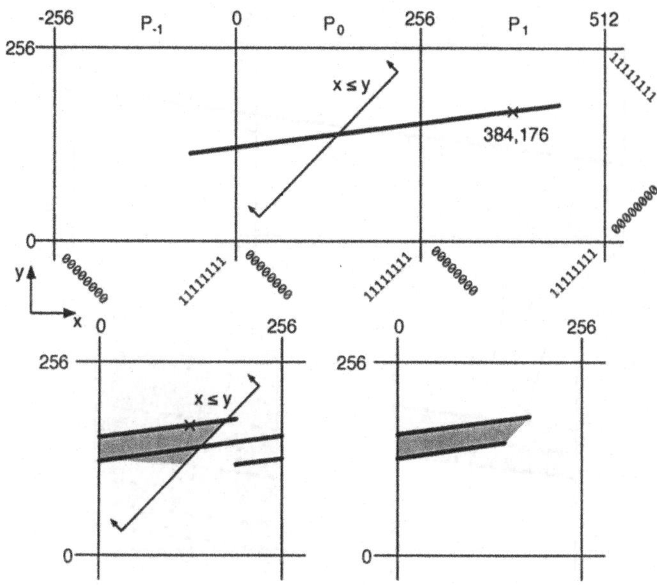

Fig. 4.4. Explicitly wrapping the possible values of x to its admissible range.

$\{\langle 384, 176\rangle\} \sqcap_P [\![x \leq y]\!]$ is empty. Hence, it is not correct to model the guard in the classic way as $P \sqcap_P [\![x \leq y]\!]$.

In order to model relational tests correctly, the values of expressions occurring on each side of a relational operator have to be wrapped to the type prescribed in a Sub C conditional. In the example, the expression y is already in the required range $[0, 255]$, whereas the range of x impinges on the two neighbouring quadrants as indicated in the upper graph of Fig. 4.4. These quadrants are obtained by partitioning the state P into $P_{-1} = P \sqcap_P [\![-256 \leq x \leq -1]\!]$, $P_0 = P \sqcap_P [\![0 \leq x \leq 255]\!]$, and $P_1 = P \sqcap_P [\![256 \leq x \leq 511]\!]$. The result of wrapping x can now be calculated by translating P_{-1} by 256 units towards positive x-coordinates and P_1 by 256 units towards negative x-coordinates, yielding $P' = P_0 \sqcup_P (P_{-1} \rhd x := x + 256) \sqcup_P (P_1 \rhd x := x - 256)$. The contribution of each partition is shown as a thick line in the lower left graph, and the grey region depicts $P' \sqcap_P [\![x \leq y]\!]$. Observe that a more precise state P'' can be obtained by intersecting each translated state separately with $[\![x \leq y]\!]$, that is, by calculating $(P_0 \sqcap_P [\![x \leq y]\!]) \sqcup_P ((P_{-1} \rhd x := x + 256) \sqcap_P [\![x \leq y]\!]) \sqcup_P ((P_1 \rhd x := x - 256) \sqcap_P [\![x \leq y]\!])$. This state, depicted as the grey area in the lower right graph, is smaller than P' since P_{-1} does not contribute at all. Indeed, this example shows that polyhedra are not meet-distributive, that is, $P \sqcap_P (P_1 \sqcup_P P_2) \neq (P \sqcap_P P_1) \sqcup_P (P \sqcap_P P_2)$. In this work, we chose to calculate the equivalent of P' in our wrapping function *wrap*, as it simplifies the presentation; implementing the refined model is mere engineering.

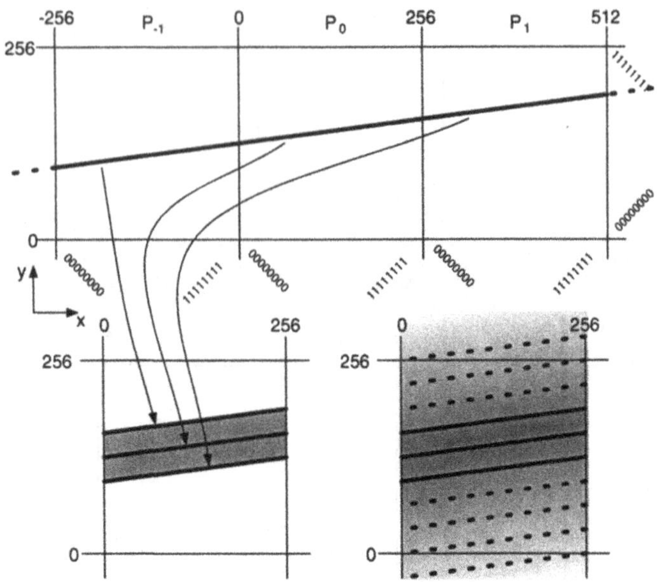

Fig. 4.5. The quest for an efficient wrapping of unbounded variables.

4.5.2 Wrapping Variables with Infinite Ranges

In the example given, it was possible to obtain a wrapped representation of the values of x and y by calculating the join of three constituent state spaces. In general, however, wrapping x and y can require the join of an infinite number of constituent state spaces, as depicted in Fig. 4.5. Here, the line in the upper graph depicts $P = [\![x + 1024 = 8y]\!]$; that is, P denotes the same linear relation as before, except that x is unbounded. Translating P by i times the range of **uint8** yields $P_i = (P \triangleright x := x + i2^8 \sqcap_P [\![0 \leq x \leq 255]\!]) \sqcup_P (P \triangleright x := x - i2^8 \sqcap_P [\![0 \leq x \leq 255]\!])$ for $i \geq 0$. A polyhedron that includes the sequence $P'_j = \bigsqcup_{0 \leq i \leq j} P_i$ can be computed using widening [62], thereby yielding the grey area in the lower right graph. In fact, this region is equivalent to $\exists_x(P) \sqcap_P [\![0 \leq x \leq 255]\!]$, as it contains neither bounds on x nor relational information between x and other variables. This suggests that, rather than wrapping unbounded variables, it is cheaper and as precise to set them to the whole range of their type. After wrapping x, it becomes apparent that y is unbounded, too, and hence needs wrapping.

4.5.3 Wrapping Several Variables

Even though the guard `x<=y` used in the example of Sect. 4.5.1 involves two variables, it was only necessary to wrap x to obtain a wrapped representation

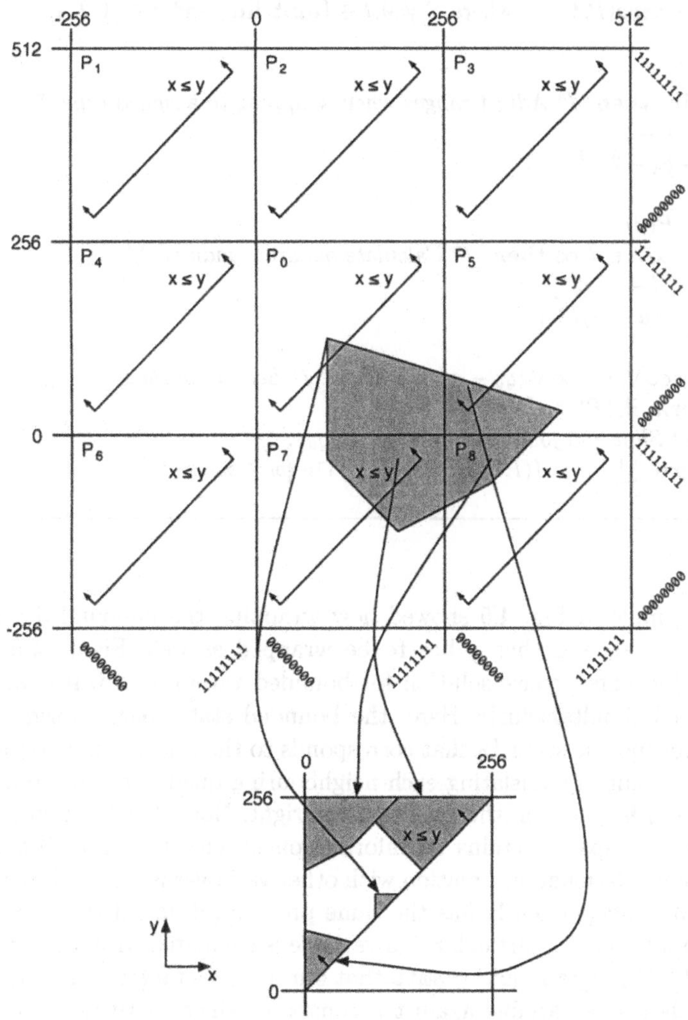

Fig. 4.6. Precise wrapping of two bounded variables.

of both x and y. The example of Sect. 4.5.2 hints at the fact that both variables might need wrapping to ensure that both sides of the guard are within range. In particular, it is not possible to translate a guard `x<=y` to the inequality $x - y \leq 0$ and to merely wrap $x - y$ to $[0, 255]$. To illustrate this, consider the simpler case `x<=42`, which is satisfied for bit sequences of `x` that fall within $[0, 42]$. In order to evaluate `x<=42`, set $x' = x - 42$ and wrap x' such that $0 \leq x' \leq 255$. The intersection with $[\![x' \leq 0]\!]$ constrains x' to 0, which implies $x = 42$ instead of $x \in [0, 42]$. Thus, both arguments to a guard `x<=y` need to be wrapped independently.

Algorithm 1 Explicitly wrapping an expression to the range of a type.

procedure *wrap*$(P, t\ s, x)$ where $P \neq \emptyset, t \in \{\mathbf{uint}, \mathbf{int}\}$ and $s \in \{1,2,4,8\}$

$b_l \leftarrow 0$
$b_h \leftarrow 2^s$
if $t = \mathbf{int}$ **then** /* *Adjust ranges when wrapping to a signed type.* */
  $b_l \leftarrow b_l - 2^{s-1}$
  $b_h \leftarrow b_h - 2^{s-1}$
end if
$[l, u] \leftarrow P(x)$
if $l \neq -\infty \wedge u \neq \infty$ **then** /* *Calculate quadrant indices.* */
  $q_l \leftarrow \lfloor (l - b_l)/2^s \rfloor$
  $q_u \leftarrow \lfloor (u - b_l)/2^s \rfloor$
end if
if $l = -\infty \vee u = \infty \vee (q_u - q_l) > k$ **then** /* *Set to full range.* */
  return $\exists_x(P) \sqcap_P [\![b_l \leq x < b_h]\!]$
else /* *Shift and join quadrants* $\{q_l, \ldots q_u\}$. */
  return $\bigsqcup_{q \in [q_l, q_u]}((P \triangleright x := x - q2^s) \sqcap_P [\![b_l \leq x < b_h]\!])$
end if

The example in Fig. 4.5 showed how wrapping the unbounded x leaves y unconstrained, and thus y has to be wrapped as well. Figure 4.6 shows a potentially more precise solution for bounded variables in which variables are wrapped simultaneously. Here, the bounded state space shown in grey expands beyond the state P_0 that corresponds to the actual range of the variables. The result of translating each neighbouring quadrant and intersecting it with $x \leq y$ is shown in the graph on the right. Note that the join of these four translated spaces retains no information on either x or y. While it is possible that relational information with other variables is retained, wrapping the variables independently has the same precision if one of the variables is within bounds and, in particular, if a variable is compared with a constant. In the 3000 LOC program `qmail-smtp` that our analysis targets, 427 out of 522 conditionals test a variable against a constant, which motivates our design choice of wrapping variables independently.

4.5.4 An Algorithm for Explicit Wrapping

Guided by the observations made in the three examples, Alg. 1 gives a procedure to wrap a polyhedral variable to the range of a given integer type. Due to the observations in the last section, we only present an algorithm to wrap one variable at a time. Hence, the evaluation of a guard requires two invocations of *wrap*, one for each expression of the condition.

The algorithm commences by calculating the maximum bounds of the type $t\ s$. A **uint8** type, for instance, will set $b_l = 0$ and $b_h = 2^8 = 256$, while an **int8** type results in the bounds $b_l = 0 - 2^{8-1} = -128$ and $b_h = 2^8 - 2^{8-1} = 128$.

Line 7 calculates the bounds of x in P. If one of these bounds is infinite, line 13 removes all information on x and restrains x to $[b_l, b_h - 1]$. In the case of finite bounds, lines 9–10 calculate the smallest and largest quadrants into which the values of x impinge. For instance, in the example of Fig. 4.4, these numbers are $q_l = -1$ (for the quadrant $[-256, -1]$) and $q_h = 1$ (for $[256, 511]$). Line 12 ensures that the linear expression is simply set to its maximum bounds if more than k quadrants have to be transposed and joined, where k is a limit that can be tuned to the required precision. Line 14 transposes each quadrant and restricts it to the bounds of the type. The correctness of *wrap* is asserted below.

Proposition 2. *Given $P \neq \emptyset$ and $P' = wrap(P, t\ s, x_i)$, the interval $P'(x_i)$ lies in the range of the type $t\ s$. Furthermore, $\gamma^{\mathbf{s}}_{\mathbf{a}}(P) \subseteq \gamma^{\mathbf{s}}_{\mathbf{a}}(P')$.*

Proof. Upon return from lines 13 and 15, x_i is restricted to lie between the bounds b_l and $b_h - 1$ of the type $t\ s$; hence $P'(x_i)$ lies in the range of type $t\ s$.

Suppose $\mathbf{a} = \langle a_1, \ldots a_n \rangle$ and $\mathbf{s} = \langle s_1, \ldots s_n \rangle$, where $a_i = addr(x_i)$ and s_i denotes the size of x_i in bytes. Let $\sigma \in \gamma^{\mathbf{s}}_{\mathbf{a}}(P)$. Then there exists $\langle v_1, \ldots v_n \rangle \in P \cap \mathbb{Z}^n$ such that $\sigma \in \gamma^{\mathbf{s}}_{\mathbf{a}}(\{\mathbf{v}\})$. We consider two behaviours of *wrap*.

- Suppose that *wrap* is exited at line 13. Observe that for any $\mathbf{b} \in \mathbb{B}^{8s_i}$ there exists $v \in \{b_l, \ldots, b_h - 1\}$ such that $bin^{8s_i}(v) = \mathbf{b}$. Hence, there exists $\mathbf{v}' = \langle v_1, \ldots v_i', \ldots v_n \rangle$ with $v_i' \in [b_l, b_h - 1] \cap \mathbb{Z}$ and $bin^{8s_i}(v_i') = bin^{8s_i}(v_i)$. Observe that $\mathbf{v}' \in P' = \exists_x(P) \sqcap_P [\![b_l \leq x < b_h]\!]$ and $\mathbf{v}' \in \mathbb{Z}^n$. Hence $bits^{\mathbf{s}}_{\mathbf{a}}(v_j) = bits^{\mathbf{s}}_{\mathbf{a}}(v_j')$ for all $j \in [1, n]$ and it follows that $\sigma \in \gamma^{\mathbf{s}}_{\mathbf{a}}(P')$.
- Suppose now that *wrap* exits at line 15. Observe that $v_i \in [l, u]$, and hence there exists $q \in [q_l, q_u]$ such that $v_i - q2^{s_1} \in [b_l, b_h - 1]$. Hence, there exists $\mathbf{v}' = \langle v_1, \ldots v_i', \ldots v_n \rangle \in P'$ such that $v_i' = v_i - q2^{s_i}$. Since $bin^{8s_i}(q2^{s_i}) = \mathbf{0}$, it follows that $bin^{8s_i}(v_i') = bin^{8s_i}(v_i - q2^{s_i}) = bin^{8s_i}(v_i)$. Thus $\sigma \in \gamma^{\mathbf{s}}_{\mathbf{a}}(P')$.

Note that the translation of quadrants using $P \rhd x := x + q2^s$ can be implemented by a potentially cheaper affine transformation [14]. However, the solution shown can be readily implemented using other polyhedral domains [169, 172] that do not directly support affine translations.

4.6 An Abstract Semantics for Sub C

This section defines the abstract semantics of Sub C by defining a set of abstract transfer functions. The idea is to calculate a single polyhedron P for each label l, where each label marks the beginning of a basic block. Starting with the unrestricted polyhedron $\mathbb{R}^{|\mathcal{X}|}$ for the first basic block and with the empty polyhedron $\emptyset \subseteq \mathbb{R}^{|\mathcal{X}|}$ for all others, the abstract transfer function for basic blocks is evaluated repeatedly until a (post-)fixpoint is reached. Once a fixpoint is reached, each state σ that may arise in the concrete program at l satisfies $\sigma \in \gamma^{\mathbf{s}}_{\mathbf{a}}(P)$, where P is the polyhedron associated with l.

Basic Blocks.
$[\![\, l : s_1; \ldots s_n; \,]\!]^\sharp_{\text{Block}} P = [\![\, lookupNext(l) \,]\!]^\sharp_{\text{Next}}([\![\, s_n \,]\!]^\sharp_{\text{Stmt}}(\ldots [\![\, s_1 \,]\!]^\sharp_{\text{Stmt}} P \ldots))$

Control Flow.
$[\![\, \textbf{jump}\ l \,]\!]^\sharp_{\text{Next}} P = \{\langle P, l\rangle\}$

$[\![\, \textbf{if}\ t\ s\ v\ op\ exp\ \textbf{then jump}\ l\ ;\ n \,]\!]^\sharp_{\text{Next}} P = \{\langle P^{then}, l\rangle\} \cup [\![\, n \,]\!]^\sharp_{\text{Next}} P^{else}$

$$\begin{aligned} \text{where } P' &= P \rhd y := [\![\, exp \,]\!]^\sharp_{\text{Expr}} \text{ where } y \in \mathcal{X}^T \text{ fresh} \\ P^{then} &= \exists_y(cond(P', t\ s, v, y, op)) \\ P^{else} &= \exists_y(cond(P', t\ s, v, y, neg(op))) \end{aligned}$$

$$cond(P, t\ s, x, y, op) = \begin{cases} (P'' \sqcap_P [\![x < y]\!]) \sqcup_P (P'' \sqcap_P [\![x > y]\!]) & \text{if } op \in \{\neq\} \\ (P'' \sqcap_P [\![x\ op\ y]\!]) & \text{otherwise} \end{cases}$$

$$\begin{aligned} \text{where } P' &= wrap(P, t\ s, x) \\ P'' &= wrap(P', t\ s, y) \end{aligned}$$

Expressions.
$[\![\, n \,]\!]^\sharp_{\text{Expr}} = n$

$[\![\, n * v + exp \,]\!]^\sharp_{\text{Expr}} = n\ v + [\![\, exp \,]\!]^\sharp_{\text{Expr}}$

Assignments.
$[\![\, s\ v = exp \,]\!]^\sharp_{\text{Stmt}} P = P \rhd v := [\![\, exp \,]\!]^\sharp_{\text{Expr}}$

Type Casts.

$$[\![\, s_1\ v_1 = t\ s_2\ v_2 \,]\!]^\sharp_{\text{Stmt}} P = \begin{cases} P' & \text{if } s_1 \leq s_2 \\ wrap(P', t\ s_2, v_1) & \text{otherwise} \end{cases}$$

where $P' = P \rhd v_1 := v_2$

Fig. 4.7. Abstract semantics of Sub C.

Specifically, given a polyhedron P that is valid at the beginning of a basic block, the first rule in Fig. 4.7 specifies how P is modified by the statements before it is fed into the evaluation of the control-flow statements of the basic block at label l. Specifically, $[\![\, lookupNext(l) \,]\!]^\sharp_{\text{Next}} P_l$ yields tuples such as $\langle P'_{l'}, l'\rangle$, indicating that $P'_{l'}$ must be joined with the existing state at l'. For instance, **jump** l merely returns the current state paired with the target label. In contrast, the conditional calculates two polyhedra, P^{then} (which is returned for the label l) and P^{else} (which is used to evaluate other control-flow instructions). The calculation of P^{else} makes use of a function *neg*, which negates a relational operator; for example, $neg('<') = '\geq'$. The auxiliary function *cond* wraps the two arguments of the relational operator *op*. Like *wrap*, this function can only wrap single polyhedral variables, which requires that *exp* be assigned to a temporary variable y that is projected out once the guard is applied.

Observe that enforcing the guard by intersecting with $[\![x\ op\ y]\!]$ has the same effect as wrapping the expression *exp* itself since $y = exp$ holds in P'.

However, if *wrap* returns from line 13 in Alg. 1, the variable y is merely set to the bounds of the type. In this case, *wrap* discards the relational information between y and *exp*, and the intersection with $[\![x\ op\ y]\!]$ does not affect the value of *exp* in P'', thereby ignoring the condition. An alternative treatment for expressions *exp* that exceed k quadrants would be to discard any previous information on variables in *exp* using projection and to modify *wrap* to intersect P with $[\![b_l \leq exp < b_h]\!]$. In this case, the information in the guard could be retained by intersecting with $[\![x\ op\ exp]\!]$ at the cost of discarding any previous bounds on the variables of *exp*.

The following proposition states the correctness of the conditional.

Proposition 3. *If* $val^{8s,t}(\sigma^s(addr^\natural(x_i)))\ op\ val^{8s,t}([\![\ exp\ ]\!]^{\natural,s}_{\text{Expr}}\sigma)$ *and* $\sigma \in \gamma^{\mathbf{s}}_{\mathbf{a}}(P)$, *then* $\langle P', l\rangle \in [\![\textbf{if}\ t\ s\ x_i\ op\ exp\ \textbf{then jump}\ l;\ n]\!]^\sharp_{\text{Next}}P$ *and* $\sigma \in \gamma^{\mathbf{s}}_{\mathbf{a}}(P')$.

Proof. Since $\sigma \in \gamma^{\mathbf{s}}_{\mathbf{a}}(P)$, there exists $\mathbf{v} \in P \cap \mathbb{Z}^n$ such that $\sigma \in \gamma^{\mathbf{s}}_{\mathbf{a}}(\{\mathbf{v}\})$. Let $\hat{P} = cond(\bar{P}, t\ s, x_i, y, op)$, where $\bar{P} = P \triangleright y := [\![\ exp\ ]\!]^\sharp_{\text{Expr}}$. Then $\langle v_1, \ldots v_n, \hat{v}\rangle \in \bar{P} \cap \mathbb{Z}^{n+1}$, where $\langle v_1, \ldots v_n\rangle = \mathbf{v}$, $\hat{v} = \mathbf{c}\cdot\mathbf{v} + d$, and $exp \equiv \mathbf{c}\cdot\mathbf{x} + d$. By Prop. 2, there exists $\mathbf{v}' = \langle v_1, \ldots v_{i-1}, v'_i, v_{i+1}, \ldots v_n, \hat{v}'\rangle \in \hat{P} \cap \mathbb{Z}^{n+1}$ such that $bin^{8s}(v_i) = bin^{8s}(v'_i) = \sigma^{8s}(addr(x_i))$. By following Prop. 1, $bin^{8s}(\hat{v}) = bin^{8s}(\hat{v}') = [\![\ exp\ ]\!]^{\natural,s}_{\text{Expr}}\sigma$. Furthermore, v'_i and $\hat{v}'$ lie in the range of $t\ s$ and thus $val^{8s,t}(\sigma^{8s}(addr(x_i))) = v'_i$ and $val^{8s,t}([\![\ exp\ ]\!]^{\natural,s}_{\text{Expr}}\sigma) = \hat{v}'$. Hence $\mathbf{v}' \in \hat{P} \sqcap_P [\![x\ op\ y]\!]$ for $op \notin \{\neq\}$. With $P' = \exists_y(\hat{P} \sqcap_P [\![x\ op\ y]\!])$, it follows that $\sigma \in \gamma^{\mathbf{s}}_{\mathbf{a}}(P')$. The argument is similar for $op \in \{\neq\}$.

The fall-through case can be shown to be correct by a similar argument.

Due to the modulo nature of $\gamma^{\mathbf{s}}_{\mathbf{a}}$, the evaluation of linear expressions and assignments resembles that of classic polyhedral analysis in that linear expressions in the program are simply reinterpreted as expressions over polyhedra variables. This holds true even for casts between different-sized variables as long as the target variable is smaller. Assigning smaller variables to larger ones, on the contrary, requires that wrapping be made explicit since a value that exceeds the range of the smaller source variable would wrap in the actual program, whereas it might not exceed the range of the larger target variable.

We conclude this section with a correctness proof for the cast statement.

Proposition 4. *Suppose* $\sigma \in \gamma^{\mathbf{s}}_{\mathbf{a}}(P)$ *and* $\sigma' = [\![\ s_1\ x_i = t\ s_2\ x_j\ ]\!]^\natural_{\text{Stmt}}\sigma$, *and let* $P' = [\![\ s_1\ x_i = t\ s_2\ x_j\ ]\!]^\sharp_{\text{Stmt}}P$. *Then* $\sigma' \in \gamma^{\mathbf{s}}_{\mathbf{a}}(P')$.

Proof. Since $\sigma \in \gamma^{\mathbf{s}}_{\mathbf{a}}(P)$, there exists $\mathbf{v} = \langle v_1, \ldots v_n\rangle \in P$ such that $\sigma \in \gamma^{\mathbf{s}}_{\mathbf{a}}(\{\mathbf{v}\})$. Let $\langle v'_1, \ldots v'_n\rangle \in P \triangleright x_i := x_j$, where $v'_i = v'_j$ and $v'_k = v_k$ for all $k \neq i$. By Lemma 1, $\sigma'^{s_k}(addr(x_k)) = \sigma^{s_k}(addr(x_k))$ for all $k \neq i$. By the definition of $[\![\ \cdot\ ]\!]^\natural_{\text{Stmt}}$, we need to show that $\sigma'^{s_1}(addr(x_i)) = bin^{8s_1}(val^{8s_2,t}(\sigma^{s_2}(addr(x_j))))$.

- Suppose $s_1 \leq s_2$. Then $bin^{8s_1}(x) = bin^{8s_1}(val^{8s_2,t}(bin^{8s_2}(x)))$. But $bin^{8s_1}(v'_i) = \sigma'^{s_1}(addr(x_i))$ and $bin^{8s_2}(v_j) = \sigma^{s_2}(addr(x_j))$; thus $\sigma' \in \gamma^{\mathbf{s}}_{\mathbf{a}}(P')$ follows.

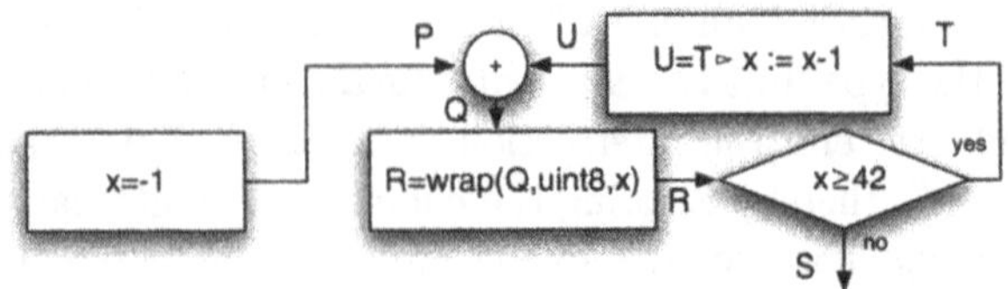

Fig. 4.8. Precision loss incurred by joining flow paths.

- Suppose now that $s_1 > s_2$. By Prop. 2, there exists $\langle v_1, \ldots \hat{v}_i, \ldots v_n \rangle \in P'$ such that $bin^{8s_2}(\hat{v}_i) = bin^{8s_2}(v_j)$ and $\hat{v}_i$ lies in the range of the type $t\ s_2$; that is, $val^{8s_2,t}(bin^{8s_2}(\hat{v}_i)) = \hat{v}_i$. But since $bin^{8s_2}(\hat{v}_i) = bin^{8s_2}(v_j) = \sigma'^{s_1}(addr(x_j))$, it follows that $\sigma'^{s_1}(addr(x_i)) = bin^{8s_1}(\hat{v}_i)$ as required.

We conclude this chapter with a discussion of the concretisation function $\gamma^{\mathbf{s}}_{\mathbf{a}}$.

4.7 Discussion

The existence of a concretisation map $\gamma^{\mathbf{s}}_{\mathbf{a}}$ begs the question of whether an abstraction map $\alpha^{\mathbf{s}}_{\mathbf{a}} : \Sigma \to \mathit{Poly}$ can be defined. For classic polyhedral analysis [62], it is well known that no best abstraction exists for certain shapes, such as a disc (see [59] or Sect. 3.3.1). In the context of our analysis, the set of concrete states Σ is finite, which implies that only a finite number of abstract states are required to represent the concrete states. However, a given set of states still has no single best abstraction. Consider $\sigma \in \Sigma$ with $\sigma^1(addr(\mathbf{x})) = 11111111$, $P_1 = [\![x = -1]\!]$, and $P_2 = [\![x = 255]\!]$. Although $\sigma \in \gamma^{\mathbf{s}}_{\mathbf{a}}(P_1) = \gamma^{\mathbf{s}}_{\mathbf{a}}(P_2)$, the polyhedra P_1 and P_2 are incomparable. As a consequence, the meet operation can only be applied after *wrap* has translated the range of $\mathbf{x}$ in P_1 to the same quadrant $[0, 255]$ as that of P_2, which makes the two polyhedra comparable.

Since different polyhedra can describe the same set of concrete states, precision can be lost when joining branches in the control-flow graph. Consider the following loop whose control-flow graph is shown in Fig. 4.8:

```
unsigned char x = (unsigned char) -1;
while (x>=42) x--;
```

The loop is entered with $P = [\![x = -1]\!]$, the largest value an unsigned variable can take. As the loop invariant $x \geq 42$ interprets x as an unsigned quantity, $Q = P \sqcup_P U$ is wrapped to $R = \mathit{wrap}(Q, \mathbf{uint8}, x) = [\![x = 255]\!]$. After evaluating the loop body, a precision loss occurs when P and $U = [\![x = 254]\!]$ are joined to obtain $P \sqcup_P U = [\![-1 \leq x \leq 254]\!]$, as x cannot fall below 42. One solution to this particular problem is to unroll the loop once, which avoids the join of the two different representations $[\![x = -1]\!]$ and $[\![x = 255]\!]$.

Observe that *wrap* does not compromise the termination behaviour of the analysis, as *wrap* is monotonic. Monotonicity is guaranteed, as the output polyhedron of *wrap* is a join of k quadrants that form a partitioning of the input polyhedron. If the input state grows in such a way that more than n quadrants need to be translated and joined, the output is a polyhedron that includes all joins of $k < n$ quadrants.

Furthermore, observe that *wrap* is idempotent and, in particular, is the identity if the variable is in range. An important consequence is that our solution is as precise as classic polyhedral analysis if all variables stay within the range of their types.

An interesting benefit of $\gamma_{\mathbf{a}}^{\mathbf{s}}$ is that the possible values of a byte x can be represented as either $[\![-128 \leq x \leq 127]\!]$ or $[\![0 \leq x \leq 255]\!]$. For example, the refinement of the analysis to handle C string buffers presented in Chap. 11 does not model individual array elements but tracks a single NUL position (a character with value zero) within the array. Even though **char** is often signed, the range $[0, 255]$ can be returned when reading a byte from the array, which can then be refined using the NUL position to $[1, 255]$ whenever the access lies in front of the NUL position (see Fig. 12.5 on p. 224 for an example). If a signed range had to be returned, it would include the NUL character since $[-128, -1] \cup [1, 127] = [-128, 127]$ is the best convex approximation. Without the ability to model an unknown byte as a strictly positive interval, it is not possible to prove that a loop iterates until the first NUL character is found.

4.7.1 Related Work

A number of works have addressed the analysis congruences [8,85,87]; that is, the inference that a variable x can only take on values such that $x \equiv b \ mod \ m$. Although affine relationships *mod* 2^n between variables can be inferred between finite integer variables [135], little work exists in the more general context of polyhedral analyses [171]. Cousot et al. use a two-tier approach [60] in the context of the Astrée analyser that is based on the definitions of the C standard [51]. For signed integers, any wrapping is erroneous. In this case, each time a variable is set, its range is checked for overflows. Overflows of unsigned integers are assumed to be intentional, as wrapping may result from bit-level operations. This approach requires a separation of signed and unsigned variables, which raises the question of how the mixing of signed and unsigned arithmetic is handled. Assuming that using a variable as both signed and unsigned integers is incorrect and thereby rejects many valid C programs. Indeed, an analysis of our example would lead to the misleading warning about converting from a signed to an unsigned integer.

As far as we are aware, wrapping of integer variables is ignored in the SLAM analyser [18], which renders it unsound with respect to conditionals. It was pointed out by Cook et al. in the talk of [52] that this deficiency is currently being addressed.

5

Overlapping Memory Accesses and Pointers

This chapter lifts the restrictions that were imposed on the *SubC* language, which was used in the last chapter to illustrate the relationship between abstract and concrete semantics. Lifting the analysis to the full expressiveness of Core C not only introduces structured memory and arrays but also casting, pointers and pointer arithmetic. This can of worms is tackled in two steps. Section 5.1 introduces a representation of memory based on fields, where the value of each field is represented by an abstract variable $x \in \mathcal{X}$. Section 5.3 enhances this model with abstract addresses (so-called l-values), which enables the analysis of pointers and pointer arithmetic.

5.1 Memory as a Set of Fields

An analysis of a language like C does not allow for a simple bijection between concrete variables in the program and abstract variables in the abstract domain. A more complex model of the program memory is required for the following three reasons.

Accesses with Incompatible Types

Programming languages commonly use types to provide a partial correctness guarantee by enforcing a certain interpretation of a memory location. In contrast, types in C merely specify the semantics of an access to memory; in particular, it is possible to access the same memory region with different types. For instance, a 4-byte memory region may be accessed as a signed or unsigned integer or as a pointer. Furthermore, it is possible to read a variable of type **uint8** from an address that was written as **uint16**. In this case, either the upper or lower 8 bits of the larger variable are read, depending on the architecture. Overlapping accesses are often used in C programs to initialise data structures efficiently by writing 4-byte quantities even though the declared elements of the underlying structure are smaller. Hence, dealing with accesses of incompatible types and sizes is not only required to ensure soundness but also important to ensure precision. Access to the same memory region with different-sized types is modelled by using one abstract variable for each access size, which requires care to update these overlapping variables consistently.

Limiting the Number of Abstract Variables

Another challenge in the analysis of generic C programs is to limit the amount of information inferred, specifically to bound the number of abstract variables $\mathcal{X}$ used at any given point. This is an important efficiency consideration in the presence of large data structures such as arrays since polyhedra are only tractable when the number of variables does not exceed a few dozen, whereas character arrays easily contain several thousand elements. Furthermore, arrays that are stored in dynamically allocated memory may only have a symbolic upper bound so that it is not clear how many abstract variables are required to describe the contents of the array. However, as the example in the introduction shows, it is not always necessary to argue about individual array elements in order to verify correct memory management, which begs the question of which parts of a memory region are relevant for verification.

Handling Dynamically Allocated Memory

A third challenge arises due to the program's allocation of a different number of memory regions depending on the input to the program. Since the number of allocated memory regions is limited only by the available memory, it is impossible to represent every dynamically allocated memory region explicitly. Our analysis summarises memory regions that were allocated at the same location in the program. Thus, an abstract dynamic memory region may relate to several (or possibly zero) concrete memory regions.

After presenting the conceptual ideas of how polyhedral variables represent the content of memory regions, Sect. 5.2 introduces access trees that model overlapping value variables, which Sect. 5.3 enriches to handle pointers.

5.1.1 Memory Layout for Core C

Motivated by the three requirements presented above, this section details how the contents of memory regions in the program are defined by the abstract variables in the corresponding numeric domain.

A concrete program operates on two kinds of memory regions, namely declared variables $\mathcal{M}$ and dynamically allocated memory. With respect to the former, observe that our analysis is fully context-sensitive in the sense that a function is reanalysed for every new call site. This approach disallows recursive function calls, as a recursive cycle can lead to an unbounded call stack in the analyser. However, the absence of recursive functions makes for a simple model in that every declared variable $m \in \mathcal{M}$ exists either once in the program, namely when the function in which it is declared is on the stack, or not at all. In contrast to these automatic variables, dynamically allocated memory regions in the concrete program are summarised into a finite set of abstract memory regions $\mathcal{D} \subseteq Label$, where $\mathcal{D} \cap \mathcal{M} = \emptyset$ and every $l \in \mathcal{D}$ denotes the location of a `malloc` instruction that allocated the concrete memory regions. In order to formalise this relationship, we introduce two mappings, namely $L : \mathcal{M} \cup \mathcal{D} \to \mathcal{A}$ and $\rho : \mathcal{A} \to \mathcal{P}([0, 2^{32} - 1])$. The former map, L, assigns an abstract address to each memory region. An address $a \in \mathcal{A}$ is called abstract since it may correspond to several concrete addresses, for instance when the abstract address denotes a dynamically allocated memory region. In order to express this property of abstract addresses, define the address map ρ to map each abstract address to a set of concrete addresses. This function is further lifted to operate on sets of abstract addresses; that is, $\rho(A) = \bigcup_{a \in A} \rho(a)$ for any $A \subseteq \mathcal{A}$. These two maps are sufficient to express the addresses at which an abstract memory region $m \in \mathcal{M} \cup \mathcal{D}$ manifests itself in a concrete store σ. What remains is to specify the content of abstract memory regions. To this end, let $F : \mathcal{M} \cup \mathcal{D} \to \mathcal{P}(\mathbb{N} \times \mathbb{N} \times \mathcal{X})$ denote the set of fields of a memory region. Given a memory region $m \in \mathcal{M} \cup \mathcal{D}$, a field $\langle o, s, x_i \rangle \in F(m)$ indicates that the abstract variable $x_i \in \mathcal{X}$ represents s bytes starting at byte offset o. As was pointed out, the bytes of a memory region do not have to be completely covered by fields; conversely, a given byte might be modelled by several abstract variables in the case of overlapping fields.

Note that the sets $\mathcal{M}$ and $\mathcal{D}$ are fully defined by the program. Furthermore, $L : \mathcal{M} \cup \mathcal{D} \to \mathcal{A}$ is initialised at the start of the analysis to assign a unique abstract address to each memory region. In contrast, the map F determines which parts of the memory regions are relevant to the verification task and which fields may be omitted for efficiency. The choice of F therefore requires insight into the behaviour of the analysed program and thus cannot be chosen up front at the beginning of the analysis. For the sake of this section, we simply assume that F is fixed at the start of the analysis by some oracle to include all fields that are relevant.

In order to illustrate the use of the maps L and F, consider the C **`struct`** declaration presented on the left of Fig. 5.1. The structure details the properties of a network socket connection, namely the port number, a 16-bit integer, and the IP address, a 4-byte value that is often accessed as four separate bytes. Suppose that the fields $F($`ip_info`$)$ are given as in the figure and that $L($`ip_info`$) = a_m$, where $a_m \in \mathcal{A}$ is the unique abstract address associated with the variable `ip_info`. Assuming that this variable is automatic (that is, locally declared in a function f), its address depends on the position of f in the current call stack. If the function is on the stack, $\rho(a_m) = \{a\}$, where $a \in [0, 2^{32} - 1]$ is the address of `ip_info` in the stack frame of f. Given a vector of possible values $\mathbf{v} \in [\![N]\!]$, the resulting stores are $mem_\rho(\mathbf{v})$, where mem_ρ is defined as follows:

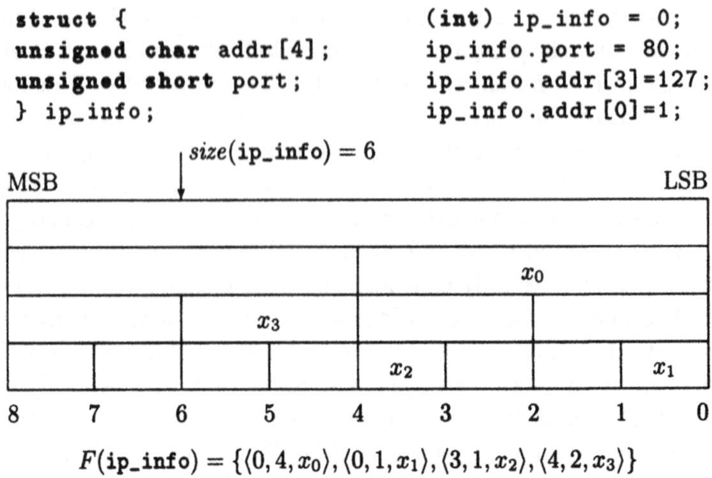

Fig. 5.1. An example of overlapping write accesses. Indexing into a memory region starts on the right to reflect the choice of a little-endian architecture. Note that the cast is not actually accepted by most C compilers. However, for the sake of readability, we chose this notation rather than the equivalent `*((int*)&ip_info)=0`.

$$mem_\rho(\mathbf{v}) = \bigcap_{m\in\mathcal{M}\cup\mathcal{D}} \left(\bigcap_{a\in\rho(L(m))} \left(\bigcap_{\langle o,s,x_i\rangle\in F(m)} bits^s_{a+o}(\pi_i(\mathbf{v})) \right)\right)$$

As before, let $\pi_i(\langle v_1, \dots v_{i-1}, v_i, v_{i+1}, \dots v_n\rangle) = v_i$ such that $bits^s_a(\pi_i(\mathbf{v}))$ creates all stores where s bytes at address a are restricted to v_i. Since no two memory regions in $\mathcal{M} \cup \mathcal{D}$ overlap, the possible stores for all memory regions can be expressed in a compositional way as implemented with the outermost intersection. Similarly, each memory region m at the abstract address $a_m \in L(m)$ gives rise to several sets of stores, one for each concrete address $a \in \rho(a_m)$ at which the memory region is currently live in the program. This is expressed by the second intersection. The innermost intersection ranges over all fields of a single memory region. In the case $F(m) = \emptyset$, we assume that the innermost intersection reduces to Σ, the neutral element. While memory regions $\mathcal{M} \cup \mathcal{D}$ never overlap, the individual fields within a single memory region may overlap if they are of different sizes. Thus, the innermost intersection can collapse to the empty set if overlapping fields are not updated in a coherent way. For instance, suppose the instructions on the upper right of Fig. 5.1 are executed by (incorrectly) updating the initial state N to $N' = N \triangleright x_0 := 0 \triangleright x_3 := 80 \triangleright x_1 := 1 \triangleright x_2 := 127$. The result $\{\mathbf{v}\} = [\![N']\!]$ implies $mem_\rho(\mathbf{v}) = \emptyset$ due to, for instance, x_0 and x_2 since $bits^4_a(0) \cap bits^1_{a+3}(127) = \emptyset$, assuming that $\{a\} = \rho(L(\texttt{ip\_info}))$. To implement the semantics of the byte-wise assignments correctly, the overlapping variable x_0 must be updated, too. This task is addressed in the next section.

5.2 Access Trees

Whenever a sequence of bits is represented by more than one abstract variable, an update of one abstract variable must update the others since the possible bit sequence in the actual store is determined by the values that all variables admit. This was illustrated in the previous section, where the variables x_0 and x_2 were not updated together, which led to an incorrect result. A simple strategy to ensure correct updates is to remove all information on all fields that overlap with the one that is being updated. However, this approach is rather imprecise. For instance, the update of line 7 sets x_2 to 127 but would have to project out the overlapping variable x_0, which means that all information about `addr[0]`, `addr[1]`, and `addr[2]` is lost. This section details a more refined update strategy, which is exact in the case of the example above. The idea is to propagate information between the different variables of fields that overlap, thereby ensuring that all variables that determine the bits of a certain memory location are set to a precise value.

Allowing arbitrary overlapping fields creates a plethora of ways to propagate information between fields, which makes the design of correct and precise algorithms a difficult task. A more elegant and principled approach is possible when restricting the possible position of fields such that their offset within a memory region can only be a multiple of their size, a requirement known as alignment. This restriction makes it possible to view possible field locations as nodes of a complete binary tree. An example of this view is depicted in Fig. 5.1, where each field of s bits is either a leaf or has exactly two children of $s/2$ bits. Suppose now that the updated information of the field $\langle 3, 1, x_2\rangle \in F(\texttt{ip\_info})$ must be propagated. The fields affected by an update of x_2 are simply all its parents and children:

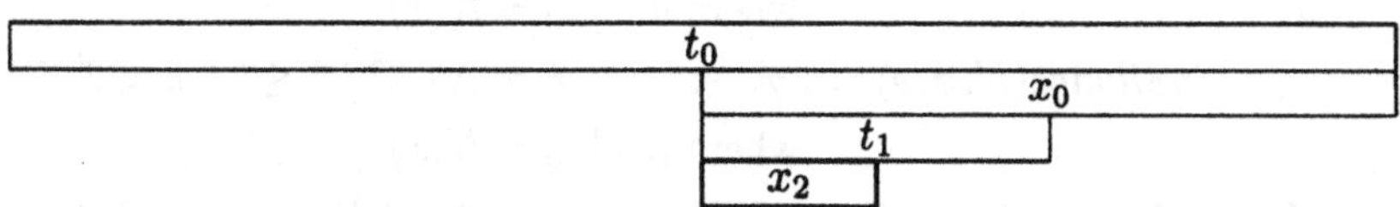

The slice of the full binary tree shown contains the field x_0 that overlaps with x_2. The idea is to propagate information from the node x_2 recursively to its parents and children in order to eventually update all overlapping fields. To this end, we populate the empty nodes of the slice with temporary variables $t_0, t_1 \in \mathcal{X}^T$. However, the temporary t_0 is irrelevant to the task of propagating information from x_2 to x_0 and can be removed, yielding the following slice:

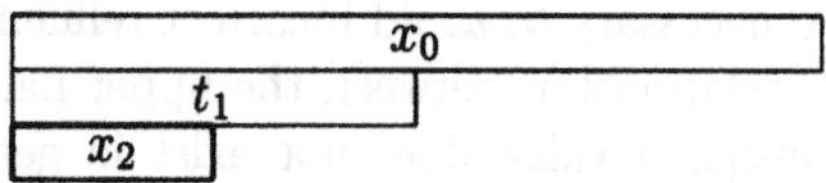

This slice is called an access tree with pivot x_2. Note that the pivot node is marked with thick borders and may have children as well as parents. Furthermore, the variable in the pivot node might be temporary. In general, the set of access trees is denoted as $\mathcal{AT}$, and we assume the existence of a function $access^F : \mathbb{N} \times \{1,2,4,8\} \times \mathcal{M} \cup \mathcal{D} \rightarrow \mathcal{AT}$ that generates an access tree from the fields $F(m)$ of a variable $m \in \mathcal{M} \cup \mathcal{D}$. The function $access^F(o, s, m)$ is defined for all aligned offsets o, that is, $o \bmod s = 0$.

Given an access tree, there are two principal ways of propagating information. Firstly, information can be propagated from the parents and children to the pivot node, and secondly, new information in the pivot node can be used to update other variables in the tree while retaining as much information as possible in the parents. We consider each direction in turn. The first approach can refine the simple strategy of projecting out overlapping fields on write accesses. Suppose that x_0 is set to zero, while x_2 is projected out to ensure that the information in x_2 does not contradict that in x_0. When x_2 is read, information from x_0 can be propagated to x_2 by passing the access tree from above to the function $prop : \{1,2,4,8\} \times \mathcal{AT} \times Num \rightarrow \mathcal{X} \times Num$ detailed in Fig. 5.2. This function returns the pivot variable x_2 and a domain in which the variables in the access tree are refined according to how they overlap. In the context of our example, let at denote the access tree; then the call $\langle x_2, N''\rangle = prop(1, at, N)$ invokes *fromUpper* recursively to calculate first $N' = N \sqcap_N \{x_0 - 65535 \leq 65536t_1 \leq x_0\}$ and then $N'' = N' \sqcap_N \{t_1 - 255 \leq 256x_2 \leq t_1\}$, yielding $N''(x_2) = [0,0]_{\equiv 64}$ as explained in detail below. In many cases, expressing linear relationships between two fields is only possible if the value of a field is within the bounds of an unsigned integer or if two adjacent fields are both within the bounds of a signed or unsigned integer. The tests to check these conditions are expressed by the following predicates:

$$\begin{aligned}
inURange(N,x,s) &= l \neq -\infty \wedge u \neq \infty \wedge 0 \leq l \wedge u < 2^{8s} \\
&\quad \text{where } [l,u]_{\equiv d} = N(x) \\
inSRange(N,x,s) &= l \neq -\infty \wedge u \neq \infty \wedge -2^{8s-1} \leq l \wedge u < 2^{8s-1} \\
&\quad \text{where } [l,u]_{\equiv d} = N(x) \\
pairInRange(N,x_1,s_1,x_2,s_2) &= inURange(x_1,s_1) \wedge inURange(x_2,s_2) \vee \\
&\quad\ inSRange(x_1,s_1) \wedge inSRange(x_2,s_2)
\end{aligned}$$

Each predicate is applied to a numeric element $N \in Num$, domain variables $x, x_1, x_2 \in \mathcal{X}$, and the size of the field in bytes. Note that *inSRange*, which checks if the given variable is in the range of a signed integer, is only used for *pairInRange*.

The predicates are necessary to avoid incorrect refinement of the domain. For instance, a linear relationship between the upper half of a signed 16-bit value and an 8-bit unsigned value does not exist in general, as the upper half of the larger variable might indicate that the values $[-1, 0]$ are possible, while the unsigned 8-bit variable indicates $[0, 255]$. Both ranges allow the bit patterns 11111111 and 00000000, but no refinement of the unsigned value is possible without explicitly wrapping it. While it is clear in this example

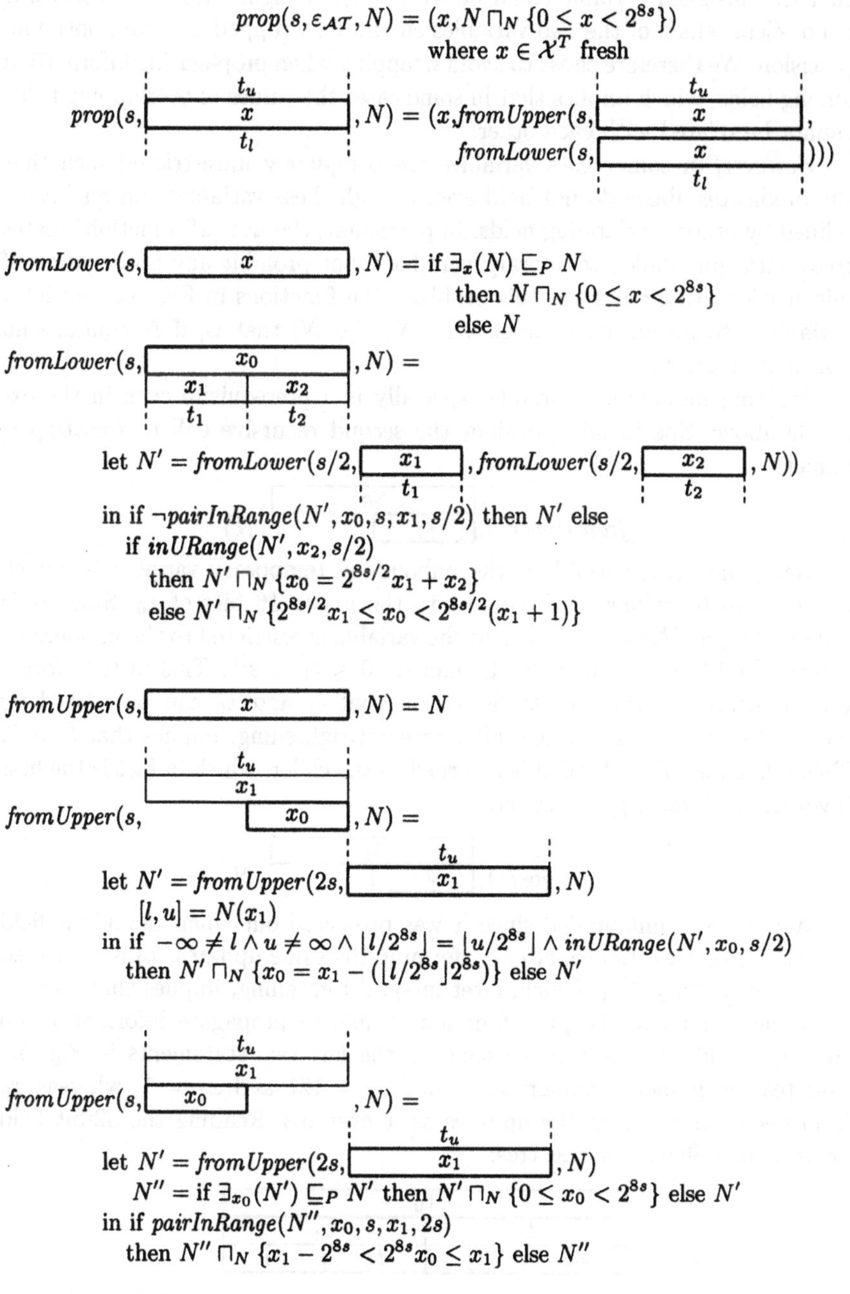

$$prop(s, \varepsilon_{AT}, N) = (x, N \sqcap_N \{0 \le x < 2^{8s}\})$$
where $x \in \mathcal{X}^T$ fresh

$$prop(s, \boxed{\begin{matrix} t_u \\ x \\ t_l \end{matrix}}, N) = (x, fromUpper(s, \boxed{\begin{matrix} t_u \\ x \end{matrix}}, fromLower(s, \boxed{\begin{matrix} x \\ t_l \end{matrix}})))$$

$$fromLower(s, \boxed{x}, N) = \text{if } \exists_x(N) \sqsubseteq_P N \text{ then } N \sqcap_N \{0 \le x < 2^{8s}\} \text{ else } N$$

$$fromLower(s, \boxed{\begin{matrix} x_0 \\ x_1 \mid x_2 \\ t_1 \mid t_2 \end{matrix}}, N) =$$

let $N' = fromLower(s/2, \boxed{\begin{matrix} x_1 \\ t_1 \end{matrix}}, fromLower(s/2, \boxed{\begin{matrix} x_2 \\ t_2 \end{matrix}}, N))$
in if $\neg pairInRange(N', x_0, s, x_1, s/2)$ then N' else
if $inURange(N', x_2, s/2)$
then $N' \sqcap_N \{x_0 = 2^{8s/2}x_1 + x_2\}$
else $N' \sqcap_N \{2^{8s/2}x_1 \le x_0 < 2^{8s/2}(x_1 + 1)\}$

$$fromUpper(s, \boxed{x}, N) = N$$

$$fromUpper(s, \boxed{\begin{matrix} t_u \\ x_1 \\ \quad x_0 \end{matrix}}, N) =$$

let $N' = fromUpper(2s, \boxed{\begin{matrix} t_u \\ x_1 \end{matrix}}, N)$
$[l, u] = N(x_1)$
in if $-\infty \ne l \wedge u \ne \infty \wedge \lfloor l/2^{8s} \rfloor = \lfloor u/2^{8s} \rfloor \wedge inURange(N', x_0, s/2)$
then $N' \sqcap_N \{x_0 = x_1 - (\lfloor l/2^{8s} \rfloor 2^{8s})\}$ else N'

$$fromUpper(s, \boxed{\begin{matrix} t_u \\ x_1 \\ x_0 \quad \end{matrix}}, N) =$$

let $N' = fromUpper(2s, \boxed{\begin{matrix} t_u \\ x_1 \end{matrix}}, N)$
$N'' = \text{if } \exists_{x_0}(N') \sqsubseteq_P N' \text{ then } N' \sqcap_N \{0 \le x_0 < 2^{8s}\} \text{ else } N'$
in if $pairInRange(N'', x_0, s, x_1, 2s)$
then $N'' \sqcap_N \{x_1 - 2^{8s} < 2^{8s}x_0 \le x_1\}$ else N''

Fig. 5.2. Read operations on access trees. The variable t_u represents a wider tree than the dotted borders suggest; t_l, t_1, and t_2 represent two trees of half-width. Each function is called with a pivot node, which is s bytes wide and drawn in bold. The empty access tree $\varepsilon_{AT} \in \mathcal{AT}$ denotes a tree with no field variables.

that the unsigned variable could be wrapped to a signed range, in general it is not clear which of the two variables should be wrapped to retain optimum precision. We therefore chose to avoid wrapping when propagating information among fields, which implies that in some cases the values of overlapping fields cannot be related with each other.

However, in some cases variables are completely unrestricted such that the predicates above do not hold even though these variables can readily be refined by other overlapping fields. In particular, the *access*^{F} function creates trees with unbounded temporary variables that prohibit any propagation of information. To circumvent this problem, the functions in Fig. 5.2 restrict a variable x to its unsigned range if $\exists_x(N) \sqsubseteq_N N$; that is, if N contains no information on x.

Treating unbounded variables specially is a prerequisite even in the example above. Specifically, consider the second recursive call to *fromUpper*, namely

fromUpper(2,

x_0	
t_1	

, N)

Here, the pivot variable is the unbounded temporary variable t_1, which is meant to be refined with respect to the upper 16 bits of x_0. Since t_1 is unbounded, $\exists_{t_1}(N) = N$ holds and the variable is restricted to the unbounded range of a 16-bit integer variable, namely $0 \leq t_1 < 2^{16}$. This in turn forces the predicate *pairInRange* to hold such that t_1 and x_0 can be related by $x_0 - 2^{16} < 2^{16} t_0 \leq x_0$, which, after integral tightening, implies that $t_1 = 0$. This refined numeric domain is returned to the caller, which in fact is the first invocation of *fromUpper*, namely

fromUpper(1,

t_1	
x_2	

, N)

Again, x_2 is unbounded since it was projected out when the 32-bit field x_0 was set to zero. Hence, after performing the same updates, x_2 is restricted to $t_1 - 2^8 < 2^8 x_2 \leq t_1$, which, after integral tightening, implies that $x_2 = 0$. Using similar tests, the *prop* function is able to propagate information up to larger fields. For instance, executing the last two statements in Fig. 5.1 updates the numeric domain such that $x_2 = 127$ and $x_1 = 1$, whereas x_0 is projected out during the updates as it overlaps. Reading the 32-bit field creates the following access tree:

x_0			
t_1		t_2	
x_2	t_3	t_4	x_1

Note that the temporaries t_3 and t_4 cannot be removed from this access tree, as they are used to define the relationship between x_2 and x_1 and their overlapping fields. In particular, the function *fromLower* is called recursively, which restricts the unbounded variables t_3 and t_4 to their unbounded ranges $[0, 255]$ such that t_1 and t_2 can be defined recursively from the variables x_2, t_3 and t_4, x_1, respectively. The final bounds of x_0 that can be inferred in this case can be expressed in hexadecimal notation as $\text{0x80000001} \leq x_0 \leq \text{0x80FFFF01}$. By projecting out the overlapping field x_0 when updating x_1 and x_2, an obvious precision loss has occurred. Instead of removing information on overlapping fields, a more precise result can be obtained by considering the second alternative of updating them.

To this end, consider the task of executing line 7 in Fig. 5.1 by updating x_2 without losing information contained in x_0. This update is performed by applying the function *update* : $\{1, 2, 4, 8\} \times \mathcal{X} \times \mathcal{AT} \times \mathit{Num} \to \mathit{Num}$ in Fig. 5.3. Specifically, the call $N' = \mathit{update}(s, y, at, N)$ updates the numeric domain N to N' such that the pivot node x_2 of the access tree at is set to the value of $y \in \mathcal{X}$. Here, s denotes the size of the pivot node in bytes.

In order to illustrate this update, reconsider the access tree for the third byte of `ip_info`:

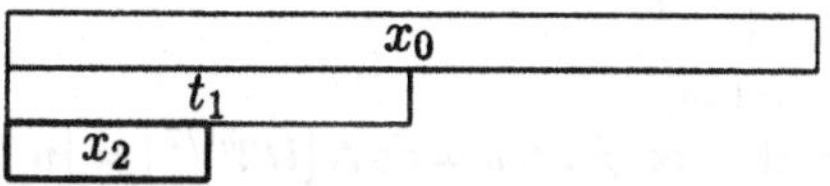

After updating x_2 to the new value y, the changed value must be propagated to the upper 8 bits of t_1. This case is handled by the last equation of *toUpper* in Fig. 5.3, which sets t_1 such that, in hexadecimal notation, $\text{0x8000} \leq t_1 \leq \text{0x80FF}$. This result is used to update x_0 to $\text{0x80000000} \leq x_0 \leq \text{0x80FF0000}$. While accessing the first four bytes of `ip_info` now results in a more precise value x_0, it is still not optimal since it is known that bits 16 to 23 are zero. Similarly to propagation to the pivot node, propagation from the pivot node suffers from loss of precision incurred by unbounded, temporary variables. In contrast, precision cannot be regained by treating unbounded variables specially. Instead, information from larger fields must be propagated to temporary variables, which can be done by calling *fromUpper* on all parents of the pivot node. In the example, we calculate

fromUpper(2, [x_0 / t_1], N)

which propagates the fact that $x_0 = 0$ to the field t_1. Given this information, the call to *toUpper* now detects that the upper 8 bits of t_1 are constant (since $\lfloor l/2^8 \rfloor = \lfloor u/2^8 \rfloor$, where $[l, u]_{\equiv d} = N(t_1)$) and hence performs the update of t_1 by subtracting $\lfloor l/2^8 \rfloor = 0$ from t_1 and adding the scaled value of x_2; that is, it calculates $N' = N \rhd t_1 := 2^8 x_2 + t_1 - 0$. The result of this calculation is used to update x_0, yielding $x_0 = \text{0x80000000}$. The strategy of propagating information to all parents of the pivot node before performing the updates is encapsulated by the function *update* in Fig. 5.3. Here the function *fromUpper* is called whenever the pivot node for which *update* is called has a parent – that is, whenever t_u is not the empty access tree $\varepsilon_{\mathcal{AT}}$. Note that no prior propagation of information is necessary for the dual case of updating children as implemented by *toLower* in Fig. 5.3.

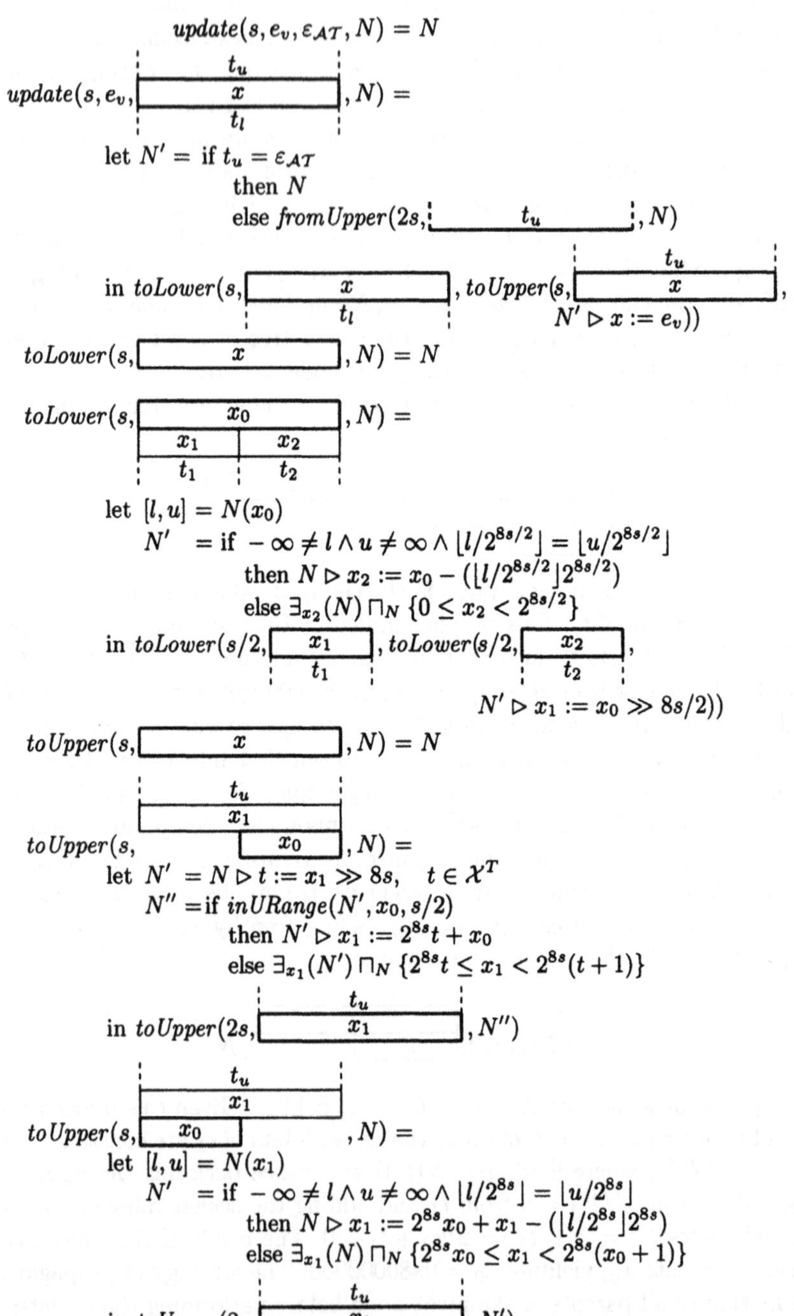

Fig. 5.3. Write operations on access trees. The variable t_u represents a wider tree than the dotted borders suggest; t_l, t_1, and t_2 represent two trees of half-width. Each function is called with a pivot node, which is s bytes wide and drawn in bold.

Thus, the tactic of updating variables that overlap is more precise than merely projecting out overlapping fields and propagating information when reading fields. Alas, the propagation functions in Fig. 5.2 are still needed in case the access tree does not contain a variable for a specific field size. For instance, consider accessing the structure `ip_info` as an array of 16-bit integers. Specifically, let the variable `i` range over $0 \ldots 2$ and consider the read access `((short*) &ip_info)[i]`. The result is the join of the pivot nodes x_3, t_1, and t_2 of the following three access trees:

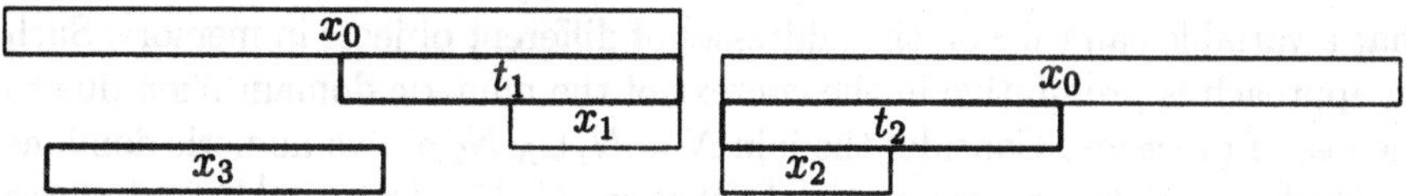

Inferring a precise result for the array access requires the *prop* function to refine the two variables t_1, and t_2 before the values of x_3, t_1 and t_2 are joined.

The upcoming Sect. 5.3 formalises accesses to several trees such as the one above. In addition, it details how to incorporate pointers into access trees.

5.2.1 Related Work

Little work has been done in the area of a detailed abstract model of the low-level memory accesses in C programs. Venet and Brat were the first to propose the automatic inference of fields that are relevant to an analysis [182]. In terms of overlapping fields, Steensgaard was the first to propose a type system to model overlapping accesses to memory [175]. His framework forms the basis of a sound points-to analysis by joining points-to sets of the different members of a structure as soon as a structure is accessed through a pointer with a non-constant offset.

Miné has extended the Astrée analyser [60] to support overlapping fields and presents a general framework in [131]. Similarly to our approach, the concrete value of a byte in the program is defined by the intersection of information of the various fields that cover this byte. Rather than defining how propagation between fields is implemented, only an example is given on how interval information can be propagated. Miné proposes to add all fields that a read or write may access, which is only possible since all memory regions in the class of programs he considers are finite. Furthermore, write operations remove overlapping fields of different sizes, which makes it necessary to treat the memory layout as an abstract domain that is modified together with the numeric state. Whenever the join of two states is computed, the memory layout is joined first and missing fields are inserted as necessary while a conservative range is inserted into the numeric domain. Section 10.2.1 will detail how conservative ranges are inserted when using a global memory map. Finally, Miné proposes an interesting extension that recovers the equality relationship between larger numeric variables and pointers when these are copied byte-wise.

5.3 Mixing Values and Pointers

Pointers in C are challenging to argue about because a pointer merely represents an address that can be used in arithmetic expressions just like integer variables. Furthermore, an abstract address $a \in \mathcal{A}$ as it is used in our analysis might correspond to an arbitrary set of addresses $\rho(a)$ in the concrete program. Thus, an abstract address can at best be represented symbolically, for instance as a polyhedral variable x_p with bounds $[0, 2^{32}]$. However, in order to summarise different runs of a program, the analysis must be able to express that a variable can take on the addresses of different objects in memory. Such an approach is prohibitive in the context of the numeric domain *Num* due to the loss of precision. Consider the join $N = N_1 \sqcup_N N_2$ of two numeric domains in which a variable x_a points to x_p^1 (that is, $N_1 \sqsubseteq_P \{x_a = x_p^1\}$) and where $N_2 \sqsubseteq_P \{x_a = x_p^2\}$. In the join N, neither points-to relationship holds, which makes it impossible to query the numeric domain as to which l-values the variable x_a may contain. This, in turn, is necessary to specify the semantics of pointer operations, where the possible l-values determine which memory regions a pointer may access. Hence, we chose to separately infer a set of possible l-values for each abstract variable in addition to the numeric information. This raises the question of how the information from the points-to domain and that of the numeric domain can be combined. For variables that correspond to sequences of 32 bits (that is, those that can contain a pointer), the answer is simply to map the abstract addresses from the variable's points-to set to concrete addresses and then add the value represented in the numeric domain as an offset. The special tag $\text{NULL} \in \mathcal{A}$, which can be part of any points-to set, is taken literally to mean that a given variable contains the address zero. This idea allows for a simple adaptation of the mem_ρ function from Sect. 5.1.1 for pointer variables. Specifically, the concrete store for each variable is composed of its value and its points-to set:

$$mem_\rho(\mathbf{v}, A) = \bigcap_{\substack{m \in \\ \mathcal{MUD}}} \left(\bigcap_{\substack{a \in \\ \rho(L(m))}} \left(\bigcap_{\substack{\langle o,s,x_i \rangle \in \\ F(m)}} \{bits^s_{a+o}(\pi_i(\mathbf{v}) + p) \mid p \in \rho(A(x_i))\} \right)\right)$$

In contrast to the previous definition of mem_ρ, the function above takes a points-to map $A \in Pts$ in addition to the vector of values $\mathbf{v} \in [\![N]\!]$ and adds, for each l-value $a \in A(x_i)$, the addresses $\rho(a)$ to the numeric value $\pi_i(\mathbf{v})$ of x_i. In order to illustrate this process, consider a program variable v with $\langle 0, 4, x_v \rangle \in F(\text{v})$ and $A(x_v) = \{\text{NULL}, a_p^1, a_p^2\}$. Given $\rho(a_p^1) = \{p_1\}$, $\rho(a_p^2) = \{p_2\}$, and $N(x_v) = [0, 4]_{\equiv 4}$, six concrete values of v are possible, namely 0, 4, p_1, $p_1 + 4$, p_2, and $p_2 + 4$. Thus, the fact that $\rho(\text{NULL}) = \{0\}$ allows a pointer-sized variable $x \in \mathcal{X}$ to be treated as a normal integer as long as $A(x) = \{\text{NULL}\}$. The view of l-values as offsets is in fact quite natural, as C pointers that are `NULL` contain the address 0. For any field that is not of pointer size, we require that its points-to set always be $\{\text{NULL}\}$. Hence, such a field can only represent values. An interesting situation arises when fields of pointer size and other fields overlap. Suppose the points-to set of the field above is reduced to $A(x_v) = \{a_p^1\}$. An overlapping field $\langle 0, 1, x'_v \rangle \in F(v)$ with $N(x'_v) = [0, 0]_{\equiv 64}$ creates stores where the first byte of v is zero, whereas the pointer-sized field creates stores where the first byte corresponds to $bin^8(p_1)$, which may not be zero. Hence, the resulting set of stores could be empty. Thus, overlapping fields have to be updated consistently with respect to the points-to sets. In particular, a field of size $s \neq 4$ that overlaps with a pointer must take on all values $v = [0, 2^{8s}]$, which ensures that $bits_a^s(v) = \Sigma$; that is, the field has no effect on the set of stores. Symmetrically, accessing a field that overlaps with a pointer-sized field with l-values other than NULL corresponds to accessing a part of a pointer. Special care has to be taken to express these operations correctly.

To this end, we define two functions, $read^{F,H}$ and $write^{F,H}$, on the abstract state $\langle N, A \rangle$ that correspond to the concrete memory access $\sigma^s(a)$ and the memory update $\sigma[a \overset{s}{\mapsto} v]$. These abstract functions make use of the *prop* and *update* functions in Figs. 5.2 and 5.3 for handling values in access tress. In order to utilise access trees for l-values, Fig. 5.4 introduces several functions to retrieve and store l-values and offsets of fields in an access tree. Specifically, the l-values of a particular memory location are returned by *getLVals*, which seeks the pointer-sized field in an access tree and returns its l-values. In contrast, the function *setLVals* is only defined when called with a pointer-sized field, in which case it merely updates its points-to set. In order to ensure that fields that overlap with pointers are always set to their maximum bounds, an offset stored in a pointer-sized field may not be propagated from and to overlapping fields, which is guaranteed by the two functions *setOfs* and *getOfs*. In particular, the *setOfs* function ensures that all overlapping fields are set to their maximum bounds. The last function for manipulating l-values is $clear^F$, which approximates a pointer with a value; that is, $clear^F$ sets the field to its maximum and resets the points-to set to {NULL}.

Treating fields in access trees as values and pointers is key to defining the semantics of the abstract memory accesses $\langle N', x, a \rangle = read^{F,H}(m, e_o, s, N, A)$ and $\langle N', A' \rangle = write^{F,H}(m, e_o, s, x, a, N, A)$ in Figs. 5.5 and 5.6, which read and write the value of $x \in \mathcal{X}$ and the points-to set $a \subseteq \mathcal{A}$ within a memory region $m \in \mathcal{M} \cup \mathcal{D}$ at offset $e_o \in Lin$. Observe that the offset e_o is not necessarily constant and hence can express an access through a pointer that might have a range of offsets. In this context, recall the notion of correct memory management, which requires that every memory access $\sigma^s(a)$ and $\sigma[a \overset{s}{\mapsto} v]$ lie within the range of *used* memory locations. In the abstract setting, we require that the access be within the bounds of the given memory region $m \in \mathcal{M} \cup \mathcal{D}$. In the case $m \in \mathcal{M}$, correctness of the access is easily checked by asserting that the offset is between zero and $size(m) - 1$. Otherwise, $m \in \mathcal{D}$ holds and the access position has to be compared with the symbolic size of the memory region. We define $H : \mathcal{D} \to \mathcal{X} \times \mathcal{X}$ to denote a map that takes

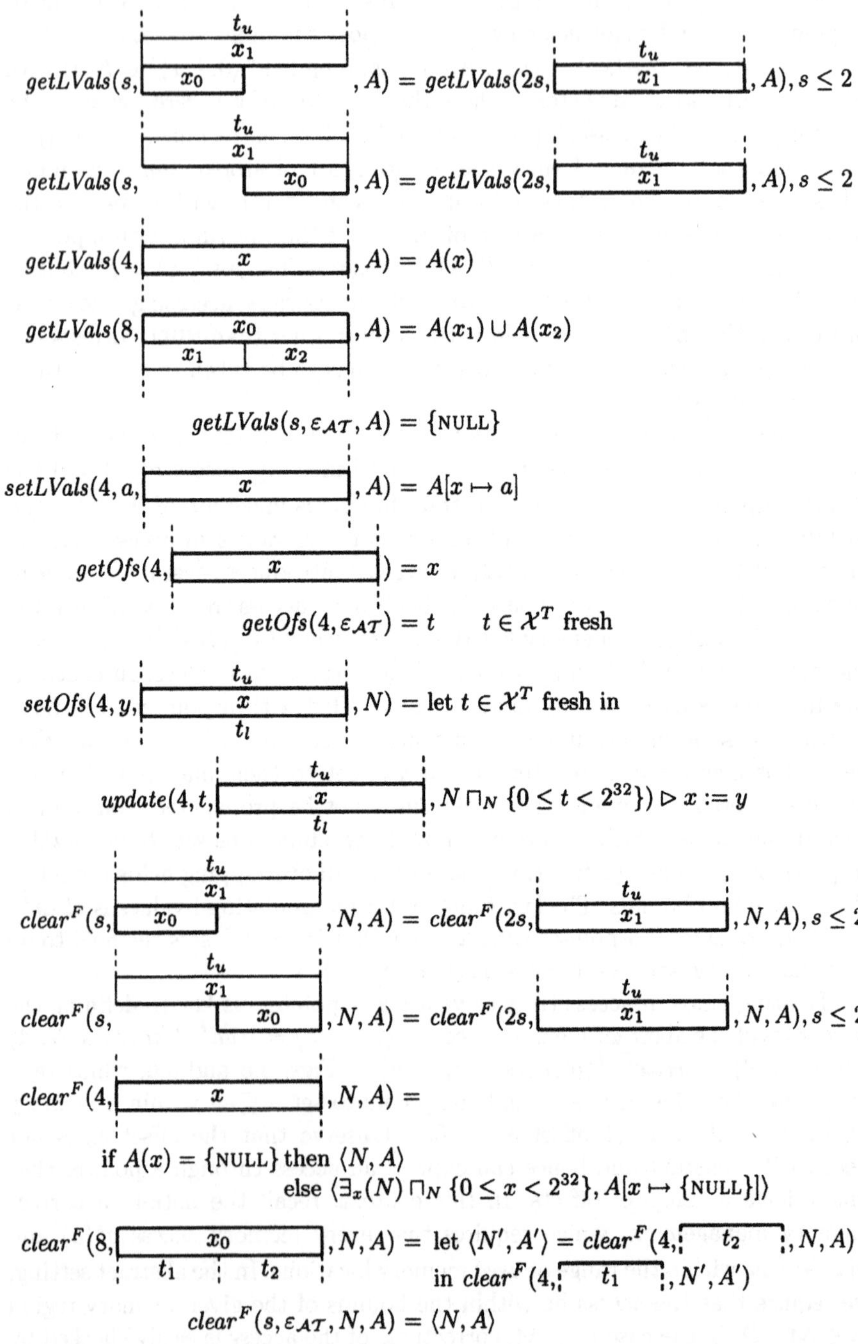

Fig. 5.4. Setting and getting l-values and their offsets. The variable t_u represents a wider tree than the dotted borders suggest; t_l represents two trees of half-width. Each function is called with a pivot node, which is s bytes wide and drawn in bold.

$$
\begin{array}{l}
\mathit{read}^{F,H}(m, e_o, s, N, A) = \textbf{if } m \in \mathcal{M} \\
\quad \textbf{then let } N_1 = N \sqcap_N \{e_o \equiv s\} \\
\qquad\qquad N_2 = N_1 \sqcap_N \{0 \leq e_o \leq \mathit{size}(m) - s\} \\
\qquad \textbf{in } \mathit{readChk}^F(m, N_2(e_o), s, N_2, A) \\
\qquad \textbf{warn } \text{“Unaligned access to } m\text{” } \textbf{if } N \not\sqsubseteq_N N_1 \\
\qquad \textbf{warn } \text{“Out-of-bounds access to } m\text{” } \textbf{if } N_1 \not\sqsubseteq_N N_2 \\
\quad \textbf{else let } \langle x_m, x_s \rangle = H(m) \\
\qquad\qquad N_1 = N \sqcap_N \{e_o \equiv s\} \\
\qquad\qquad N_2 = N_1 \sqcap_N \{0 \leq e_o \leq x_s - s\} \\
\qquad\qquad N_3 = N_2 \sqcap_N \{x_m \geq 1\} \\
\qquad\qquad \langle N_4, x, a \rangle = \mathit{readChk}^F(m, N_3(e_o), s, N_3, A) \\
\qquad \textbf{in if } N_3 \sqsubseteq_N \{x_m \leq 1\} \textbf{ then } \langle N_4, x, a \rangle \\
\qquad\qquad\qquad\qquad\qquad\qquad\qquad \textbf{else } \langle N_4 \rhd x_t := N_4(x), x_t, a \rangle,\ x_t \in \mathcal{X}^T \text{ fresh} \\
\qquad \textbf{warn } \text{“Unaligned access to memory allocated at } m\text{” } \textbf{if } N \not\sqsubseteq_N N_1 \\
\qquad \textbf{warn } \text{“Out-of-bounds access to memory allocated at } m\text{” } \textbf{if } N_1 \not\sqsubseteq_N N_2 \\
\qquad \textbf{warn } \text{“Access to freed region allocated at } m\text{” } \textbf{if } N_2 \not\sqsubseteq_N N_3 \\
\mathit{readChk}^F(m, [l, u]_{\equiv d}, s, N, A) = \langle N', x, a' \rangle \\
\quad \textbf{where } \{at_1, \ldots at_n\} = \bigcup_{o \in \{l, l+d, \ldots u\}} \mathit{access}^F(o, s, m) \\
\qquad\quad a = \bigcup_{i=1}^{n} \mathit{getLVals}(s, at_i, A) \\
\qquad\quad x \in \mathcal{X}^T \text{ fresh} \\
\qquad\quad \langle N', a' \rangle = \textbf{if } a = \{\text{NULL}\} \textbf{ then } \langle \bigsqcup_{i=1}^{n} N_n \rhd x := x_i, a \rangle \textbf{ where} \\
\qquad\qquad\qquad\qquad\qquad N_0 = N \\
\qquad\qquad\qquad\qquad\qquad \langle N_i, x_i \rangle = \mathit{prop}(s, at_i, N_{i-1}) \textbf{ for } i = 1, \ldots n \\
\qquad\qquad\qquad\quad \textbf{else if } s = 4 \textbf{ then } \langle \bigsqcup_{i=1}^{n} N \rhd x := \mathit{getOfs}(at_i), a \rangle \\
\qquad\qquad\qquad\quad \textbf{else } \langle N \sqcap_P \{0 \leq x < 2^{8s}\}, \{\text{NULL}\} \rangle
\end{array}
$$

Fig. 5.5. Reading abstract memory regions.

each heap region $m \in \mathcal{D}$ to two abstract variables $\langle x_m, x_s \rangle = H(m)$, where x_m denotes the number of dynamic memory regions that m summarises, whereas x_s represents the possible size of these regions. Given the symbolic size of summarised memory regions on the heap, correct memory management is checked on the access offsets in both $\mathit{read}^{F,H}$ and $\mathit{write}^{F,H}$. Specifically, both functions are split into a wrapper that checks and restricts the access position, and a part that infers and evaluates the access trees of the memory region.

For instance, the call $\mathit{read}^{F,H}(m, e_o, s, N, A)$ as implemented in Fig. 5.5 first checks that reading s bytes at the offsets $N(e_o)$ constitutes a valid access to the memory region $m \in \mathcal{M}$ by defining N_1 and N_2. Specifically, $\mathit{read}^{F,H}$ enforces that a declared variable $m \in \mathcal{M}$ is at least accessed at multiples of the size s, yielding N_1, and then enforces that the locations accessed at offset e_o lie within $[0, \mathit{size}(m) - 1]$, yielding N_2. The warnings shown are emitted as side effects of the function whenever the conditions shown hold. The idea is that no undefined behaviour can occur in the concrete transfer function if none of these warnings are triggered. The possible values of e_o in

N_2 are passed to $readChk^F$ as a set of access positions $[l, u]_{\equiv d}$, which then returns a variable $x \in \mathcal{X}$ representing the extracted value and a points-to set $a \subseteq \mathcal{A}$. The second case of $read^{F,H}$ handles accesses to dynamically allocated memory. The calculated states N_1, N_2, and N_3 guarantee that the access position is aligned, that it is within bounds, and that the abstract memory region represents at least one region, respectively. Warnings are emitted if these requirements are not met. As in the case of $m \in \mathcal{M}$, $readChk^F$ is called. However, its result is only returned if x_m is at most one; that is, if m represents no more than one memory region. If m represents several memory regions, the possible values of the result $N_4(x)$ are assigned to a new temporary x_t, which is then returned as the result. Note that $N_4(x)$ is an interval and that no linear relationship exists between x_t and x. This is important to ensure that the summary memory region m faithfully approximates all summarised regions. Suppose that x is returned, rather than x_t, and that the result is used in a test. Evaluating the condition involving x will restrict the field of all memory regions that are summarised by m, which is incorrect, as each read access in the program can only read one of the summarised memory regions. In order to ensure correctness, it is sufficient to copy the value of x to x_t without creating a linear relationship between the two variables, an operation known as "expand" [80].

The actual result of the memory access is calculated by creating access trees for all possible access positions. Note that $access^F$ returns an empty tree $\varepsilon_{\mathcal{AT}} \in \mathcal{AT}$ whenever a position in m is accessed that is not covered by any field. Since $\{at_1, \ldots at_n\}$ is a set and hence contains no duplicates, n is bounded by the number of fields $F(m)$ rather than the number of access positions $\{l, l+d, \ldots u\}$. In particular, n is always finite since $|F(m)|$ is finite. Hence, an approximation to the read value can be calculated effectively by calling *prop* to extract a result x_i for every access tree at_i as done in the case $a = \{\text{NULL}\}$, thereby refining N_{i-1} to N_i in every call of *prop*. These variables are then assigned to a fresh temporary variable x in the refined domain N_n, and the join of all of these assignments $\bigsqcup_{i=0}^{n} N_n \triangleright x := x_i$ summarises the result of all access trees in the variable x. Note that if a memory region m contains no fields in the accessed range, then $access^F(o, s, m) = \{\varepsilon_{\mathcal{AT}}\}$ so that $\langle N_1, x_1 \rangle = prop(s, \varepsilon_{\mathcal{AT}}, N_0)$ sets the variable x_1 to all possible values, which are then assigned to x and returned. Thus, $readChk^F$ always returns $N' \neq \bot_N$, even if the accessed range covered no fields in $F(m)$.

The condition $a = \{\text{NULL}\}$ ensures that propagating values among overlapping fields is only performed when the accessed region contains no pointers. If some of the accessed fields contain a points-to set other than $\{\text{NULL}\}$, then the access size determines if a pointer offset is read ($s = 4$) or if a part of a pointer is read ($s \neq 4$). In the first case, *getOfs* is called for each access tree to gather a summary of all offsets in x, which is then returned together with the joined points-to sets of all accessed fields. In the case $s \neq 4$, the bytes that would be read in the concrete memory contain parts of the addresses $\rho(a) \subseteq [0, 2^{32} - 1]$, which cannot be represented as abstract addresses

$$
\begin{array}{l}
\mathit{write}^{F,H}(m, e_o, s, e_v, a, N, A) = \textbf{if } m \in \mathcal{M} \\
\quad \textbf{then let } N_1 = N \sqcap_N \{e_o \equiv s\} \\
\qquad\qquad N_2 = N_1 \sqcap_N \{0 \leq e_o \leq \mathit{size}(m) - s\} \\
\qquad \textbf{in } \mathit{writeChk}^F(m, N_2(e_o), s, e_v, a, N_2, A) \\
\qquad \textbf{warn } \text{“Unaligned access to } m\text{”} \textbf{ if } N \not\sqsubseteq_N N_1 \\
\qquad \textbf{warn } \text{“Out-of-bounds access to } m\text{”} \textbf{ if } N_1 \not\sqsubseteq_N N_2 \\
\quad \textbf{else let } \langle x_m, x_s \rangle = H(m) \\
\qquad\qquad N_1 = N \sqcap_N \{e_o \equiv s\} \\
\qquad\qquad N_2 = N_1 \sqcap_N \{0 \leq e_o \leq x_s - s\} \\
\qquad\qquad N_3 = N_2 \sqcap_N \{x_m \geq 1\} \\
\qquad\qquad \langle N', A' \rangle = \mathit{writeChk}^F(m, N_3(e_o), s, e_v, a, N_3, A) \\
\qquad \textbf{in if } N_3 \sqsubseteq_N \{x_m \leq 1\} \textbf{ then } \langle N', A' \rangle \textbf{ else } \langle N_3 \sqcup_N N', A \sqcup_A A' \rangle \\
\qquad \textbf{warn } \text{“Unaligned access to memory allocated at } m\text{”} \textbf{ if } N \not\sqsubseteq_N N_1 \\
\qquad \textbf{warn } \text{“Out-of-bounds access to memory allocated at } m\text{”} \textbf{ if } N_1 \not\sqsubseteq_N N_2 \\
\qquad \textbf{warn } \text{“Access to freed region allocated at } m\text{”} \textbf{ if } N_2 \not\sqsubseteq_N N_3
\end{array}
$$

$$
\begin{array}{l}
\mathit{writeChk}^F(m, [l, u]_{\equiv d}, s, e_v, a, N, A) = \langle \bigsqcup_{i=1}^n N_i, \bigsqcup_{i=1}^n A_i \rangle \\
\quad \text{where } \{at_1, \ldots at_n\} = \bigcup_{o \in \{l, l+d, \ldots u\}} \mathit{access}^F(o, s, m) \\
\qquad \langle N_i, A_i \rangle = \mathit{writeTree}(s, e_v, a, at_i, N, A) \quad \forall i \in [1, n]
\end{array}
$$

$$
\begin{array}{l}
\mathit{writeTree}(s, e_v, a, at, N, A) = \\
\quad \textbf{if } s = 4 \textbf{ then } \langle N', \mathit{setLVals}(s, a, at, A) \rangle \textbf{ where} \\
\qquad\quad N' = \textbf{if } a = \{\text{NULL}\} \textbf{ then } \mathit{update}(s, e_v, at, N) \textbf{ else } \mathit{setOfs}(s, e_v, at, N') \\
\quad \textbf{else } \langle \mathit{update}(s, e_v, at, N'), A' \rangle \textbf{ where} \\
\qquad\quad \langle N', A' \rangle = \mathit{clear}^F(s, at, N, A)
\end{array}
$$

Fig. 5.6. Writing abstract memory regions.

$a \subseteq \mathcal{A}$. As a consequence, an arbitrary value is returned as the result of the read operation by restricting the temporary variable x with $\{0 \leq x < 2^{8s}\}$. An arbitrary value in the concrete program also includes any possible address such that the points-to set can be refined to $\{\text{NULL}\}$.

Analogously to $\mathit{read}^{F,H}$, a call to $\mathit{write}^{F,H}(m, x_o, s, x, a, N, A)$ as implemented in Fig. 5.6 first checks that accessing s bytes of the memory region $m \in \mathcal{M} \cup \mathcal{D}$ at offset $x_o \in \mathcal{X}$ is within bounds before writing the value $x \in \mathcal{X}$ and the points-to set $a \subseteq \mathcal{A}$. The checks shown are identical to the $\mathit{read}^{F,H}$ case except when writing to a dynamically allocated memory region $m \in \mathcal{D}$ with $H(m) = \langle x_m, x_s \rangle$, where x_m may be greater than one. The latter indicates that the abstract region m represents more than one concrete memory region. Given that a write in the concrete program will only modify one of the regions summarised by $m \in \mathcal{D}$, the summarised region itself has to be changed to the new value (to reflect the update to the memory region that was written) and has to retain its previous values (to represent the regions that remain unchanged). This is achieved by joining the new state $\langle N', A' \rangle$ with the original state $\langle N, A \rangle$ whenever $x_m > 1$ is satisfiable in N.

The update itself is performed by the functions $writeChk^F$ and *writeTree*. The former creates the access trees corresponding to the accessed range and calls *writeTree* on each of them. Similar to the updates of dynamically allocated memory regions, the concrete memory access updates only one position in the given range, and hence the results of calling *writeTree* are joined to cater to any possible access position the concrete program can use. Updating a single access tree as implemented by *writeTree* distinguishes between three cases. The first two cases update a pointer-sized field. Depending on the set of l-values $a \subseteq \mathcal{A}$ to be written, the value of the field is set either by *update* or by *setOfs*. The former case is chosen if a pure value is written, in which case all fields that overlap with the written positions are updated by *update*. In contrast, if a points-to set other than {NULL} is written, overlapping fields are set to their maximum range by *setOfs*, which ensures that only the written pointer-sized field determines the value of the corresponding concrete memory. In both cases, the points-to set is updated by calling *setLVals*. The third case handles non-pointer fields by merely calling *update*. Special care has to be taken when writing to such a field, as it may overlap with a pointer-sized field that has an l-value set different from {NULL}. Suppose a byte-sized value 0 is written over a pointer field $x_p \in \mathcal{X}$, where $A(x_p) = \{a_p\}$. In the concrete program, a part of an address $p \in \rho(a_p)$ is overwritten. Calling the function $clear^F$ ensures that the pointer-sized field is set to its maximum range, which represents a conservative approximation to a partly overwritten pointer. Note that $clear^F$ has no effect if an overlapping pointer-sized field is a pure value.

The basic abstract functions for reading and writing to memory are key to a concise abstract semantics of Core C. Before Chap. 6 presents this semantics, we detail the general relationship between concrete and abstract states.

5.4 Relating Concrete and Abstract Domains

The relationship between the concrete and the abstract domains forms the basis of any correctness proof of an analysis. While we give no formal proof of the analysis presented, the abstraction relation is still helpful in the construction and understanding of the abstract semantics. Furthermore, formulating an abstraction map for a full-fledged imperative programming language is an interesting exercise in itself, not least in that it requires certain design choices.

Before we elaborate on possible alternatives, we present the abstraction relation of our analysis, which is based on the function γ_ρ defined as follows:

$$\gamma_\rho(N, A) = \{mem_\rho(\mathbf{v}, A) \mid \mathbf{v} \in [\![N]\!]\}$$

The stores $\gamma_\rho(N, A) \subseteq \Sigma$ that correspond to an abstract state $\langle N, A\rangle$ depend on the address map ρ, which maps (sets of) abstract addresses to sets of concrete addresses. As such, ρ determines the position of each function-local variable and each heap-allocated variable, whose positions in the concrete program are determined by the memory allocation function $fresh_\theta^\delta$. In order to prove correct memory management, the abstraction must be independent of how $fresh_\theta^\delta$ is implemented. Since it is difficult to define ρ for all possible implementations of $fresh_\theta^\delta$, we choose to define a relational abstraction that specifies when an abstract state $\langle N, A\rangle$ is an abstraction of a concrete state $\langle \sigma, \theta, \delta\rangle$. The idea is to extract concrete addresses from the stack θ and the set of dynamically allocated memory regions δ to synthesise ρ, which is used to define the set of possible concrete stores $\gamma_\rho(N, A)$. To this end, let $\delta_m^a = \{p \mid \langle l, p, s\rangle \in \delta \wedge l = m\}$ denote the addresses of all memory regions allocated at location m in the program and let $\delta_m^s = \{s \mid \langle l, p, s\rangle \in \delta \wedge l = m\}$ denote their sizes. Suppose that $\theta = \langle l_0, A_0\rangle \cdots \langle l_s, A_s\rangle$. Then define ρ as follows:

$$\rho(a) = \begin{cases} \{p\} & \text{if } \exists m \in \mathcal{M}, p \in [0, 2^{32}-1] \,.\, L(m) = a \wedge \langle m, p\rangle \in \bigcup_{i=0}^{s} A_i \\ \delta_m^a & \text{if } \exists m \in \mathcal{D} \,.\, L(m) = a \wedge |\delta_m^a| \geq 1 \\ \{p_f\} & \text{if } a \in \mathcal{A}^{\mathcal{F}} \text{ represents the address } p_f \text{ of the function } f \\ \{0\} & \text{if } a = \text{NULL} \\ [0, 2^{32}-1] & \text{otherwise} \end{cases}$$

The first case matches if the abstract address $a \in \mathcal{A}$ is that of a declared variable $m \in \mathcal{M}$, which is contained in a stack frame as $\langle m, p\rangle$. Note that this case only applies if the variable is present in exactly one stack frame. In particular, a variable cannot appear in several stack frames, which thereby prohibits the analysis of recursive functions. While the definition of ρ can easily be adapted to cater to several invocations of the same function, the latter would also require changes to $write^{F,H}$, which currently assumes that the given memory region is always overwritten, which is incorrect if it exists several times. In contrast, $write^{F,H}$ performs weak updates to dynamically allocated memory regions m when its abstract address $L(m)$ corresponds to several concrete addresses δ_m^a. Note that if $|\delta_m^a| = 0$, then m corresponds to no concrete memory region, which is handled by the last case. The third case maps the abstract address $a_f \in \mathcal{A}$ of every function f to the address p_f of the function itself. With respect to the fourth case, as pointed out earlier, the tag NULL represents the address 0. The last case applies if an abstract memory region does not currently exist in the concrete program, which can happen in two cases: firstly, when a local variable $m \in \mathcal{M}$ resides in a function that is currently not on the call stack; and secondly, when a dynamic memory region $m \in \mathcal{D}$ has not yet been allocated or is already freed. Variables that contain pointers to these memory regions are assumed to take on all possible addresses, which is why the set of all addresses $[0, 2^{32}-1]$ is returned. In this context, observe that a free memory region must exist somewhere since otherwise $\rho(m) = \emptyset$ and concretising the variable containing the address of m forces the result of mem_ρ on p. 100 to be empty since $\{bits_{a+o}^s(\pi_i(\mathbf{v}) + p) \mid p \in \rho(m)\} = \emptyset$. Indeed, the last case in the definition of ρ forces mem_ρ to create all possible bit patterns at all possible addresses for each non-existent variable, so that the concretisation of non-existent variables does not affect the concrete states. Note that the question of mapping non-existent memory regions is purely technical and independent of the analysis, which emits a warning whenever a pointer is dereferenced that corresponds to an address of a freed memory region or to a local variable that has gone out of scope.

Once an address relation ρ is derived from θ and δ, the relationship between an abstract state $\langle N, A\rangle$ and the concrete state $\langle \sigma, \theta, \delta\rangle$ is surprisingly concise:

$$\begin{aligned}\langle N, A\rangle \propto \langle \sigma, \theta, \delta\rangle \text{ iff } & \sigma \in \gamma_\rho(N, A) \wedge \\ & \forall m \in \mathcal{D} \,.\, \langle x_m, x_s\rangle = H(m) \wedge \delta^s_m \subseteq N(x_s) \wedge |\delta^a_m| \in N(x_m)\end{aligned}$$

Thus, an abstract and a concrete state are in $\propto$ relation whenever the concrete store σ is in the set of stores created by γ_ρ. The second line enforces additional requirements on each memory region $m \in \mathcal{D}$, which may represent several concrete memory regions in δ. Specifically, the number of memory regions $|\delta^a_m|$ that are summarised by m must be a possible value of x_m, and the size of each of these memory regions $s \in \delta^s_m$ must be a possible value of x_s. Using the relation above, it is possible to show that executing a Core C statement on the concrete state and on the abstract state results in two modified states that are in $\propto$ relation if the original states were in $\propto$ relation. Correctness of a fixpoint computation on the abstract semantics follows if $\langle \Sigma, \subseteq\rangle$ is a lower cone, [29] which is trivially true since $\langle \Sigma, \subseteq\rangle$ together with $\cup$ and $\cap$ is a complete lattice.

We conclude this chapter with a discussion of abstraction frameworks; that is, of ways to formalise a static analysis. This discussion is relevant since certain frameworks allow an assessment of the quality of the analysis results. Yet the relational abstraction chosen in this book carries no such quality guarantees, and the next section therefore justifies its use.

5.4.1 On Choosing an Abstraction Framework

The relational abstraction $\propto$ used above is rather weak in that it merely allows a correctness proof but no reasoning about the quality of an abstraction. For instance, a best abstraction is a function that maps every concrete state to the most precise abstract state, which is a desirable property for abstract domains. In order to discuss why this relational abstraction was chosen, let $S^\natural$ and $S^\sharp$ denote the sets of concrete and abstract states, respectively, and let $\preceq^\natural$ and $\preceq^\sharp$ impose an order on each set. A classic approach in data-flow analysis is the use of a Galois connection; that is, a pair of adjoint functions $\alpha : S^\natural \to S^\sharp$ and $\gamma : S^\sharp \to S^\natural$ such that

$$\forall s^\natural \in S^\natural \,.\, \forall s^\sharp \in S^\sharp \,.\, s^\natural \preceq^\natural \gamma(s^\sharp) \iff \alpha(s^\natural) \preceq^\sharp s^\sharp$$

The property above requires both domains, $S^\natural$ and $S^\sharp$, to be complete lattices [122]; that is, each (finite or infinite) subset of $S \subseteq S^\natural$ must have a supremum $\bigcup S$ and an infimum $\bigcap S$ in $S^\natural$. This requirement renders a

Galois connection unsuitable to analyse a standard denotational semantics that uses a flat cpo (complete partial order) to describe values at a particular point in the program. A cpo is a partial order $\langle S, \preceq \rangle$ in which every (infinite) chain $s_1 \preceq s_2 \preceq \ldots, s_i \in S$ has a supremum $\bigcup_{i \in \mathbb{N}} s_i \in S$; in a flat cpo, the order $\preceq$ distinguishes only two classes of elements. A flat cpo for Core C is $S^\natural = \Sigma \cup \bot$ such that $s_1 \preceq^\natural s_2$ iff $s_1 = \bot \vee (s_1 \in \Sigma \wedge s_2 \in \Sigma)$. In general, given $s_1, s_2 \in S^\natural$ with $s_1 \neq s_2$ and $\alpha(s_1) = \alpha(s_2)$, the value $\gamma(\alpha(s_1))$ may not exist in $S^\natural$. One solution is to lift the concrete domain to sets of objects; this was presented in Sect. 2.5 as the collecting semantics. In order to avoid a more complex correctness proof with respect to the collecting semantics (which itself is subject to correctness concerns), it is possible to only use an abstraction map $\alpha : S^\natural \to S^\sharp$ and a complete abstract lattice $\langle S^\sharp, \preceq^\sharp \rangle$. The latter requires that α map a concrete state to the most precise abstract state [56], which implies that the abstract domain is a cpo. However, some abstract domains, such as convex polyhedra, are not cpos since they are devoid of some (infinite) elements, such that they cannot represent a best abstraction for certain concrete objects [59,122]. In this case, a single function $\gamma : S^\sharp \to \mathcal{P}(S^\natural)$ can be used, although it maps to a set of concrete states, which, as argued above, comes at the cost of additional complexity. In order to avoid arguing correctness with respect to a set of states, it is possible to use a relational abstraction $\propto \subseteq S^\sharp \times S^\natural$, which relates individual elements of each domain [122], thereby allowing proofs on individual elements. In fact, we require the relational framework since $\propto$ is defined in terms of γ_ρ, which maps to a set of concrete states, whereas ρ is synthesised from a single concrete state.

Another interesting aspect of the analysis is the quality of the abstract transfer functions. To this end, note that, in the context of a Galois connection, soundness can be asserted by proving $\alpha \circ f^\natural \preceq^\sharp f^\sharp \circ \alpha$ or, equivalently, $f^\natural \circ \gamma \preceq^\natural \gamma \circ f^\sharp$, where $(g \circ f)(x) = g(f(x))$ and $\preceq^\sharp$, $\preceq^\natural$ are lifted to functions. Interestingly, if the transfer functions fulfill these relationships as equalities, a statement about the quality of the transfer functions is possible. In particular, if $\alpha \circ f^\natural = f^\sharp \circ \alpha$, then $f^\sharp$ is α-complete, meaning that it is as precise as possible within the expressiveness of the abstract domain [56,57,79]. In the context of the analysis presented, Sect. 4.7 already pointed out that no abstraction map α exists, and hence the precision of the transfer functions can only be assessed in special circumstances, such as by assuming that α maps a bit pattern to the lowest positive values. On the contrary, if $f^\natural \circ \gamma = \gamma \circ f^\sharp$ holds, $f^\sharp$ is γ-complete, which is an orthogonal property to α-completeness [78]. This property always holds unless the abstract states themselves are mere equivalence classes; that is, if several abstract states map to the same concrete state [156]. In the context of polyhedra, Sect. 3.3.1 presented the lattice of $\mathbb{Z}$-polyhedra, which are the smallest polyhedra in each equivalence class of rational polyhedra that contain the same integral points. In this context, evaluating linear expressions in the polyhedral domain is not γ-complete. Suppose the interval $[0, 1.6]$, which includes the integral values 0 and 1, is multiplied by 2, resulting in $[0, 3.2]$, which implies that the largest integral value of the result may be 3 rather than $1 \times 2 = 2$. Integral tightening may improve precision, but it is unclear if and how γ-completeness can be achieved. Note that all completeness results rely on an abstraction that is formulated as a Galois connection or at least one in which the domains involved are complete partial orders [149]. Hence it is not clear how the completeness properties presented map to our more general setting, in which no unique α exists and in which the abstract domain does not constitute a complete partial order.

6

Abstract Semantics

The functions *wrap*, $read^{F,H}$, and $write^{F,H}$ presented in the last chapter facilitate the definition of the abstract transfer functions; that is, the semantic equations that specify how an abstract state is converted by executing Core C programs. In contrast to the concrete semantics, the abstract semantics naturally summarises several runs of a concrete program such that a lifting to sets of states, as done in the form of the collecting semantics, is not necessary to calculate a fixpoint. Thus, rather than calculating a set of states for each program point, a fixpoint calculation on the abstract domains updates single stores in a map $\psi : Label^* \to Num \times Pts$, where $Label^*$ denotes sequences of labels that specify the current position in the program. Specifically, a sequence $\langle l_0 \cdots l_s \cdot l \rangle$ is mapped to the pair of abstract domains $\psi(\langle l_0 \cdots l_s \cdot l \rangle) = \langle N, A \rangle$ that represents the program state at label l in a function that was invoked through calls residing at the labels $l_0, \ldots l_s$. Here, l_s denotes the immediate caller of the function in which l resides. A fixpoint of the program can be calculated by repeatedly evaluating the abstract transfer function $[\![\, lookupBlock(l) \,]\!]^\sharp_{\mathsf{Block}} \psi(l_0 \cdots l_s \cdot l)$ for all $l_0 \cdots l_s \cdot l$ in the pre-image of ψ. The result of each call is a set of triples, where each triple $\langle N, A, l_0 \cdots l_s \rangle$ denotes a new state $\langle N, A \rangle$ for the position $l_0 \cdots l_s$ in the program. These positions of ψ are then updated by joining the new state element-wise for each domain $N \in Num$ and $A \in Pts$ with the previous state using the operators $\sqcup_N$ and $\sqcup_A$, respectively. This process is repeated until the abstract states at all positions in ψ are stable, which is checked using the operators $\sqsubseteq_N$ and $\sqsubseteq_A$. Let ψ^{init} denote the initial map where $\psi(l_1 \cdots l_s) = \langle \bot_A, \bot_N \rangle$ for all $l_1 \cdots l_s \in Label^*$ with $\bot_A(x) = \emptyset$ for all $x \in \mathcal{X}$. Given a Core C program $V^g \; s_0; \ldots s_n; F$, where V^g denotes the set of global variables, s_i the sequence of initialisation statements, and F the set of function declarations that include a **main** function, the evaluation of

$$[\![\, \mathbf{main}() \,]\!]^\sharp_{\mathsf{Block}}([\![\, s_n \,]\!]^\sharp_{\mathsf{Stmt}} \cdots ([\![\, s_1 \,]\!]^\sharp_{\mathsf{Stmt}} \langle A_{init}, N_{init}, l \rangle) \ldots)$$

returns a first state $\langle A, N, l_1 \cdots l_s \rangle$, which is used to update ψ. Here, the map A_{init} is defined such that $A(x) = \{\text{NULL}\}$ for all $x \in \mathcal{X}$ and, given the set of dynamically allocated memory regions $\mathcal{D} = \{m_1, \ldots m_n\}$ with $H(m_i) = \langle x^i_m, x^i_s \rangle$ for all $i \in [1, n]$, let $N_{init} = \{x^1_m = 0, x^1_s = 0, \ldots x^n_m = 0, x^n_s = 0\}$. Thus, all variables are set to contain only values initially, and all dynamically allocated memory regions represent no concrete memory regions. Setting the size of these memory regions avoids a loss of precision that otherwise occurs in the convex hull calculation but has no effect otherwise since x^i_s is not evaluated if $x^i_m = 0$. The label $l \in \mathit{Label}$ that defines the call context of **main** may not occur in the program. Whenever the return statement of the **main** function is evaluated, the transfer function for control statements is called as $[\![\ \mathit{lookupNext}(l)\ ]\!]^\sharp_{\mathsf{Next}}$, which we assume returns an empty set such that ψ is not updated, thereby ignoring the exit state of the program.

During normal evaluation of the program, basic blocks and the control-flow statements are evaluated by the abstract transfer functions presented in Fig. 6.1. The first rule for a sequence of statements resembles that of the concrete semantics: After evaluating the semantics of every statement in the sequence, the control-flow statements of that block are looked up and evaluated. The notation of a nested function application turns out to be rather unwieldy for more complex iterative calculations, which is why the abstract semantics of a function call in the next rule features a different way of formulating iteration. Here, the n arguments of the function f are assigned to the formal parameters by defining an equation that is parameterised by i, which is executed for all i in the given interval. The initial domains N_0 and A_0 are set to the incoming domains N and A, whereas the resulting domains N_n and A_n are used to evaluate the first basic block of the called function. Note that the return value of $[\![\cdot]\!]^\sharp_{\mathsf{Block}}$ is that of $[\![\cdot]\!]^\sharp_{\mathsf{Next}}$, which in turn returns sets of triples as explained above. This feature is used in the third case, where a function is called through a pointer variable. The four bytes of the pointer v are read by the call to $\mathit{read}^{F,H}$, and the resulting pointer offset x is restricted to zero in order to check that the function pointers have no offset. The extracted points-to set a is restricted to the set $\mathcal{A}^{\mathcal{F}}$ of function addresses, resulting in the abstract addresses $\{a_{f_1}, \ldots a_{f_k}\}$, where each a_{f_i} corresponds to the function f_i. The actual invocation of the functions is deferred by calling $[\![\cdot]\!]^\sharp_{\mathsf{Block}}$ for each function f_i. The resulting triples of these invocations are joined and returned. Note that the functions are invoked on the state $N''' = \exists_{\mathcal{X}^T}(N'')$. Here, the notation $\exists_X(N)$ abbreviates the sequence of projection operations $\exists_{x_n}(\ldots \exists_{x_1}(N)\ldots)$, where $\{x_1, \ldots x_n\} = X$. Specifically, all temporary variables $x \in \mathcal{X}^T$ are projected out in order to remove variables that were introduced when populating access trees during the evaluation of the call to $\mathit{read}^{F,H}$. Note that $\mathit{read}^{F,H}$ may return temporary variables as a result, and hence these variables may only be removed once they are no longer needed to express relationships between non-temporary variables in fields.

The semantics of basic blocks is tightly coupled with the evaluation of control-flow statements. For instance, the **return** statement complements the semantics for function calls in that it evaluates the control-flow statements

Basic Blocks.

$[\![\, l : s_1; \ldots s_n;\,]\!]^{\sharp}_{\mathsf{Block}}\langle N, A, l_0 \cdots l_s\rangle = [\![\, \mathit{lookupNext}(l_s)\,]\!]^{\sharp}_{\mathsf{Next}}\langle N', A', l_0 \cdots l_s\rangle$

where $\langle N', A'\rangle = [\![\, s_n\,]\!]^{\sharp}_{\mathsf{Stmt}}(\ldots([\![\, s_1\,]\!]^{\sharp}_{\mathsf{Stmt}}\langle N, A\rangle)\ldots)$

$[\![\, l : f_i(a_1, \ldots a_n);\,]\!]^{\sharp}_{\mathsf{Block}}\langle N, A, l_0 \cdots l_s\rangle = [\![\, \mathit{lookupBlock}(l_t)\,]\!]^{\sharp}_{\mathsf{Block}}\langle N', A', l_0 \cdots l_s \cdot l\rangle$

where $\langle\, \langle p_1, \ldots p_n\rangle, \langle v_1, \ldots v_k\rangle, l_t\rangle = \mathit{lookupFunc}(f_i)$
$s_i = \mathit{size}(p_i)$ for all $i \in [1, n]$
$N_0 = N,\ A_0 = A$
$\langle N_i, A_i\rangle = [\![\, \textbf{structure}\ s_i\ p_i.0 = a_i;\,]\!]^{\sharp}_{\mathsf{Stmt}}\langle N_{i-1}, A_{i-1}\rangle$ for all $i \in [1, n]$
$N' = N_n,\ A' = A_n$

$[\![\, l : *v(a_1, \ldots a_n);\,]\!]^{\sharp}_{\mathsf{Block}}\langle N, A, l_0 \cdots l_s\rangle =$
$\bigcup_{i=1}^{k}\Big([\![\, l : f_i(a_1, \ldots a_n);\,]\!]^{\sharp}_{\mathsf{Block}}\langle N''', A, l_1 \cdots l_s\rangle\Big)$

where $\langle N', x, a\rangle = \mathit{read}^{F,H}(v, 0, 4, N, A)$
$N'' = N' \sqcap_P \{x = 0\}$
$N''' = \exists_{\mathcal{X}^T}(N'')$
$\{a_{f_1}, \ldots a_{f_k}\} = a \cap \mathcal{A}^{\mathcal{F}}$
warn "Function pointer has an offset." **if** $N' \not\sqsubseteq_N N''$
warn "Call to a non-function pointer." **if** $a \not\subseteq \{a_{f_1}, \ldots a_{f_k}\}$

Control Flow.

$[\![\, \textbf{return}\,]\!]^{\sharp}_{\mathsf{Next}}\langle N, A, l_0 \cdots l_{s-1} \cdot l_s\rangle = [\![\, \mathit{lookupNext}(l_s)\,]\!]^{\sharp}_{\mathsf{Next}}\langle \exists_X(N), A, l_0 \cdots l_{s-1}\rangle$

where f: the currently executed function
M: variables and in stack frames $l_0, \ldots l_{s-1}$
$\langle P, V, l_t\rangle = \mathit{lookupFunc}(f)$
$X = \{x \mid \langle o, s, x\rangle \in F(m) \wedge m \in P \cup V\}$
warn "Returning address of local variable."
if $\exists a \in A. a \in \{L(m) \mid m \in P \cup V\} \wedge$
$a \in \{A(x) \mid \langle o, 4, x\rangle \in F(m), m \in M\}$

$[\![\, \textbf{jump}\ l\,]\!]^{\sharp}_{\mathsf{Next}}\langle N, A, l_0 \cdots l_s\rangle = \{\langle N, A, l_0 \cdots l_s \cdot l\rangle\}$

$[\![\, \textbf{if}\ t\ s\ v.o\ \mathit{op}\ \mathit{exp}\ \textbf{then jump}\ l\ ;\ \mathit{nxt}\,]\!]^{\sharp}_{\mathsf{Next}}\langle N, A, l_0 \cdots l_s\rangle =$
$\{\langle \exists_{\mathcal{X}^T}(N^{then}), A^{then}, l_0 \cdots l_s \cdot l\rangle\} \cup [\![\, \mathit{nxt}\,]\!]^{\sharp}_{\mathsf{Next}}\langle \exists_{\mathcal{X}^T}(N^{else}), A^{else}, l_0 \cdots l_s\rangle$

where $\langle N', x, a_x\rangle = \mathit{read}^{F,H}(v, o, s, N, A)$
$\langle N'', e, a_y\rangle = [\![\, \mathit{exp}\,]\!]^{\sharp, s}_{\mathsf{Expr}}\langle N', A\rangle$
$N''' = N'' \triangleright y := e$ where $y \in \mathcal{X}^T$ fresh
$\langle N^{then}, A^{then}\rangle = \mathit{cond}(N''', A, t\ s, x, a_x, y, a_y, \mathit{op})$
$\langle N^{else}, A^{else}\rangle = \mathit{cond}(N''', A, t\ s, x, a_x, y, a_y, \mathit{neg}(\mathit{op}))$

Fig. 6.1. Abstract transfer functions for basic blocks.

lookupNext(l_s), which are skipped during the invocation of a function at the call site l_s. Besides redirecting the control flow, the **return** statement removes all information pertaining to local variables and function arguments, thereby guaranteeing that out-of-scope variables are unrestricted, as required by the abstraction relation (see Sect. 5.4). The next rule defines the **jump** statement, which merely returns the current state $\langle N, A\rangle$ together with the target of the jump appended to the current calling context $l_1 \cdots l_s$. This simple case depicts how the returned triple can be used to update the ψ map for the target of the jump $l_1 \cdots l_s \cdot l$. More involved is the definition of the conditional, which naturally creates one triple if the condition holds and other triples for the remaining control-flow statements if the condition does not hold. The states $\langle N^{then}, A^{then}\rangle$ and $\langle N^{else}, A^{else}\rangle$ are created by calling the helper function *cond* on the results of the evaluation of the variable and expression that form the arguments of the condition. As before, the function *neg* maps a relational operator to its opposite;, e.g., $neg('\leq') = '>'$. A conditional statement is the only operation that restricts the state space represented by the abstract domains; its implementation crucially determines the precision of the analysis. While many different cases are possible, we present the most important ones in sequence. The most basic case occurs when two values are compared. If the operation is not $\neq$, the resulting variable x and the expression e are translated directly to a meet operation on the numeric domain:

$$
\begin{aligned}
&cond(N, A, t\ s, x, \{\text{NULL}\}, y, \{\text{NULL}\}, op) = \langle N'' \sqcap_N \{x\ op\ y\}, A\rangle \\
&\quad \text{where } N' = wrap(N, t\ s, x) \\
&\qquad\qquad N'' = wrap(N', t\ s, y)
\end{aligned}
$$

Note that the pattern above only matches whenever the points-to sets of both operands are $\{\text{NULL}\}$; that is, neither side contains a pointer. The expression $e \in Lin \times \mathbb{Z}$ may contain a linear expression over $\mathcal{X}$ and a constant. The *wrap* function is applied to each argument to ensure that the relation expressed on the variables in the polyhedron corresponds to the finite program variables as described in Sect. 4.5. Note that the *wrap* function over $N \in Num$ can be derived straightforwardly from the implementation over $P \in Poly$ on p. 82. A special case is needed to implement the disequality operator $\neq$:

$$
\begin{aligned}
cond(N, A, t\ s, x, \{\text{NULL}\}, y, \{\text{NULL}\}, \neq) = \langle\, & N'' \sqcap_N \{x < y\} \sqcup_N \\
& N'' \sqcap_N \{x > y\}, A\rangle
\end{aligned}
$$

$$
\begin{aligned}
&\text{where } N' = wrap(N, t\ s, x) \\
&\qquad\quad N'' = wrap(N', t\ s, y)
\end{aligned}
$$

The operation above may in practice be unable to restrict the input polyhedron N if neither argument to the join $\sqcup_N$ is unsatisfiable. However, even in this case, useful information can sometimes be inferred, as was shown in Fig. 3.6 on p. 60, where the first state is intersected with $x_2 \neq 5$, thereby restricting x_1 from $1 \leq x_1 \leq 5$ to $2 \leq x_1 \leq 4$. Comparisons of pointers deviate only slightly from those of values. Consider operations other than $\neq$ on points-to sets that contain a single l-value:

$$\begin{aligned}
&\mathit{cond}(N, A, t\ s, x, \{a_1\}, y, \{a_2\}, op) = \langle N'' \sqcap_N \{x\ op\ y\}, A\rangle \text{ iff } a_1 = a_2 \land \\
&\qquad\qquad\qquad\qquad\qquad\qquad\qquad\qquad\qquad\qquad \forall m \in \mathcal{D}.\ L(m) \neq a_1 \\
&\quad \text{where } N' = N \sqcap_N \{-2^{12} \leq x < 2^{30}\} \\
&\quad\qquad N'' = N' \sqcap_N \{-2^{12} \leq y < 2^{30}\} \\
&\quad\qquad \textbf{warn } \text{“Pointer variable on lhs has excessive offset.” } \textbf{if } N \not\sqsubseteq_N N' \\
&\quad\qquad \textbf{warn } \text{“Pointer expression on rhs has excessive offset.” } \textbf{if } N' \not\sqsubseteq_N N''
\end{aligned}$$

Note that wrapping the operands of the test to a signed or unsigned integer of 32 bits would be incorrect since the values represent offsets to the addresses $\rho(a_1)$. However, wrapping cannot occur if the pointer offset is larger than -2^{12}, as the lowest 4 KB are in $\mathit{reserved}_\theta$, and similarly offsets smaller than 2^{30} cannot wrap since the upper 1 GB of memory is reserved for the operating system. While the C standard insists that a pointer is invalid once its offset exceeds the size of the underlying memory region, this assumption is quite strong and we are not aware of any C implementation where for any pointer `p` the expression `p==v+p-v` would not hold for any value `v`. In fact, a small negative offset allows the implementation of Pascal-like arrays where indexing starts at 1. On the contrary, restricting offsets to less than 1 GB will generate false warnings whenever a program is capable of handling data structures larger than 1 GB. In practice, the upper bound can be extended to offsets up to 3 GB, although a correctness argument is more difficult, as the sum of the pointer's base address and the pointer offset may become larger than 4 GB and hence could wrap.

As a consequence of limiting the value of pointer offsets, the function above may emit warnings whenever the 3-GB limit is exceeded. Furthermore, note that the comparison above is only valid when both points-to sets only contain one l-value. For instance, if both operands contain the points-to set $\{a_1, a_2\}$, it is possible that in the concrete program the addresses $\rho(a_1)$ will be compared with the addresses $\rho(a_2)$ and vice versa. Even if the points-to sets are singleton elements, a comparison cannot be made when the single abstract address corresponds to several concrete addresses $\rho(a_1)$. This case can only occur if a_1 refers to a dynamically allocated memory region, which is disallowed by requiring $\forall m \in \mathcal{D}\ .\ L(m) \neq a_1$. The case above can be refined if the operator is an equality test:

$$\begin{aligned}
&\mathit{cond}(N, A, t\ s, x, a_x, y, a_y, =) = \langle N'' \sqcap_N \{x = y\}, A[x \mapsto (a_x \cap a_y)]_{x \in X}\rangle \\
&\quad \text{where } N' = N \sqcap_N \{-2^{12} \leq x < 2^{30}\} \\
&\quad\qquad N'' = N' \sqcap_N \{-2^{12} \leq y < 2^{30}\} \\
&\quad\qquad X = \{y \in \mathcal{X} \mid N'' \sqsubseteq_N \{y = x\}\} \\
&\quad\qquad \textbf{warn } \text{“Pointer variable on lhs has excessive offset.” } \textbf{if } N \not\sqsubseteq_N N' \\
&\quad\qquad \textbf{warn } \text{“Pointer expression on rhs has excessive offset.” } \textbf{if } N' \not\sqsubseteq_N N''
\end{aligned}$$

Here, the offsets of the two operands are restricted as before. However, if the equality test holds, the two operands must indeed be equal, which implies that the offsets are equal and that both operands must contain the same l-values, which therefore restricts the possible l-values to $a_x \cap a_y$. The points-to domain is updated to this set for all variables that are equal to x (including x itself), thereby propagating information from the numeric domain to the points-to domain. Note that the case above is particularly useful when testing a pointer variable for `NULL`. While the test above is able to restrict the domain considerably, the opposite test for $\neq$ is rather benign:

$$
\begin{aligned}
&\mathit{cond}(N, A, t\ s, x, a_x, y, \{a\}, \neq) = \\
&\quad \text{if } N'' \sqsubseteq_N \{x = y\} \text{ then } \langle N'', A[x \mapsto a_x \setminus \{a\}]_{x \in X}\rangle \text{ else } \langle N'', A\rangle \\
&\quad \text{where } N' = N \sqcap_N \{-2^{12} \leq x < 2^{30}\} \\
&\qquad\qquad N'' = N' \sqcap_N \{-2^{12} \leq y < 2^{30}\} \\
&\qquad\qquad X = \{z \in \mathcal{X} \mid N'' \sqsubseteq_N \{z = x\}\} \\
&\qquad\qquad \textbf{warn } \text{“Pointer variable on lhs has excessive offset.” } \textbf{if } N \not\sqsubseteq_N N' \\
&\qquad\qquad \textbf{warn } \text{“Pointer expression on rhs has excessive offset.” } \textbf{if } N' \not\sqsubseteq_N N''
\end{aligned}
$$

This test is able to remove a single l-value contained in the right operand from the variable x whenever the offset is equal. The test is rather specific to the test if a given pointer is not equal to `NULL`; however, it is the most important test on pointers in the analysis. Observe that the choice to treat a pointer as an l-value set with an offset shows its limitations in the test above in that a variable to which a non-zero offset is added cannot be tested for `NULL`. This problem is further discussed in Chap. 13. The cases presented must be tested in the order presented and the first matching case is to be evaluated. If none of the cases match, the input state is returned verbatim, which is a conservative approximation of any conditional in the program. Thus, define $\mathit{cond}(N, A, t\ s, x, a_x, y, a_e, op) = \langle N, A\rangle$ as a catch-all case.

We now detail the semantics of expressions and simple assignments.

6.1 Expressions and Simple Assignments

Figure 6.2 presents the semantics of Core C expressions and simple assignments. The function $[\![\ e\ ]\!]^{\sharp,s}_{\mathrm{Expr}}$, which evaluates an expression e where each variable is s bytes wide, inductively evaluates each term in turn. The function expects a points-to set as the third argument, which is added by the first line if it is missing. This points-to set contains the l-values of the variables encountered so far during the structural induction over the expression. The base case of a constant is handled by the second case, which merely returns the numeric domain N together with the value n as an offset to the passed-in points-to set. The third case details how the sum of a term and a remaining expression is evaluated. The idea is to evaluate the remaining expression and

Expressions.

$[\![\ exp\ ]\!]^{\sharp,s}_{\text{Expr}}\langle N, A\rangle = [\![\ e\ ]\!]^{\sharp,s}_{\text{Expr}}\langle N, A, \{\text{NULL}\}\rangle$

$[\![\ n\ ]\!]^{\sharp,s}_{\text{Expr}}\langle N, A, a\rangle = \langle N, n, a\rangle$

$[\![\ n * v.o + exp\ ]\!]^{\sharp,s}_{\text{Expr}}\langle N, A, a\rangle =$
let $\langle N', x, a'\rangle = read^{F,H}(v, o, s, N, A)$
in if $a' = \{\text{NULL}\}$ then
let $\langle N'', e, a''\rangle = [\![\ exp\ ]\!]^{\sharp,s}_{\text{Expr}}\langle N', A, a\rangle$ in $\langle N'', nx + e, a''\rangle$
else if $n = 1 \wedge a = \{\text{NULL}\}$ then
let $\langle N'', e, a''\rangle = [\![\ exp\ ]\!]^{\sharp,s}_{\text{Expr}}\langle N', A, a'\rangle$ in $\langle N'', x + e, a''\rangle$
else if $n = -1 \wedge (a \setminus \{\text{NULL}\}) = (a' \setminus \{\text{NULL}\})$ then
let $\langle N'', e, a''\rangle = [\![\ exp\ ]\!]^{\sharp,s}_{\text{Expr}}\langle N', A, \{\text{NULL}\}\rangle$ in $\langle N'', -x + e, a''\rangle$
warn "Subtracting pointers to different objects." if $|a| > 1$
warn "Positive pointer can be NULL." if $\text{NULL} \in a$
warn "Negative pointer can be NULL." if $\text{NULL} \in a'$
else let $t \in \mathcal{X}^T$ fresh in $\langle N \sqcap_N \{0 \le t < 2^{8s}\}, t, \{\text{NULL}\}\rangle$

Assignments.

$[\![\ s\ v.o = exp\ ]\!]^{\sharp}_{\text{Stmt}}\langle N, A\rangle = \langle \exists_{\mathcal{X}^T}(N''), A'\rangle$
where $\langle N', e, a\rangle = [\![\ exp\ ]\!]^{\sharp,s}_{\text{Expr}}\langle N, A\rangle$
$\langle N'', A'\rangle = write^{F,H}(v, o, s, e, a, N', A)$

$[\![\ s\ v \to o = exp\ ]\!]^{\sharp}_{\text{Stmt}}\langle N, A\rangle = \langle \exists_{\mathcal{X}^T}(N'''), A'\rangle$
where $\langle N', e, a\rangle = [\![\ exp\ ]\!]^{\sharp,s}_{\text{Expr}}\langle N, A\rangle$
$\langle N'', x_o, a'\rangle = read^{F,H}(v, 0, 4, N', A)$
$\{m_1, \ldots m_n\} = \{m \in \mathcal{M} \cup \mathcal{D} \mid L(m) \in a'\}$
$\langle N_i, A_i\rangle = write^{F,H}(m_i, x_o + o, s, e, a, N'', A)$ for all $i \in [1, n]$
$N''' = \bigsqcup_{i=1}^{n} N_i, A' = \bigsqcup_{i=1}^{n} A_i$
warn "Dereferencing a NULL pointer." if $\text{NULL} \in a'$
warn "Dereferencing a function pointer." if $a' \cap \mathcal{A}^{\mathcal{F}} \neq \emptyset$

$[\![\ s\ v_1.o_1 = v_2 \to o_2\ ]\!]^{\sharp}_{\text{Stmt}}\langle N, A\rangle = \langle \exists_{\mathcal{X}^T}(N'''), A\rangle$
where $t \in \mathcal{X}^T$ fresh
$\langle N', x_o, a\rangle = read^{F,H}(v_2, 0, 4, N, A)$
$\{m_1, \ldots m_n\} = \{m \in \mathcal{M} \cup \mathcal{D} \mid L(m) \in a\}$
$\langle N_i, x_i, a_i\rangle = read^{F,H}(m_i, x_o + o_2, s, N', A)$ for all $i \in [1, n]$
$N'' = \bigsqcup_{i=1}^{n} N_i \rhd t := x_i$
$\langle N''', A'\rangle = write^{F,H}(v_1, o_1, s, t, (a_1 \cup \ldots \cup a_n), N'', A)$
warn "Dereferencing a NULL pointer." if $\text{NULL} \in a$
warn "Dereferencing a function pointer." if $a \cap \mathcal{A}^{\mathcal{F}} \neq \emptyset$

Fig. 6.2. Abstract semantics for expressions and assignments.

to add the result to the outcome of this calculation. In the simplest case, the term $n * v\ .o$ contains a value such that the result is merely that of the remaining expression to which the term is added. Slightly more interesting are the occurrences of variables that contain l-values. In the case with $n = 1$ and a passed-in points-to set $a = \{\text{NULL}\}$, the new points-to set a' is passed to the recursive call $[\![\ exp\ ]\!]^{\sharp,s}_{\text{Expr}}$ and the offset of the pointer variable is added on return. This new points-to set is passed down until the base case is reached or until a coefficient of -1 is encountered in front of a variable that has the same points-to set. In this case, the difference of the two pointers is calculated and the difference between them is returned as the value; that is, with the points-to set $\{\text{NULL}\}$. The final case conservatively approximates the result when a pointer is scaled by a constant unequal to 1 and -1 or the points-to sets of positive and negative terms do not coincide. Note that expressions such as $-p+q$, where both p and q have the same points-to set, are not recognised, although $q-p$ is identified as the pointer difference. This limitation simplifies the presentation of $[\![\cdot]\!]^{\sharp,s}_{\text{Expr}}$ and is not present in the actual implementation.

The remaining three transfer functions in Fig. 6.2 are assignments of base types between variables and through pointers. The first case assigns the value e of the expression and the points-to set a to the memory region v at offset o. The second case assigns an expression to the memory regions pointed to by v at the offset o. In order to infer the possible set of memory regions, the four bytes of the pointer v are read into $x_o \in \mathcal{X}$ and $a' \subseteq \mathcal{A}$. The points-to set is mapped to memory regions $\{m_1, \ldots m_n\}$, to which the result is written at offset $x_o + o$; that is, at the sum of the pointer offset x_o and the constant displacement o stemming from the statement itself. Since the concrete program only writes to one of these memory regions, the resulting n domain pairs $\langle N_1, A_1\rangle, \ldots \langle N_n, A_n\rangle$ are joined and returned. The case of reading through a pointer is built up analogously. In order to approximate a value that contains the values from all pointed-to memory regions, the result of each read is assigned to the temporary t. After joining all domains that contain the resulting value in t, an approximation of the values from all accessed memory regions is available in t, which is then written to the variable v_1. The join of the points-to set is calculated by simple set union $a_1 \cup \ldots \cup a_n$. Note that using access trees guarantees a precise result even if the pointer offset $x_o + o$ is a range. Less precision is attainable when assigning whole structures.

6.2 Assigning Structures

The abstract transfer functions that assign whole structures cannot make use of access trees since only the total size of a structure is known rather than the size and offsets of individual fields. Since assigning whole structures at a range of offsets is rather difficult, the transfer functions for assigning whole structures in Fig. 6.4 only distinguish between two cases: Either the two structures

Create bounds on access to memory region.

$$check^H(m, e_o, s) = \text{if } m \in \mathcal{M} \text{ then } \{0 \leq e_o \leq size(m) - s\}$$
$$\text{else let } \langle x_m, x_s\rangle = H(m) \text{ in } \{0 \leq e_o \leq x_s - s, x_m \geq 0\}$$

Removing all information pertaining to a memory region.

$$clearMem(m, e_o, s, N, A) = \langle N^n, A^n\rangle$$

$$\begin{aligned}
\text{where } & N' = N \sqcap_N check^H(m, e_o, s) \\
& [l, u]_{\equiv d} = N'(e_o) \\
& \{\langle x_1, s_1\rangle, \ldots \langle x_n, s_n\rangle\} = \{\langle x, s'\rangle \mid \forall \langle o', s', x\rangle \in F(m) \,. \\
& \qquad [l, u + s] \cap [o', o' + s'] \neq \emptyset\} \\
& N_0 = N',\ N_i = \exists_{x_i}(N_{i-1}) \sqcap_N \{0 \leq x_i < 2^{8s_i}\} \text{ for all } i \in [1, n] \\
& A_0 = A,\ A_i = A_{i-1}[x_i \mapsto \{\text{NULL}\}] \text{ for all } i \in [1, n] \\
& \textbf{warn } \text{“Illegal write access to structure.” } \textbf{if } N \not\sqsubseteq_N N'
\end{aligned}$$

Copying whole memory regions.

$$copyMem(m_1, o_1, m_2, o_2, s, N, A) = \langle N_m, A_m\rangle$$

$$\begin{aligned}
\text{where } & F_i = \{\langle o - o_i, s, x\rangle \mid \langle o, s, x\rangle \in F(m_i) \wedge [o, s] \cap [o_i, s_i] \neq \emptyset\} \text{ for } i = 1, 2 \\
& \{\langle x_1^l, x_1^r\rangle, \ldots \langle x_n^l, x_n^r\rangle\} = \{\langle x^l, x^r\rangle \mid \forall \langle o, s, x^l\rangle \in F_1 \,. \\
& \qquad \exists x^r \in \mathcal{X} \,.\, \langle o, s, x^r\rangle \in F_2\} \\
& \{\langle s_{n+1}, x_{n+1}\rangle, \ldots \langle s_m, x_m\rangle\} = \{\langle s, x^l\rangle \mid \forall \langle o, s, x^l\rangle \in F_1 \,. \\
& \qquad \forall x^r \in \mathcal{X} \,.\, \langle o, s, x^r\rangle \notin F_2\} \\
& N_0 = N \sqcap_N check^H(m_1, o_1, s) \sqcap_N check^H(m_2, o_2, s),\ A_0 = A \\
& N_i = N_{i-1} \rhd x_i^l := x_i^r \text{ for all } i \in [1, n] \\
& A_i = A_{i-1}[x_i^l \mapsto A(x_i^r)] \text{ for all } i \in [1, n] \\
& N_i = \exists_{x_i}(N_{i-1}) \sqcap_N \{0 \leq x_i < 2^{8s_i}\} \text{ for all } i \in [n+1, m] \\
& A_i = A_{i-1}[x_i \mapsto \{\text{NULL}\}] \text{ for all } i \in [n+1, m] \\
& \textbf{warn } \text{“Illegal access to structure.” } \textbf{if } N \not\sqsubseteq_N N_0
\end{aligned}$$

Fig. 6.3. Helper functions to copy and clear memory regions.

in the assignment are located at constant offsets, in which case the individual fields of the access trees can simply be copied, or an assignment through a pointer has a range of offsets, in which case all affected fields in the target memory region are set to their maximum bounds. To this end, Fig. 6.3 defines three helper functions. The first one restricts an access position to the range of the given memory region. This function is used in $clearMem(m, e_o, s, N, A)$, which clears all fields in $m \in \mathcal{M} \cup \mathcal{D}$ in the range $[e_o, e_o + s - 1]$, where $e_o \in Lin \times \mathbb{Z}$ may itself be a range of offsets. On the contrary, the third helper function, $copyMem(m_1, o_1, m_2, o_2, s, N, A)$, copies s bytes starting at the constant offsets o_1 in m_1 to the fields starting at the constant offset o_2 of memory region m_2. Both functions extract fields that overlap with the accessed range. While *clearMem* resets all fields to their maximal value and the points-to sets to {NULL}, the *copyMem* function distinguishes between fields in the source memory region that match fields in the destination region and the remaining fields in the destination. Specifically, for both regions m_1 and m_2, *copyMem* extracts those fields of m_i that overlap with the accessed region, relocates

Assignment of Structures.

$[\![\textbf{structure } s\ v_1.o_1 = v_2.o_2]\!]^\sharp_{\text{Stmt}}\langle N, A\rangle = \mathit{copyMem}(v_1, o_1, v_2, o_2, s, N, A)$

$[\![\textbf{structure } s\ v_1 \to o_1 = v_2.o_2]\!]^\sharp_{\text{Stmt}}\langle N, A\rangle = \langle \exists_{\mathcal{X}^T}(\bigsqcup_{i=1}^{n} N_i), \bigsqcup_{i=1}^{n} A_i\rangle$
where $\langle N', x_o, a\rangle = \mathit{read}^{F,H}(v_1, 0, 4, N, A)$
$\{m_1, \ldots m_n\} = \{m \in \mathcal{M} \cup \mathcal{D} \mid L(m) \in a\}$
$[l, u]_{\equiv d} = N'(x_o)$
if $l = u$
then $\langle N_i, A_i\rangle = \mathit{copyMem}(m_i, l + o_1, v_2, o_2, s, N', A)$ for all $i \in [1, n]$
else $\langle N_i, A_i\rangle = \mathit{clearMem}(m_i, x_o + o_1, s, N', A)$ for all $i \in [1, n]$
warn "Dereferencing `NULL` pointer." **if** NULL $\in a$
warn "Dereferencing function pointer." **if** $a \cap \mathcal{A}^{\mathcal{F}} \neq \emptyset$

$[\![\textbf{structure } s\ v_1.o_1 = v_2 \to o_2]\!]^\sharp_{\text{Stmt}}\langle N, A\rangle = \langle \exists_{\mathcal{X}^T}(N'''), A'\rangle$
where $\langle N', x_o, a\rangle = \mathit{read}^{F,H}(v_2, 0, 4, N, A)$
$\{m_1, \ldots m_n\} = \{m \in \mathcal{M} \cup \mathcal{D} \mid L(m) \in a\}$
$N'' = N' \sqcap_N (\bigsqcup_{i=1}^{n} \mathit{check}^H(m_i, x_o + o_2, s))$
$[l, u]_{\equiv d} = N''(x_o)$
if $u = l$
then $\langle N_i, A_i\rangle = \mathit{copyMem}(v_1, o_1, m_i, l + o_2, s, N'', A)$ for all $i \in [1, n]$
$\langle N''', A'\rangle = \langle \bigsqcup_{i=1}^{n} N_i, \bigsqcup_{i=1}^{n} A_i\rangle$
else $\langle N''', A'\rangle = \mathit{clearMem}(v_1, o_1, s, N'', A)$
warn "Dereferencing `NULL` pointer." **if** NULL $\in a$
warn "Dereferencing function pointer." **if** $a \cap \mathcal{A}^{\mathcal{F}} \neq \emptyset$
warn "Illegal write access to structure." **if** $N' \not\sqsubseteq_N N''$

Fig. 6.4. Abstract transfer functions for assignments of structures.

them by removing the offset o_i, and finally stores them in F_i. All fields that have the same offset and size are then assigned in N by calculating $N_1, \ldots N_n$ and $A_1, \ldots A_n$, whereas all fields in F_1 that have no matching source field are set to their maximum bounds by calculating $N_{n+1}, \ldots N_m$ and $A_{n+1}, \ldots A_m$.

These helper functions are then used to define assignments of structures in Fig. 6.4, which in its simplest case merely calls *copyMem*. If a structure is written through a pointer, the possible target memory regions are extracted. Depending on whether the offset of the pointer is constant or not, each memory region is either copied to or overwritten. The resulting pairs of domains $\langle N_i, A_i\rangle$ are joined and returned as the result. Note that checking that all memory accesses are within bounds is done within the helper functions. Reading a structure through a pointer that has a range of offsets will merely clear all fields in the target structure without ever reading any fields from the source variable. Hence, check^H is used to constrain the read offset such that the access to each source region is within bounds.

The next section details abstract transfer functions for casts, address-of operators, and memory allocation, and thereby completes the abstract semantics.

6.3 Casting, &-Operations, and Dynamic Memory

Figure 6.5 presents the abstract transfer functions for the remaining Core C statements. An interesting case is the cast between two integer types. Not surprisingly, the result of the assignment is the maximum range of the target type if the source can contain a pointer. Otherwise, two cases can occur. Firstly, the type on the left may be smaller than the type on the right. As pointed out in Chap. 4, no special conversion has to be done, as wrapping will be made explicit when needed – for instance, when comparing the target variables. In contrast, *wrap* needs to be called if the target type is larger than the source type since otherwise a value might be assigned to the target that is larger than that of the maximum range of the source type, which is impossible in the concrete program.

The cases for the address-of operators need little explanation, nor does the assignment of a string constant that merely clears the memory, as the analysis does not track individual elements of an array for efficiency reasons.

The last two statements concern the allocation and deallocation of dynamic memory. The **malloc** function increments the number of concrete memory regions x_m that the abstract memory region l summarises. Analogously, the **free** function decrements this counter by 1 by calling *freeMem*^{H} on every memory region that the passed-in pointer may point to. Note that the main purpose of x_m is to distinguish if $l \in \mathcal{D}$ corresponds to some concrete heap regions ($x_m > 0$) or if l represents a freed memory region ($x_m = 0$). Each time a memory region l is freed, the counter is not only decremented but also explicitly set to zero in order to indicate that l might from now on refer to a freed region. A subtle consequence is that x_m merely tracks the maximum number of memory regions that the abstract address l summarises rather than tagging each allocated memory region with a number. This is important in that the **free** function in the concrete program can free any of the regions that are summarised by l. Thus, if two memory regions of sizes s_1 and s_2 are both summarised by the same abstract memory region $l \in \mathcal{D}$, freeing one of them will leave one memory region of size s_1 or s_2. Since it is not known which memory region is freed, it is important that no linear relationship between x_m and x_s exist, as this would inevitably restrict the value of x_s whenever **free** decrements x_m. The implementation of **malloc** guarantees that no linear relationship exists between x_s and x_m whenever $x_m > 0$ by first incrementing x_m by calculating $N'' = N' \triangleright x_m := x_m + 1$ and afterwards setting the new size x_s for all $x_m > 0$.

Given the abstract semantics of Core C, the last section of this chapter will detail some refinements necessary to create a fully automated analysis.

Type Casts.

$$[\![\ s_1\ v_1.o_1 = t\ s_2\ v_2.o_2\ ]\!]^\sharp_{\text{Stmt}}\langle N, A\rangle = \langle \exists_{\mathcal{X}^T}(N'''), A'\rangle$$

where $\langle N', x, a\rangle = \mathit{read}^{F,H}(v_2, o_2, s_2, N, A)$

$x_t \in \mathcal{X}^T$ fresh

$$\langle N'', x'\rangle = \begin{cases} \langle N' \sqcap_N \{0 \leq x_t < 2^{8s_1}\}, x_t\rangle & \text{if } a \neq \{\text{NULL}\} \wedge t = \textbf{uint} \\ \langle N' \sqcap_N \{-2^{8s_1-1} \leq x_t < 2^{8s_1-1}\}, x_t\rangle & \\ & \text{if } a \neq \{\text{NULL}\} \wedge t = \textbf{int} \\ \langle N', x\rangle & \text{if } s_1 \leq s_2 \\ \langle \mathit{wrap}(N', t\ s_2, x), x\rangle & \text{otherwise} \end{cases}$$

$\langle N''', A'\rangle = \mathit{write}^{F,H}(v_1, o_1, s_1, x', \{\text{NULL}\}, N'', A)$

Address-Of Operators.

$$[\![\ v_1.o_1 = \&v_2.o_2\ ]\!]^\sharp_{\text{Stmt}}\langle N, A\rangle = \mathit{write}^{F,H}(v_1, o_1, 4, o_2, \{L(v_2)\}, N, A)$$

$$[\![\ v_1.o_1 = \&f\ ]\!]^\sharp_{\text{Stmt}}\langle N, A\rangle = \mathit{write}^{F,H}(v_1, o_1, 4, 0, \{a_f\}, N', A)$$

where $a_f \in \mathcal{A}^{\mathcal{F}}$ corresponds to address p_f of function f

String Constants.

$$[\![\ v = \text{"}c_0c_1\ldots c_{k-1}\text{"}\ ]\!]^\sharp_{\text{Stmt}}\langle N, A\rangle = \mathit{clearMem}(v, 0, k, N, A)$$

Dynamic Memory Allocation

$$[\![\ l:\ v_1 = \textbf{malloc}(v_2)\ ]\!]^\sharp_{\text{Stmt}}\langle N, A\rangle = \langle \exists_{\mathcal{X}^T}(N''''), A''\rangle$$

where $\langle N', x_v, a\rangle = \mathit{read}^{F,H}(v_2, 0, 4, N, A)$

$N'' = \mathit{wrap}(N', \textbf{uint32}, x_v)$

$\langle x_m, x_s\rangle = H(l)$

$N'' = N' \triangleright x_m := x_m + 1$

$N''' = N'' \sqcup_N (N'' \sqcap_N \{x_m > 0\} \triangleright x_s := x_v)$

$\langle N'''', A''\rangle = \mathit{write}^{F,H}(v_1, 0, 4, 0, \{l, \text{NULL}\}, N''', A')$

warn "Parameter contains pointer." **if** $a \neq \{\text{NULL}\}$

$$[\![\ \textbf{free}(v)\ ]\!]^\sharp_{\text{Stmt}}\langle N, A\rangle = \langle \exists_{\mathcal{X}^T} N''', A\rangle$$

where $\langle N', x_v, a\rangle = \mathit{read}^{F,H}(v, 0, 4, N, A)$

$N'' = N' \sqcap_N \{x_v = 0\}$

$\{m_1, \ldots m_n\} = \{m \in \mathcal{M} \cup \mathcal{D} \mid L(m) \in a\}$

$N''' = \bigsqcup_{i=1}^{n} \mathit{freeMem}^H(m_i, N'')$

warn "Freed pointer has an offset." **if** $N' \not\sqsubseteq_N N''$

warn "Freeing function pointer." **if** $a \cap \mathcal{A}^{\mathcal{F}} \neq \emptyset$

$$\mathit{freeMem}^H(m, N) = (N' \triangleright x_m := x_m - 1) \sqcup_N (N' \triangleright x_m := 0 \triangleright x_s := 0)$$

where $\langle x_m, x_s\rangle = H(m)$

$N' = N \sqcap_N \{x_m > 0\}$

warn "Repeatedly freeing region allocated at m" **if** $N \not\sqsubseteq_N N'$

Fig. 6.5. Abstract transfer functions for type casts, address-of operators, string constants, and dynamic memory allocation.

6.4 Inferring Fields Automatically

The analysis presented so far has a major drawback: It is not fully automatic since the fields in the map F on which information is inferred have to be set manually. Populating all fields with domain variables is not feasible for performance reasons, as every 8 bytes can have 15 overlapping fields, which is unacceptable even for programs with a small memory footprint. Furthermore, fully populating access trees is wasteful since arrays of sufficient size are always accessed within loops such that accesses will occur through a pointer with a range of offsets. However, write accesses to a range of array elements cannot induce any information on individual elements, as a write to several elements has to retain the previous value of each element. Hence, populating memory regions that are accessed through a pointer that has a range of offsets is of no benefit and a waste of resources. This observation leads to a heuristic for populating F automatically with fields that are of interest to the analysis. The idea is to add a field $\langle o, s, x\rangle$ to $F(m)$ each time s bytes are written to the memory region m at the constant offset o. The number of fields added to F in this way is bounded by the number of write operations in the program since a repeated analysis of a write operation will feature either the same constant offset, in which case the field in F is reused, or a range of offsets, in which case no new field is added. In particular, a single write operation can never be evaluated with two different constant offsets since the state space grows monotonically; that is, the state during the second evaluation has to include the offset of the first evaluation. Adding new fields should also be avoided when writing through a pointer that contains several l-values, as no target can be updated without retaining the previous value. By following these principles, the analysis only adds fields that can be set to a precise value and can thus become relevant for verifying the program at hand.

While this heuristic yields a field map F that only contains fields that can contribute to a more precise analysis result, it turns out that adding fields on-the-fly is not without drawbacks, as illustrated by the following code fragment:

```
if (rand()) { (int) ip_info=0; } else {
  ip_info.addr[0]=0; ip_info.addr[1]=0;
  ip_info.addr[2]=0; ip_info.addr[3]=0; }
```

The fragment above sets the fields of the `ip_info` structure in Fig. 5.1 to zero either by writing the 4-byte field x_0 or by writing four byte-sized fields that overlap with x_0. Assuming that the then-branch of the conditional is analysed first, x_0 will be present when analysing the else-branch, so that it is updated correctly when writing the individual bytes. However, when joining the two branches, the information on the byte-sized fields is lost since these four fields are unbounded in the then-branch. Irritatingly, the lost information cannot easily be recovered: Reading a single byte in the joined state can only propagate information through the generated access tree if the byte-sized fields are in the range $[0, 255]$. However, since these fields are completely unbounded in the then-branch, their value is unbounded in the join and propagation is disabled since the values are not in $[0, 255]$. Chapter 10 will detail a workaround that mostly avoids a precision loss in this situation.

An orthogonal aspect is to make the analysis more versatile by adapting it to different platforms. The abstract semantics and the examples are based on little-endian, 32-bit machines that can operate on integers with at most 64 bits. Thus, the widest fields in F can be 8 bytes, thereby limiting the height of any given access tree. Furthermore, all pointer-sized accesses are encoded as 4-byte accesses. While changes to these parameters are obvious, the change to a big-endian machine is straightforward but somewhat more intriguing. Changing the endianness assumption of the underlying machine entails that writing a single byte at the address of a 4-byte integer will overwrite not the lower bits but rather the upper bits. In fact, analysing a program for a big-endian machine merely requires that the children in each access tree be swapped recursively before calling *prop* and *update*.

Finally, we point out that on rare occasions the alignment assumption of our model might not fit the actual alignment of program variables. Consider the following two declarations of function-local variables:

```
int i;
struct {
  int j;
  double k;
} l;
```

Assuming that the integer `i` is located at offset o with $o \ mod \ 8 = 0$, an optimising compiler may place the elements of the structure at $o+4$ for `j` and $o+8$ for `k`. In this case, the beginning of the memory region represented by `l` is not aligned at an 8-byte boundary, which is the assumption when overlaying complete binary trees over a memory region starting from offset 0. The effect is that accessing `k` will trigger a false warning about incorrect alignment. A workaround in the form of access trees that are offset by 4 bytes is certainly possible but was not found necessary for the C programs of interest.

Part II

Ensuring Efficiency

7

Planar Polyhedra

The analysis presented in the previous chapters relies on an efficient implementation of the underlying polyhedral operations in order to attain acceptable performance. However, common implementations of convex polyhedra [14, 27, 93, 119] suffer from inherent scalability problems that mainly relate to the calculation of the join operation, which corresponds to the convex hull in the context of polyhedra. The classic approach for calculating the convex hull of two polyhedra is to convert the half-space representation using inequalities into the generator representation consisting of vertices, rays, and lines. Vertices are points in the polyhedron that cannot be represented by a convex combination of other points. Rays and lines are vectors that represent unidirectional and bidirectional trajectories, respectively, towards which the polyhedron extends to infinity. The convex hull of two input polyhedra can be calculated by converting both polyhedra into their generator representations, joining their sets of vertices, rays, and lines, and converting these three sets back into the half-space representation. In order to illustrate the problems using this approach, consider Fig. 7.1. Here, the shown polyhedra $P_1 = [\![\{1 \leq x_1 \leq 7, 2 \leq x_2 \leq 8\}]\!]$ and $P_2 = [\![\{1 \leq x_1 \leq 7, 10 \leq x_2 \leq 16\}]\!]$

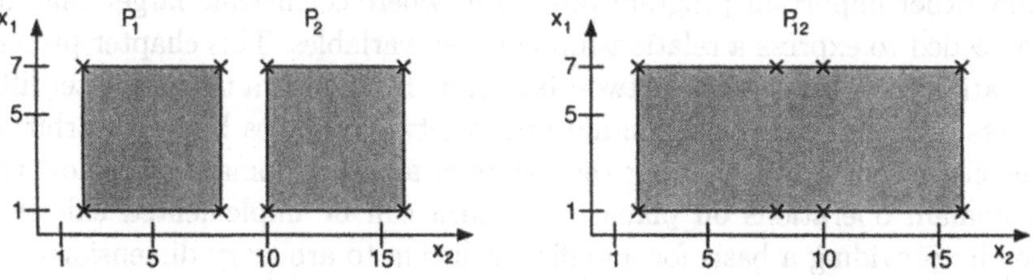

Fig. 7.1. Calculating the convex hull $P_{12} = P_1 \sqcup_P P_2$ of planar polyhedra using the generator representation.

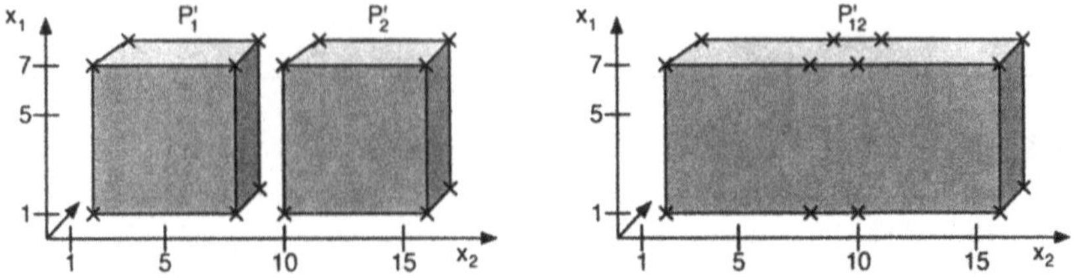

Fig. 7.2. Calculating the convex hull $P'_{12} = P'_1 \sqcup_P P'_2$ of three dimensional polyhedra using the generator representation.

contain neither rays nor lines, as they are both bounded. The sets of vertices are shown as crosses. These vertices are included in the resulting convex hull $P_{12} = P_1 \sqcup_P P_2$, which is shown in the right graph of the figure. A similar example in three dimensions is shown in Fig. 7.2, which depicts the convex hull $P'_{12} = P'_1 \sqcup_P P'_2$ of the polyhedra $P'_1 = [\![\{1 \leq x_1 \leq 7, 2 \leq x_2 \leq 8, 0 \leq x_3 \leq 3\}]\!]$ and $P'_2 = [\![\{1 \leq x_1 \leq 7, 10 \leq x_2 \leq 16, 0 \leq x_3 \leq 3\}]\!]$. While in the two-dimensional case each input polyhedron can be described by four inequalities or, equivalently, four vertices, each input in the three-dimensional case is described by six inequalities or, equivalently, eight vertices. In general, calculating the convex hull of two d-dimensional hypercubes requires $2d$ inequalities to represent each input polyhedron or, equivalently, 2^d vertices. Thus, even though input and output polyhedra can be described by a small number of inequalities, the intermediate representation using generators can be exponential. While circumventing this exponential blowup is possible by approximating the convex hull operation [169], the most compelling example of large-scale program analysis that uses a relational domain [30] is based on a sub-class of polyhedra, namely the Octagon domain [128]. The Octagon domain only allows inequalities with at most two variables per inequality, where the coefficients of these variables have to be one or minus one. These inequalities are too weak to express relationships between overlapping fields, and many other important program properties where coefficients larger than one are needed to express a relationship between variables. This chapter presents operations on planar (that is, two-dimensional) polyhedra that are later lifted to sets of inequalities that contain at most two variables but with arbitrary coefficients, thereby extending the Octagon abstract domain. We show that all domain operations on planar polyhedra can be implemented efficiently, thereby providing a basis for an efficient lifting to arbitrary dimensions. The key observation for implementing efficient algorithms is that inequalities in two dimensions can be sorted by angle. The chapter therefore commences by presenting basic properties of inequalities in planar space. The notation introduced here is then used to define the various domain operations on planar polyhedra.

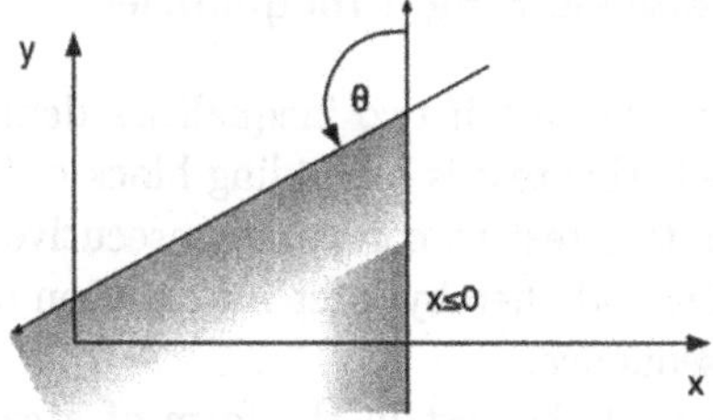

Fig. 7.3. The angle of a planar inequality is measured relative to $x \leq 0$.

7.1 Operations on Inequalities

For the sake of this chapter, let $\mathcal{X} = \{x, y\}$ denote the set of polyhedral variables that correspond to the axes of the two-dimensional Euclidian space. Observe that the vector $\langle a, b\rangle$ is orthogonal to the line $ax + by = c$ and points away from the induced half-space $[\![ax+by \leq c]\!]$. This vector induces an ordering on half-spaces via the orientation mapping θ. This map $\theta : \mathit{Ineq} \to [0, 2\pi)$ is defined such that $\theta(ax + by \leq c) = \psi$, where $\cos(\psi) = a/\sqrt{a^2+b^2}$ and $\sin(\psi) = b/\sqrt{a^2+b^2}$. The mapping θ corresponds to the counterclockwise angle through which the half-space of $x \leq 0$ has to be turned to coincide with that of $ax + by \leq c$, as illustrated in Fig. 7.3. In the context of this work, θ is mainly used to compare the orientations of two half-spaces, which is key to sorting a set of inequalities. For the sake of efficiency and numeric stability, it is desirable to implement this comparator without a recourse to trigonometric functions [158]. To this end, define the function *class* : *Ineq* $\to \{1, 2, \ldots, 8\}$ as follows:

$$\mathit{class}(ax + by \leq c) = \begin{cases} 7 - \mathit{sign}(b) & : \; a < 0 \\ 1 & : \; a = 0 \wedge b \leq 0 \\ 5 & : \; a = 0 \wedge b > 0 \\ 3 + \mathit{sign}(b) & : \; a > 0 \end{cases}$$

Here $\mathit{sign} : \mathbb{Z} \to \{-1, 0, 1\}$ is the function that returns -1 if the given number is negative, 1 if it is positive, and zero otherwise. A comparison between the angles of $\iota_1 \equiv a_1x+b_1x \leq c_1$ and $\iota_2 \equiv a_2x+b_2x \leq c_2$ can now be implemented as follows:

$$\begin{aligned} \theta(\iota_1) \leq \theta(\iota_2) \iff & \mathit{class}(\iota_1) \leq \mathit{class}(\iota_2) \vee \\ & \mathit{class}(\iota_1) = \mathit{class}(\iota_2) \wedge a_1b_2 \leq a_2b_1 \end{aligned}$$

Furthermore, define the angular difference $\iota_1 \measuredangle \iota_2$ between two inequalities ι_1 and ι_2 as the counterclockwise angle between $\theta(\iota_1)$ and $\theta(\iota_2)$. More precisely, $\iota_1 \measuredangle \iota_2 = (\theta(\iota_2) - \theta(\iota_1)) \; \mathit{mod} \; 2\pi$. This function is used to test if two inequalities are less than π apart. As above, this test can be implemented without recourse to trigonometric functions.

7.1.1 Entailment between Single Inequalities

A recurring function is the test if two inequalities define a sub-space of another inequality. In fact, this test is a building block of the upcoming domain operation that applies this test to compare consecutive elements of a sorted sequence of inequalities and thereby infer information on an inequality with respect to the whole sequence.

We give a definition of the test in the form of a case distinction on the coefficients of the inequalities involved. Let $\iota_i \equiv a_i x + b_i y \leq c_i$ for $i = 1, 2$ and $\iota \equiv ax + by \leq c$. Assume $\iota_1 \measuredangle \iota_2 \leq \pi$; otherwise exchange ι_1 and ι_2. We define the following predicates, which are explained below:

$$\{\iota_1\} \sqsubseteq \iota \iff \begin{cases} \mathit{false} & \text{if } a_1 b - a b_1 \neq 0 \\ \mathit{false} & \text{else if } a_1 a < 0 \\ \mathit{false} & \text{else if } b_1 b < 0 \\ \frac{a}{a_1} c_1 \leq c & \text{else if } a_1 \neq 0 \\ \frac{b}{b_1} c_1 \leq c & \text{else if } b_1 \neq 0 \\ (c < 0 \wedge a = 0 \wedge b = 0) \Rightarrow c_1 < 0 & \text{otherwise} \end{cases}$$

$$\{\iota_1, \iota_2\} \sqsubseteq \iota \iff \begin{cases} \{\iota_1\} \sqsubseteq e \vee \{\iota_2\} \sqsubseteq \iota & \text{if } d = a_1 b_2 - a_2 b_1 = 0 \\ \mathit{false} & \text{else if } \lambda_1 = (a b_2 - a_2 b)/d < 0 \\ \mathit{false} & \text{else if } \lambda_2 = (a_1 b - a b_1)/d < 0 \\ \lambda_1 c_1 + \lambda_2 c_2 \leq c & \text{otherwise} \end{cases}$$

Intuitively, the predicate $\{\iota_1\} \sqsubseteq \iota$ holds iff $[\![\iota_1]\!] \subseteq [\![\iota]\!]$ and, analogously, $\{\iota_1, \iota_2\} \sqsubseteq \iota$ holds iff $[\![\{\iota_1, \iota_2\}]\!] \subseteq [\![\iota]\!]$. The reasoning behind the definitions above is as follows.

Inclusion between two single inequalities never holds if they are not parallel – that is, if the determinant of their coefficients $a_1 b - a b_1$ is non-zero. Furthermore, the inclusion cannot hold if ι_1 and ι are anti-parallel, which is the case if the coefficients for x have different signs, and similarly for y. Otherwise, the intersection points with the y-axis of ι_1 and ι are calculated and compared. In particular, the subset relation $\{x \mid a_1 x \leq c_1\} \subseteq \{x \mid ax \leq c\}$ implies that $x \leq \frac{c}{a}$ if $x \leq \frac{c_1}{a_1}$, assuming that $a_1 > 0$ and $a > 0$. The latter is equivalent to $\frac{c_1}{a_1} \leq \frac{c}{a}$, and since a is positive, $\frac{a}{a_1} c_1 \leq c$ follows. Now assume $a_1 < 0$ and $a < 0$. From $x \geq \frac{c}{a}$, if $x \geq \frac{c_1}{a_1}$, then $\frac{c_1}{a_1} \geq \frac{c}{a}$ follows. Multiplying by $a < 0$ yields $\frac{a}{a_1} c_1 \leq c$. If $a_1 = 0$ but $b_1 \neq 0$, the intersection points with the x-axis can be calculated and compared. If $b_1 = 0$, too, then ι_1 is tautologous or unsatisfiable, which is handled by the implication that formed the last case.

The second test, $\{\iota_1, \iota_2\} \sqsubseteq \iota$, reduces to the first test whenever ι_1 and ι_2 are parallel – that is, if the determinant of the coefficients $a_1 b_2 - a_2 b_1$ is zero. Otherwise, a linear combination of ι_1 and ι_2 is calculated, yielding an inequality that is parallel to ι. Specifically, λ_1 and λ_2 are calculated such that $\lambda_1 a_1 + \lambda_2 a_2 = a$ and $\lambda_1 b_1 + \lambda_2 b_2 = b$. If either λ_1 or λ_2 is negative, the resulting half-space of the parallel inequality faces the opposite direction and the inequality ι is not entailed. If both λ_1 and λ_2 are positive, entailment can

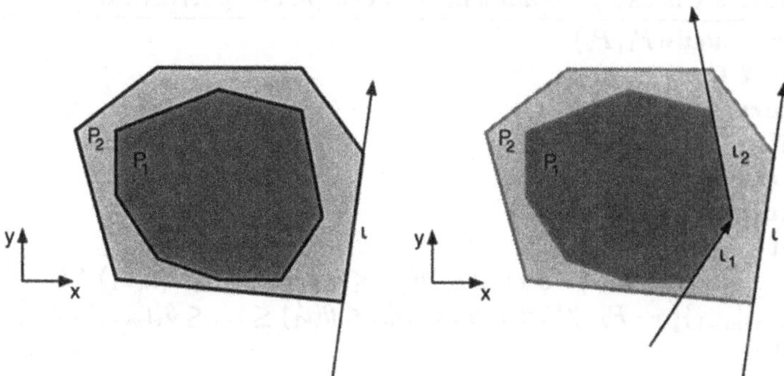

Fig. 7.4. The entailment check $P_1 \sqsubseteq_P P_2$ is reduced to checking entailment $\{\iota_1, \iota_2\} \sqsubseteq \iota$ of each inequality ι of P_2 with respect to some ι_1, ι_2 of P_1.

be determined by comparing the constant of the parallel inequality, namely $\lambda_1 c_1 + \lambda_2 c_2$, with the constant of ι.

Due to the ability to sort inequalities by angle, this constant-time entailment check between three inequalities can be lifted to a linear-time entailment check between two planar polyhedra, as presented in the next section.

7.2 Operations on Sets of Inequalities

In the following, we present operations on planar polyhedra that exploit the fact that inequalities can be sorted by angle. Given this order on inequalities, entailment, redundancy removal, the convex hull, and a linear programming algorithm can all be implemented efficiently.

7.2.1 Entailment Check

By traversing the inequalities so that their angles are increasing, a linear-time entailment check between two planar polyhedra can be implemented. Consider the task of checking whether $P_1 \sqsubseteq_P P_2$; that is, if $P_1 \subseteq P_2$ holds. It is sufficient to show that each inequality ι that defines a facet of P_2 contains the polyhedron P_1, as shown on the left of Fig. 7.4. Specifically, it is sufficient to find the two adjacent inequalities ι_1, ι_2 of P_1 that are angle-wise no larger and strictly larger than ι and ensure that $\{\iota_1, \iota_2\} \sqsubseteq \iota$. All inequalities in P_1 that are not tested in this way will only make the inner polyhedron P_1 smaller and hence cannot affect the outcome of the entailment check. Algorithm 2 formalises this idea. Here, the trivial cases of $P_1 = \emptyset$ and $P_2 = \emptyset$ are checked before the inequalities that constitute the two polyhedra are extracted with indices that increase with the angle. For each inequality ι_i in the facet list of P_2, the inner loop in lines 12–15 finds two adjacent inequalities $\iota_{l \bmod n}, \iota_{u \bmod n}$ in P_1 that

Algorithm 2 Checking entailment between planar polyhedra.

procedure *entails*(P_1, P_2)

if $P_1 = \emptyset$ **then**
  return *true*
end if
if $P_2 = \emptyset$ **then**
  return *false*
end if
$[\![\{\iota_0, \ldots \iota_{n-1}\}]\!] \leftarrow P_1$ /* *such that* $\theta(\iota_0) \leq \theta(\iota_1) \leq \ldots \leq \theta(\iota_{n-1})$ */
$[\![\{\iota'_0, \ldots \iota'_{m-1}\}]\!] \leftarrow P_2$ /* *such that* $\theta(\iota'_0) \leq \theta(\iota'_1) \leq \ldots \leq \theta(\iota'_{m-1})$ */
$u \leftarrow 0$
$l \leftarrow n - 1$
for $i \in [0, m-1]$ **do**
  while $u < n \wedge \theta(\iota_u) < \theta(\iota'_i)$ **do**
    $l \leftarrow u$
    $u \leftarrow u + 1$
  end while
  if $\{\iota_{l \bmod n}, \iota_{u \bmod n}\} \not\sqsubseteq \iota'_i$ **then**
    return *false*
  end if
end for
return *true*

enclose the inequality ι_i angle-wise. If the entailment $\{\iota_{l \bmod n}, \iota_{u \bmod n}\} \sqsubseteq \iota_i$ holds for all ι_i and corresponding $\iota_{l \bmod n}, \iota_{u \bmod n}$, then P_2 is entailed by P_1. Note that the total number of times the inner loop iterates is $|I_2|$, where $[\![I_2]\!] = P_2$. The whole algorithm runs in $O(|I_1| + |I_2|)$, where $[\![I_1]\!] = P_1$ and its input is thereby linear in size.

A slightly more complicated iteration strategy is necessary to remove redundant inequalities from a system. This task is the topic of the next section.

7.2.2 Removing Redundancies

This section presents an algorithm to remove redundant inequalities. Although most algorithms presented in this chapter take and return non-redundant systems of inequalities, redundancy removal is important for adding new inequalities to a system as required to implement the meet operation $\sqcap_P$ of the polyhedral domain. Specifically, define the function *nonRedundant* $(\{\iota_1, \ldots, \iota_m\})$, which takes a set of inequalities $\{\iota_1, \ldots, \iota_m\}$ that are sorted by angle. We use the notation $\langle \iota_1, \ldots, \iota_m \rangle$ to describe a sorted sequence of elements that in practice may be implemented as a doubly linked list. In this representation, assigning a sequence $\langle \iota_m, \iota_1, \ldots \iota_{m-1} \rangle$ to I is a constant-time operation that rotates the original sequence $I = \langle \iota_1, \ldots \iota_m \rangle$ one position to the right.

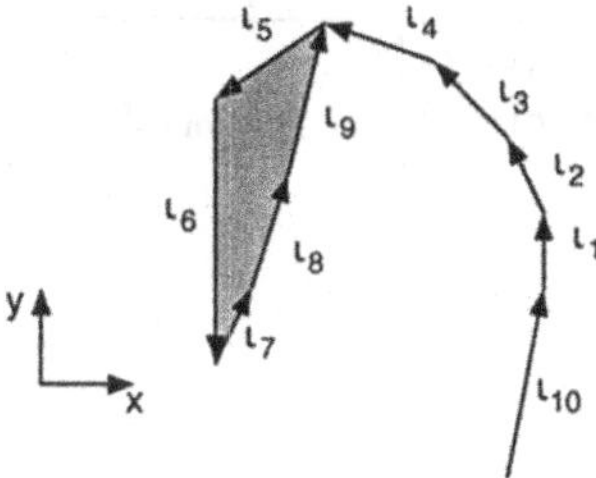

Fig. 7.5. A chain of inequalities $\iota_1, \ldots \iota_4$ that are non-redundant with respect to their neighbours but are redundant with respect to ι_9 and ι_5.

The key to removing redundant inequalities lies in the observation that the inequality ι_i is redundant if $\{\iota_{(i-1)\ mod\ m}, \iota_{(i+1)\ mod\ m}\} \sqsubseteq \iota_i$. Checking each inequality once against its neighbours is not sufficient to determine that no such pair of inequalities exists since each time an inequality is removed, the previously separate neighbours become adjacent, which might make one of them redundant, too. This is illustrated in Fig. 7.5. Here, the feasible space of the polyhedron is shown in grey, and hence the inequalities $\iota_{10}, \iota_1, \ldots \iota_4$ are redundant. Consider a redundancy removal function that starts checking each inequality from ι_1 onwards. Since $\{\iota_{10}, \iota_2\} \not\sqsubseteq \iota_1$, the inequality ι_1 is non-redundant with respect to its two neighbours and iteration proceeds to infer $\{\iota_1, \iota_3\} \not\sqsubseteq \iota_2$, $\{\iota_2, \iota_4\} \not\sqsubseteq \iota_3$, and so forth. It is not until iteration ten when $\{\iota_9, \iota_1\} \sqsubseteq \iota_{10}$ that ι_{10} is discarded. If iteration stops here, the redundant inequalities $\iota_1, \ldots \iota_4$ are not detected as such and an incorrect result is returned. In a correct implementation, iteration has to proceed until a fixpoint is reached – that is, until each inequality is found to be non-redundant.

This strategy is implemented as Alg. 3. The input inequalities are stored in I in increasing angular order. The variable *todo* tracks the number of inequalities that still need to be examined and is initially set to the size of the sequence I. Lines 4–5 stop the loop short when the size of the system is so small that there is no neighbour to test against which, in turn, would lead to the incorrect removal of the single remaining inequality. Otherwise, $|I| \geq 2$ and the conditional in line 8 tests if the first inequality in I is redundant with respect to its two neighbours. If ι_1 is redundant, lines 9–10 remove this inequality from I and the *todo* counter is reset to enforce that every inequality in I is checked once more. Lines 12–13 deal with the case where ι_1 is non-redundant, in which case the sequence I is rotated in order to check the next inequality. In this case, the *todo* counter is merely decremented, with the effect that the loop eventually stops when all remaining inequalities are non-redundant. With respect to the running time of the algorithm, observe that Fig. 7.4 constitutes the worst-case scenario, in which the loop iterates over a maximum chain of inequalities until the last inequality ι_{10} is found to be redundant. At this point, *todo* is repeatedly reset to $|I|$ until ι_4 is removed

Algorithm 3 Removal of redundant inequalities.

procedure *nonRedundant*($\{\iota_1, \ldots \iota_n\}$)
$I \leftarrow \langle \iota_1, \ldots \iota_n \rangle$ /* $\theta(\iota_1) \leq \theta(\iota_2) \leq \ldots \leq \theta(\iota_n)$ */
$todo \leftarrow |I|$
while $todo > 0$ **do**
  if $|I| \leq 1$ **then**
    return I
  end if
  $\langle \iota_1, \ldots \iota_m \rangle \leftarrow I$
  if $\{\iota_m, \iota_2\} \sqsubseteq \iota_1$ **then**
    $I \leftarrow \langle \iota_m, \iota_2, \iota_3, \ldots \iota_{m-1} \rangle$
    $todo \leftarrow |I|$
  else
    $I \leftarrow \langle \iota_2, \iota_3, \ldots \iota_{m-1}, \iota_m, \iota_1 \rangle$
    $todo \leftarrow todo - 1$
  end if
end while
return I

and the loop iterates further until ι_9, only to find that all inequalities are non-redundant. In summary, the algorithm runs for at most two complete iterations such that it is in $O(n)$, where n is the number of input inequalities. Note that the first inequality in the returned set I does not necessarily have the smallest angle. Rather than sorting the resulting set, it can be rotated until the smallest inequality is at the beginning of the sequence.

A special case arises when the input to *nonRedundant* is an unsatisfiable set of inequalities. The algorithms will terminate with either $I = \{\iota_0, \iota_1\}$, where $\iota_0 \measuredangle \iota_1 = \pi$, or with $I = \{\iota_0, \iota_1, \iota_2\}$, where $\iota_i \measuredangle \iota_{(i+1) \bmod 3} < \pi$. In the former case, the coefficients of the inequalities need to be compared in order to detect that their intersection is empty. In the latter case, the boundaries of the half-spaces $[\![\iota_0]\!]$ and $[\![\iota_1]\!]$ intersect in a point $\langle v_x, v_y \rangle$. The system is unsatisfiable if $\langle v_x, v_y \rangle \notin [\![\iota_2]\!]$ (i.e. if $av_x + bv_y > c$) where $\iota_2 \equiv ax + by \leq c$.

This completes the description of the redundancy removal algorithm that forms the basis for the meet operation. We now proceed to define the join operation, which turns out to be the most intricate.

7.2.3 Convex Hull

In 1972, Graham published the first sub-quadratic algorithm to compute the convex hull of a set of points in planar space [84]. Since then, numerous improvements [2, 4, 5, 36, 113] and extensions to polytopes [146, 179] have been proposed. An overview can be found in [147, 160]. Interestingly, some of these improvements turned out to be incorrect, [88, 179, 180] which suggests that geometric algorithms are difficult to construct correctly. While the convex

Algorithm 4 Calculating an inequality from two points.

procedure *connect*$(\langle x_1, y_1\rangle, \langle x_2, y_2\rangle)$
  return $(y_2 - y_1)x + (x_1 - x_2)y \leq (y_2 - y_1)x_1 + (x_1 - x_2)y_1$

hull of polytopes (bounded polyhedra) can be calculated straightforwardly by taking the convex hull of their extreme points, calculating the convex hull of unbounded polyhedra turns out to be more subtle due to a large number of geometric configurations. Even for planar polyhedra, the introduction of rays makes it necessary to handle polyhedra such as a single half-space, a single ray, a single line, two facing (not coinciding) half-spaces, etc., all of which require special handling in a point-based algorithm. The problem is exacerbated by the number of ways these special polyhedra can be combined. In order to simplify the correctness argument, we present a direct reduction of the convex hull problem of planar polyhedra to the classic convex hull problem for a set of points. The idea of the algorithm presented is to confine vertices of the input polyhedra to a box and to use the rays to translate these points outside the box. A linear pass around the convex hull of all these points is then sufficient to determine the resulting polyhedron. This approach inherits the time complexity of the underlying convex hull algorithm, which is in $O(n \log n)$ [84]. Our algorithm follows the standard tactic for calculating the convex hull of polyhedra that are represented as sets of inequalities, namely to convert the input into an intermediate ray and vertex representation. Two approaches to the general (n-dimensional) conversion problem are the double description method [132] (also known as the Chernikova algorithm [44, 117]) and the vertex enumeration algorithm of Avis and Fukuda [6]. While our approach is linear, the Chernikova method leads to a cubic time solution for calculating the convex hull of planar polyhedra [117], whereas the method of Avis and Fukuda is quadratic.

Before we detail the algorithm itself, we define a few auxiliary functions. We then give an explanation in the context of an example in which the join of a bounded polyhedron and a polyhedron with two rays is calculated. A note on degenerated input polyhedra and their treatment completes the description of the algorithm.

Auxiliary Functions

The auxiliary function *intersect*$(a_1x + b_1y \leq c_1,\ a_2x + b_2y \leq c_2)$ calculates the set of intersection points of the two lines $a_1x + b_1y = c_1$ and $a_2x + b_2y = c_2$. In practice, an implementation of this function only needs to be partial since it is only applied when the resulting set contains a single point. Algorithm 4 presents *connect*, which generates an inequality from two points subject to the following constraints: the half-space induced by *connect*(p_1, p_2) has p_1 and p_2 on its boundary and, if p_1, p_2, p_3 are sorted counterclockwise,

Algorithm 5 Algorithm to calculate the generators of a planar polyhedron.

procedure $extreme(\{\iota_0, \ldots \iota_{n-1}\})$ where $\theta(\iota_0) \leq \theta(\iota_1) \leq \ldots \theta(\iota_{n-1})$
$\langle V, R\rangle \leftarrow \langle \emptyset, \emptyset\rangle$
if n=1 **then**
  $ax + by \leq c \leftarrow \iota_0$
  $R \leftarrow \langle -a/\sqrt{a^2+b^2}, -b/\sqrt{a^2+b^2}\rangle$
end if
for $i \in [0, n-1]$ **do**
  $ax + by \leq c \leftarrow \iota_i$
  $d_{pre} \leftarrow \iota_{(i-1) \text{ mod } n} \measuredangle \iota_i \geq \pi \vee n = 1$
  $d_{post} \leftarrow \iota_i \measuredangle \iota_{(i+1) \text{ mod } n} \geq \pi \vee n = 1$
  if d_{pre} **then**
    $R \leftarrow R \cup \{\langle b/\sqrt{a^2+b^2}, -a/\sqrt{a^2+b^2}\rangle\}$
  end if
  if d_{post} **then**
    $R \leftarrow R \cup \{\langle -b/\sqrt{a^2+b^2}, a/\sqrt{a^2+b^2}\rangle\}$
  else
    $V \leftarrow V \cup intersect(\iota_i, \iota_{(i+1) \text{ mod } n})$
  end if
  if $d_{pre} \wedge d_{post}$ **then**
    $V \leftarrow V \cup \{v\}$ where $v \in \{\langle x, y\rangle \mid ax + by = c\}$
  end if
end for
return $\langle V, R\rangle$

then p_3 is in the feasible space. The notation $\overline{p_1, p_2}$ is used to abbreviate $connect(p_1, p_2)$. Furthermore, the predicate $saturates(p, \iota)$ holds whenever the point p is on the boundary of the half-space defined by the inequality ι; that is, $saturates(\langle x_1, y_1\rangle, ax + by \leq c)$ iff $ax_1 + by_1 = c$. Finally, the predicate $inBox(s, p)$ holds whenever the point p is strictly contained within a square of width $2s$ that is centred on the origin; specifically, $inBox(s, \langle x, y\rangle)$ iff $|x| < s \wedge |y| < s$.

A Typical Run of the Algorithm

The algorithm divides into a decomposition and a reconstruction phase. The *hull* function, presented as Alg. 6, decomposes the input polyhedra into their corresponding ray and vertex representations by calling the function *extreme* in lines 4 and 5, which is defined as Alg. 5. The remainder of the *hull* function reconstructs a set of inequalities whose half-spaces enclose both sets of rays

Algorithm 6 Calculating the convex hull of planar polyhedra.

```
procedure hull(I_1, I_2) where each I_i satisfiable, sorted by angle, non-redundant
if I_1 = ∅ ∨ I_2 = ∅ then
   return ∅
end if
⟨P_1, R_1⟩ ← extreme(I_1)
⟨P_2, R_2⟩ ← extreme(I_2)
P ← P_1 ∪ P_2
R ← R_1 ∪ R_2 /* Note: |R| ≤ 8 */
s ← 1 + max{|x|, |y| | ⟨x, y⟩ ∈ P}
Q ← P
for ⟨⟨x, y⟩, ⟨a, b⟩⟩ ∈ P × R do
   Q ← Q ∪ {⟨x + 2√2sa, y + 2√2sb⟩}
end for
if Q = {⟨x_1, y_1⟩} then /* result is zero dimensional (a point) */
   return {x ≤ x_1, y ≤ y_1, −x ≤ −x_1, −y ≤ −y_1}
end if
q_p ← ⟨Σ_{⟨x,y⟩∈Q} x/|Q|, Σ_{⟨x,y⟩∈Q} y/|Q|⟩ /* q_p is feasible, but not a vertex */
⟨q_0, ..., q_{n−1}⟩ ← sort(q_p, Q) /* sort points by angle with q_p */
   /* identify the vertices q_{k_i} where 0 ≤ k_0 < ... < k_{m−1} < n */
⟨q_{k_0}, ..., q_{k_{m−1}}⟩ ← scan(⟨q_0, ..., q_{n−1}⟩)
I_res ← ∅
for i ∈ [0, m − 1] do
   ⟨x_1, y_1⟩ ← q_{k_i}
   ⟨x_2, y_2⟩ ← q_{k_{(i+1) mod m}}
   ι ← connect(⟨x_1, y_1⟩, ⟨x_2, y_2⟩) /* add ι to I_res if q_{k_i} or q_{k_{i+1}} is in the box */
   add ← inBox(s, ⟨x_1, y_1⟩) ∨ inBox(s, ⟨x_2, y_2⟩) ∨ m = 2
   j ← (k_i + 1) mod n
   while ¬add ∧ j ≠ k_{i+1} do /* ...or any point on ι is in the box */
      add ← saturates(q_j, ι) ∧ inBox(s, q_j)
      j ← (j + 1) mod n
   end while
   if m = 2 ∧ inBox(s, ⟨x_1, y_1⟩) then
      if y_1 = y_2 then
         I_res ← I_res ∪ {sgn(x_1 − x_2)x ≤ sgn(x_1 − x_2)x_1}
      else
         I_res ← I_res ∪ {sgn(y_1 − y_2)y ≤ sgn(y_1 − y_2)y_1}
      end if
   end if
   if add then
      I_res ← I_res ∪ {ι}
   end if
end for
return I_res
```

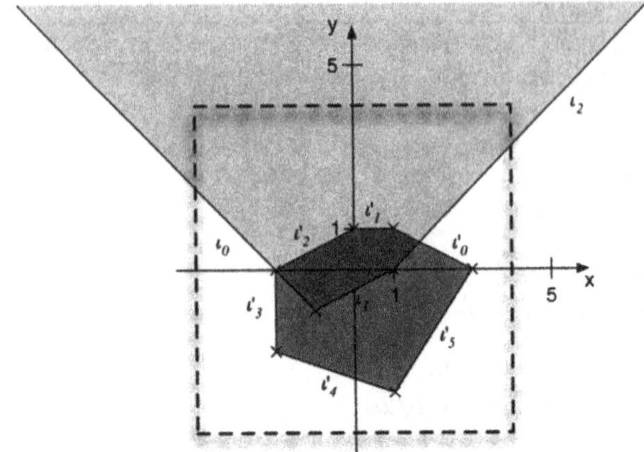

Fig. 7.6a. Calculate a square around the origin that includes all vertices.

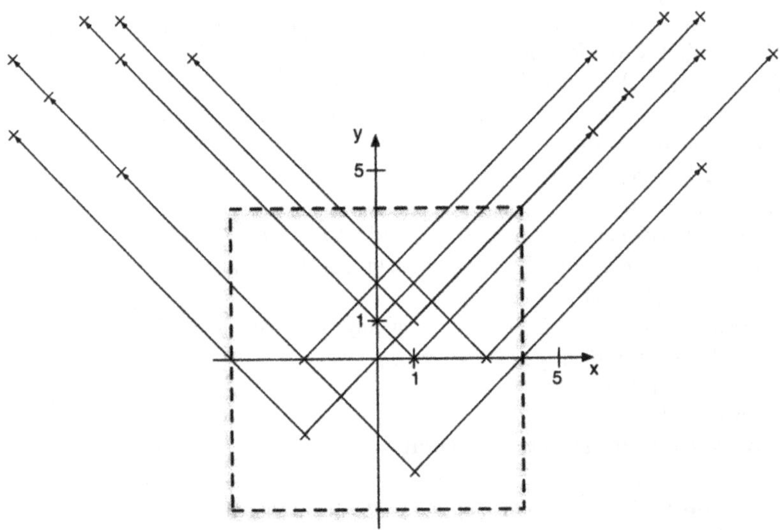

Fig. 7.6b. Translate all vertices along the rays such that they lie outside the square.

and points. The algorithm requires that the input polyhedra be non-redundant and sorted; its output is also non-redundant and sorted.

In order to illustrate the algorithm, consider Fig. 7.6a. The polyhedron $I = \{\iota_0, \iota_1, \iota_2\}$ and the polytope $I' = \{\iota'_0, \ldots, \iota'_5\}$ constitute the inputs to the *hull* function. They are passed to the function *extreme* at lines 4 and 5. Note that we assume that the set of inequalities is sorted by angle such that their indices increase with the angle. The loop at lines 6–22 examines the

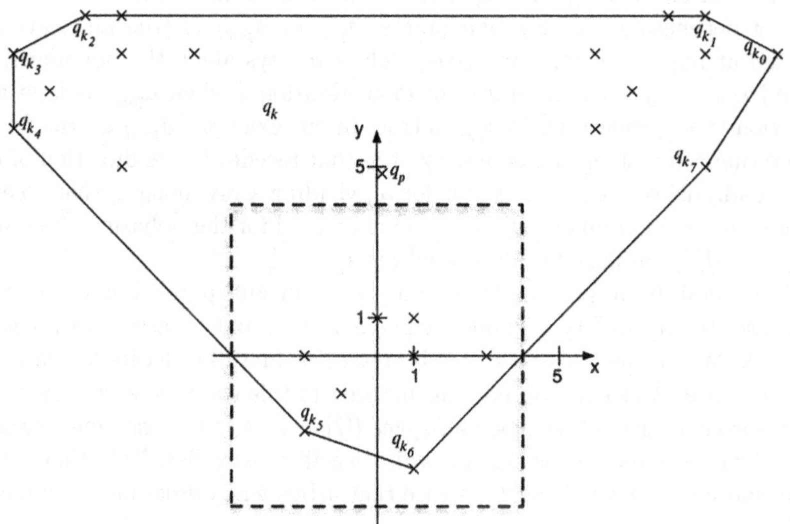

Fig. 7.6c. Calculate the convex hull of all points.

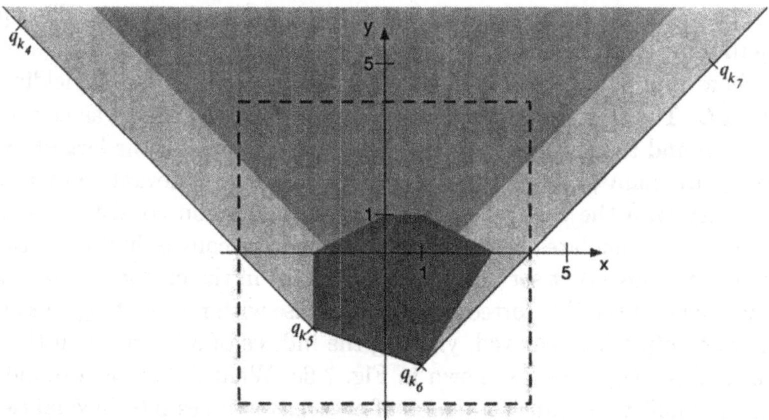

Fig. 7.6d. The three inequalities $\overline{q_{k_4}, q_{k_5}}$, $\overline{q_{k_5}, q_{k_6}}$ and $\overline{q_{k_6}, q_{k_7}}$ define a polyhedron that includes the two polyhedra $I = \{\iota_0, \iota_1, \iota_2\}$ and $I' = \{\iota'_0, \ldots \iota'_5\}$ from Fig. 7.6a.

relationship of each inequality with its two angular neighbours. If d_{post} is false, the intersection point $intersect(\iota_i, \iota_{(i+1) \bmod n})$ is a vertex, which is added at line 16. In the example, two vertices are created for I, namely v_1 and v_2, where $\{v_1\} = intersect(\iota_0, \iota_1)$ and $\{v_2\} = intersect(\iota_1, \iota_2)$. Six vertices are created for I'. Conversely, if d_{post} is true, the intersection point is degenerate; that is, either I contains a single inequality or the angular difference between the current inequality and its successor is greater than or equal to π. For instance, $intersect(\iota_2, \iota_0)$ is degenerated and thus is not added to V. In the case of degenerated intersection points, d_{pre} or d_{post} is true and rays are created at line 11 or 14, respectively. The two rays along the boundaries of ι_i and $\iota_{(i+1)\ mod\ n}$ are generated in loop iteration i when d_{post} is true and iteration $(i + 1)\ mod\ n$ when d_{pre} is true. In our example, d_{post} is true for ι_2, generating a ray along the boundary of ι_2 that recedes in the direction of the first quadrant, whereas d_{pre} is true for ι_0, yielding a ray along ι_0 that recedes towards the second quadrant. No rays are created for the polytope I' because d_{post} and d_{pre} are false for all inequalities $\iota'_0, \ldots \iota'_5$.

In general, both flags might be true – e.g., for anti-parallel half-spaces. In this case, the inequality ι_i cannot define a vertex, and an arbitrary point on the boundary of the half-space of ι_i is created at line 19 to fix its representing rays in space. Another case not encountered in this example arises when the polyhedron consists of a single half-space ($|I| = 1$). In this case, line 4 creates a third ray to indicate the side on which feasible space lies. Note that R has never more than four elements, a case that arises when describing two facing half-spaces.

The remainder of the *hull* function is dedicated to the reconstruction phase. The point and ray sets, returned by *extreme*, are merged at lines 6 and 7. At line 8, the size of a square is calculated, which includes all points in P. The square has $\langle s, s\rangle$, $\langle -s, s\rangle$, $\langle s, -s\rangle$, $\langle -s, -s\rangle$ as its corners. The square in the running example is depicted at all stages with a dashed line. Figure 7.6b shows how each point $p \in P$ is then translated by each ray $r \in R$ yielding the point set Q. The translated points are always outside the square since rays are normalised and then scaled by $2\sqrt{2}s$, which corresponds to the largest extent of the square, namely its diagonal. Lines 13–14 are not relevant to this example, as they trap the case where the output polyhedron consists of a single point. Line 16 calculates a feasible point q_p of the convex hull of Q that is not a vertex. This point serves as the pivot point in the classic Graham scan. Firstly, the point set Q is sorted counterclockwise with respect to q_p. Secondly, all interior points are removed, yielding the indices of all vertices in the case of the example $k_0, \ldots k_7$, as shown in Fig. 7.6c. What follows is a round-trip around the hull that translates pairs of adjacent vertices into inequalities by calling *connect* at line 23. Whether this inequality actually appears in the result depends on the state of the *add* flag. In our particular example, the *add* flag is only set at line 24. Whenever it is set, it is because one of the two vertices lies within the square. The resulting polyhedron is shown in Fig. 7.6d and consists of the inequalities $\overline{q_{k_4}, q_{k_5}}$, $\overline{q_{k_5}, q_{k_6}}$, and $\overline{q_{k_6}, q_{k_7}}$, which is a correct solution for this example.

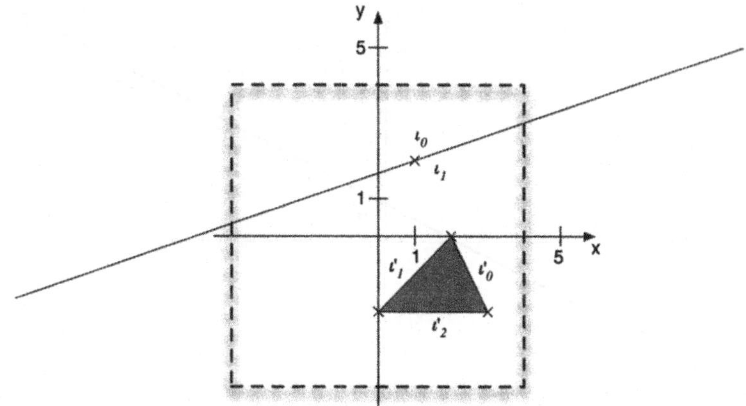

Fig. 7.7a. Creating a point in the box for each line.

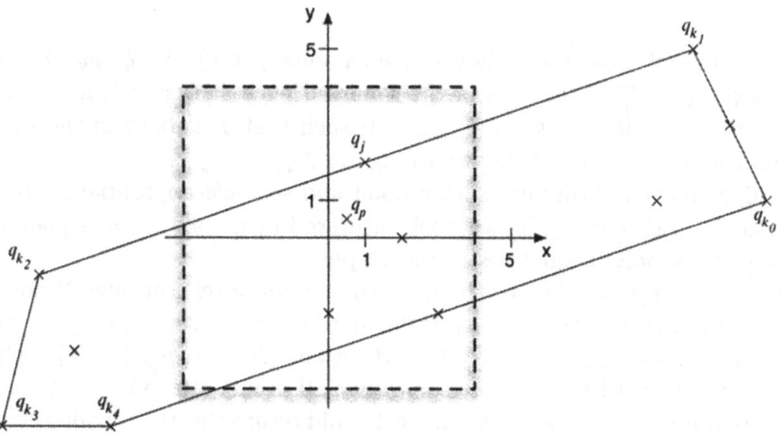

Fig. 7.7b. Creating inequalities if one of the points lies in the box.

Pathological Configurations

The reconstruction phase has to consider certain anomalies that mainly arise in outputs of lower dimensionality. One subtlety in the two-dimensional case is the handling of polyhedra that contain lines. This is illustrated in Fig. 7.7a, where the two inequalities ι_0, ι_1 are equivalent to one equation that defines a space that is a line or, equivalently, two opposing rays. The result of translating the vertices by the two rays and their convex hull is shown in Fig. 7.7b. Observe that no point in the square is a vertex in the hull of Q. Therefore, the predicate *inBox* does not hold for the two vertices q_{k_1} and q_{k_2} and the desired inequality $\overline{q_{k_1}, q_{k_2}}$ is not emitted. The same holds for q_{k_4} and q_{k_0}.

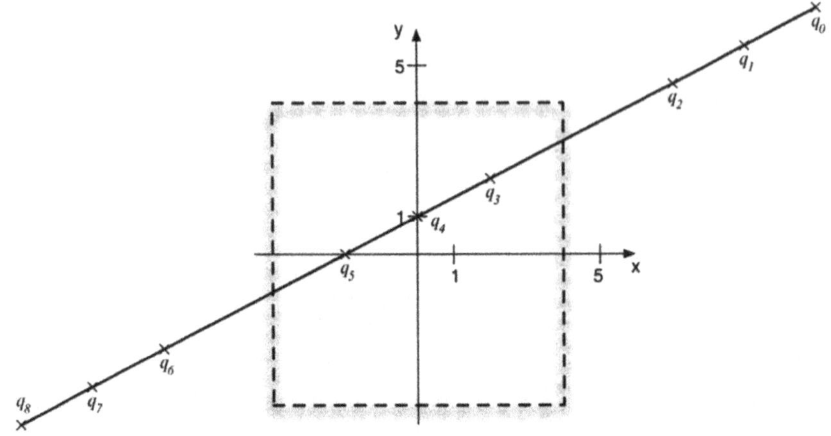

Fig. 7.8. Handling the one-dimensional case.

However, in such cases there always exists a point $q_j \in Q$ with $\overline{q_p, q_{k_i}} \measuredangle \overline{q_p, q_j} < \overline{q_p, q_{k_i}} \measuredangle \overline{q_p, q_{k_{(i+1)\ mod\ m}}}$ that lies in the square. Hence, it is sufficient to search for an index $j \in [k_i + 1, k_{(i+1)\ mod\ m} - 1]$ such that q_j is both in the square and on the line connecting the vertices q_{k_i} and $q_{k_{(i+1)\ mod\ m}}$. The inner loop at lines 26–29 tests if Q contains such a point and sets *add* appropriately. In the example, $\overline{q_{k_1}, q_{k_2}}$ and $\overline{q_{k_4}, q_{k_5}}$ are each saturated by a point in the square and are in fact the only inequalities in the output.

The one-dimensional case is handled by the $m = 2$ tests at lines 24 and 30. Figure 7.8 illustrates the necessity of the first test. Suppose I_1 and I_2 are given such that $extreme(I_1) = \langle\{q_4, q_5\}, \emptyset\rangle$ and $extreme(I_2) = \langle\{q_3\}, \{r, -r\}\rangle$, where r is any ray parallel to $\overline{q_4, q_5}$. Observe that all points are collinear; thus the pivot point is on the line and a stable sort could return the ordering depicted in the figure. The correct inequalities for this example are $I_{res} = \{\overline{q_0, q_8}, \overline{q_8, q_0}\}$. The Graham scan will identify $q_{k_0} = q_0$ and $q_{k_1} = q_8$ as vertices. Since there exists $j \in [k_0+1, k_1-1]$ such that $inBox(s, q_j)$ holds, $\overline{q_0, q_8} \in I_{res}$. In contrast, although there are boundary points between q_8 and q_0, the loop is not aware of them since sorting the points removed all points between q_8 and q_0. In this case, the $m = 2$ test sets *add* and thereby forces $\overline{q_8, q_0} \in I_{res}$.

Another complication arises when generating line segments, as shown in Fig. 7.9. Observe that the output polyhedron must include q_{k_i} as a vertex whenever $inBox(s, q_{k_i})$ holds. If $inBox(s, q_{k_i})$ holds, the algorithm generates the inequalities $\iota_{i-1} = \overline{q_{k_{(i-1)\ mod\ m}}, q_{k_i}}$ and $\iota_i = \overline{q_{k_i}, q_{k_{(i+1)\ mod\ m}}}$. If $\iota_{i-1} \measuredangle \iota_i < \pi$, then $\{q_{k_i}\} = intersect(\iota_{i-1}, \iota_i)$ and the vertex q_{k_i} is realised. However, if $m = 2$, then $\iota_{i-1} \measuredangle \iota_i = \pi$, which requires an additional inequality to define the vertex q_{k_i}. This is the role of the inequalities generated on lines 32 and 34.

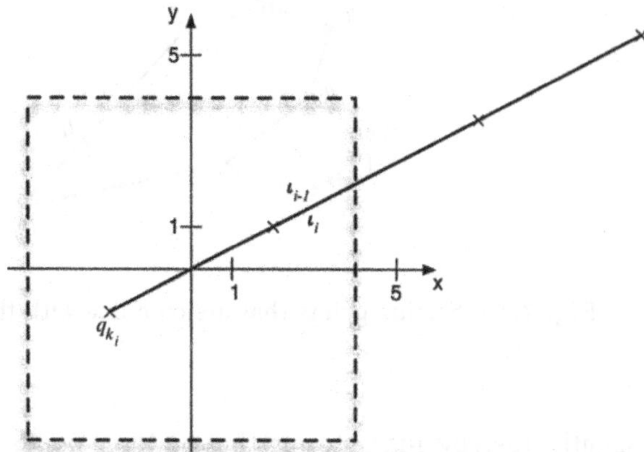

Fig. 7.9. Joining a vertex that lies on a ray with the ray itself.

This new inequality ι obeys $\iota_{i-1}\measuredangle\iota < \pi$ and $\iota\measuredangle\iota_i < \pi$ and thus suffices to define q_{k_i}.

The last special case to be considered is a result that is zero-dimensional. This case can only occur when both input polyhedra consist of the same single point v. Line 13 traps this case and returns a set of inequalities describing $\{v\}$.

Even though some subtle problems arise in dealing with the pathological cases that occur, note that the zero- and one-dimensional cases only require minute changes to the general two-dimensional case. We show in [168] that these modifications are indeed sufficient to ensure that the algorithm is correct on all possible inputs.

Special care has to be taken when implementing the actual convex hull algorithm. The pivot point calculated in line 16 of Alg. 6 is likely to have rational coordinates with large numerators and denominators, thereby slowing down the algorithm. The original algorithm of Graham [84] creates an interior point as the pivot point by choosing two arbitrary points q_1, q_2 and searching the point set for a point q_i that does not saturate the line $\overline{q_1, q_2}$. The center of the triangle q_1, q_2, q_i is also guaranteed to be an interior point of Q. While the pivot point so found may have a smaller representation, generating inequalities from the resulting sequence of vertices does not guarantee that the first inequality generated will have the smallest angle of all inequalities. The disadvantage is that the resulting sequence of inequalities needs to be rotated until the sequence is in increasing order. Another pitfall is when two or more points lie on a line with the pivot point and thereby compare as equal even though the points have different coordinates. Graham suggests retaining only the point that lies farthest away from the pivot point since only the outermost point can ever become a vertex of the convex hull. However, the removal of points is also at odds with our algorithm, as the input point set may not be modified. A common way to address this problem is to perturb the input point

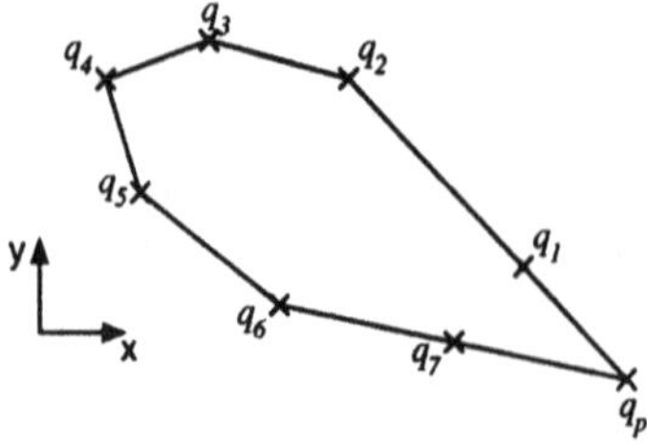

Fig. 7.10. Sorting points that are on a line with the pivot point q_p.

set slightly, thereby guaranteeing that no three points lie on a line [68]. For a sound analysis, the perturbation would have to be removed after the convex hull algorithm finishes, which complicates the algorithm further.

Our approach to circumventing these problems is to pick one point from the input set that is a definite vertex and use this point as the pivot point (as proposed in [4]). Specifically, we choose the vertex with the largest x-coordinate and the smallest y-coordinate (in that order). Creating inequalities starting with this vertex is guaranteed to generate a sequence with strictly increasing angles. As before, complications arise when ordering points that lie on a line with the pivot point. For instance, $q_1, \ldots q_7$ in Fig. 7.10 are to be returned in increasing sequence. However, the pairs q_1, q_2 and q_6, q_7 compare as equal since they have the same angle to the pivot point q_p. By enhancing the comparison function to lexicographically sort by angle, then by larger x-coordinates, and then by smaller y-coordinates, the points q_6 and q_7 are sorted in the correct order, thereby leaving the other pair in the incorrect order q_2, q_1. To ensure that the points q_1 and q_2 appear in increasing sequence, all points at the beginning of the sequence that lie on a line with q_p are reversed.

Finally, observe that the loop at lines 26–29 can often be skipped: If the line between q_{k_i} and $q_{k_{(i+1)\ mod\ m}}$ does not intersect with the square, *inBox*(s, q) cannot hold for any $q \in Q$. In this case, *add* cannot set in line 27, and the loop has no effect. Hence, if all corners of the box lie in the feasible region of the potential inequality $\overline{q_{k_i}, q_{k_{(i+1)\ mod\ m}}}$, the loop can safely be skipped. These special cases complete the description of the convex hull on planar polyhedra.

7.2.4 Linear Programming and Planar Polyhedra

Determining the set of fields that a pointer may access requires the minimum and maximum offsets that the pointer variable can take on at the given program point. In general, possible values of an expression $\mathbf{a} \cdot \mathbf{x}$ with respect to a given polyhedron $P \in Poly$ can be inferred by running a linear program *minExp* twice, once for an upper bound and once for a lower bound on the expression. In fact, finding the tightest bounds $[l, u]$ such that $l \leq \mathbf{a} \cdot \mathbf{x} \leq u$

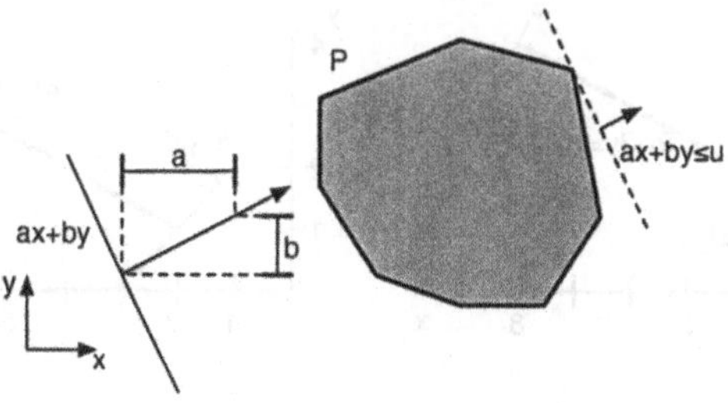

Fig. 7.11. Calculating the maximum of a linear expression in a planar polyhedron.

holds in P can be implemented more simply in planar space: Finding the upper bound u amounts to finding a minimal u such that $P \sqcap_P [\![ax+by \leq u]\!] = P$. To this end, Fig. 7.11 depicts the line $ax + by = c$ and its coefficient vector $\langle a, b\rangle$ as a vector that points towards the direction of larger values of c. The minimal value of u can be found by increasing u until $P \sqcap_P [\![ax+by \leq u]\!] = P$ holds. In practice, an inequality ι_i of P can be found in $O(\log n)$ time such that $\theta(\iota_i) \leq \theta(ax + by \leq u) < \theta(\iota_{(i+1) \ mod \ n})$ by performing a binary search on the sorted set of inequalities with $\theta(ax + by \leq u)$ as the key. If $\theta(\iota_i) = \theta(ax + by \leq u)$, then $u = c_i$, where $\iota_i \equiv a_i x + b_i y \leq c_i$. Otherwise, if $\iota_i \measuredangle \iota_{(i+1) \ mod \ n} < \pi$, then the intersection point $\langle x', y'\rangle$ of the two inequalities yields the minimal value of u, namely $u = ax' + by'$. Otherwise, no upper bound exists and $u = \infty$. The lower bound l can be inferred in a similar way using the angle $\theta(-ax - by \leq -l)$ as key.

Inferring the bounds of a single variable x is a special case that reduces to finding the vertex with the smallest and largest x-coordinates. Incidentally, the projection operator $\exists_y$ for planar polyhedra can be implemented by inferring the bounds on x and returning the corresponding interval. For instance, the projection operation $P' = \exists_x(P)$ yields $P' = [\![\{l \leq y \leq u\}]\!]$ iff $[l, u] = P(y)$ and $l \neq -\infty$, $u \neq \infty$. In practice, our implementation stores the upper and lower bounds of each variable explicitly so that a projection onto one variable is unnecessary.

The explicit representation of bounds also impacts on the way widening is implemented, which is the next and final planar operation.

7.2.5 Widening Planar Polyhedra

Calculating the fixpoint of a loop using polyhedra may result in an infinite chain of iterates. By removing inequalities that describe changing facets of a polyhedron, the fixpoint calculation is accelerated and in fact forced to converge, a process known as widening [59,62]. Suppose that two consecutive loop iterates $I_1 = \{y \geq 1, 2x + y \leq 20, 2x - 3y = 1\}$ and $I_2 = \{x + y \geq 3, x \leq 8,$

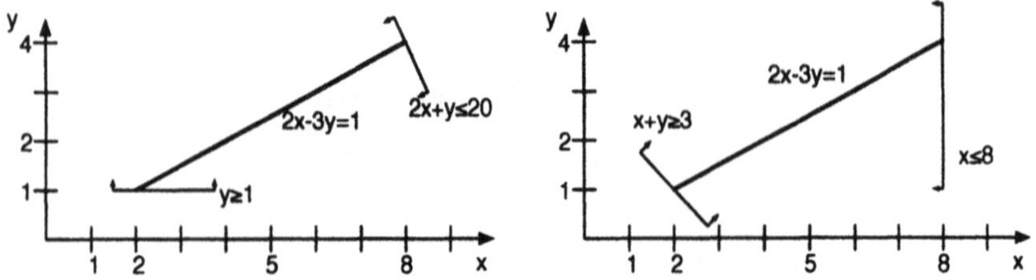

Fig. 7.12. Planar polyhedra may have several representations if they extend in less than two dimensions (degrees of freedom).

$2x - 3y = 1\}$ are given as shown in Fig. 7.12. Both systems describe the same set of points $[\![I_1]\!] = [\![I_2]\!]$ and hence $[\![I_1]\!]\nabla[\![I_2]\!] = [\![I_1]\!]$ since the iterates are stable. Thus, since $I_1 \neq I_2$, widening cannot generally be implemented as a syntactic operation and has to be defined semantically; that is, in terms of entailment [62]. In particular, the original widening is defined such that $[\![I_1]\!]\nabla[\![I_2]\!] = [\![I']\!]$, where $[\![I']\!] = \emptyset$ if $[\![I_1]\!] = \emptyset$ and otherwise $\iota \in I'$ if $\iota \in I_1$and $[\![I_2]\!] \sqsubseteq_P [\![\iota]\!]$. While this operation can be implemented similarly to the entailment check on planar polyhedra, an even simpler implementation is possible in the context of the TVPI domain described in the next chapter, where each planar polyhedron has a unique representation. Given a unique representation of a planar polyhedron, widening can be implemented in a purely syntactic way, as it reduces to a simple set-difference operation.

The next chapter elaborates on how to use the algorithms presented on planar polyhedra to implement polyhedra of arbitrary dimension where each inequality has at most two non-zero coefficients.

8

The TVPI Abstract Domain

This chapter presents the abstract domain of polyhedra, where each facet can be described by an inequality that has at most two non-zero variables. These so-called TVPI polyhedra form a proper subset of general convex polyhedra. For instance, consider the inequality set $\{x \geq 0, y \geq 0, z \geq 0, x+y+z \leq 1\}$. The resulting state space is depicted on the left of Fig. 8.1. This system can be approximated with TVPI inequalities by replacing the inequality $x+y+z \leq 1$ with three inequalities of the forms $x+y \leq c_{xy}, x+z \leq c_{xz}$, and $y+z \leq c_{yz}$. The constant c_{xy} can be determined by inserting the bounds for z into $x+y+z \leq 1$, yielding $x+y+[0,1] \leq 1$. Moving the interval to the right yields $x+y \leq 1-[-1,0]$; that is, $x+y \leq 1+[0,1]$. Thus, the tightest bound that can be inferred for $x+y$ is $c_{xy}=1$, and similarly $c_{xz}=c_{yz}=1$. The resulting space is depicted on the right of the figure. Note that no TVPI approximation exists if the variables are not bounded from below; for instance, the polyhedron $[\![x+y+z \leq 1]\!]$ has no TVPI approximation. Thus, TVPI polyhedra are a strict subset of general polyhedra.

An interesting property of TVPI inequalities is that they are closed under projection. Consider projecting $I = \{2x+3y \leq 4, -2y+2z \leq 2\}$ onto the x,z-plane by applying Fourier-Motzkin variable elimination [104] on y. This is carried out by scaling the first inequality by 2 and the second by 3 and adding them to yield $2x+3z \leq 7$, which describes all possible x,z-values of the original polyhedron $[\![I]\!]$. The observation is that projecting out variables of TVPI inequalities removes a common variable and thereby yields an inequality with at most two variables.

Interestingly, when the coefficients of inequalities are normalised to their lowest common denominator, the number of inequalities that can be added through projection is polynomial in the size of the input system [138]. The process of calculating all projections is called closure. In a closed system, the set of inequalities containing the variables $x_i, x_j \in \mathcal{X}$ expresses all information that is available with respect to these variables. In fact, the key idea of the TVPI domain is to apply the planar operations on each x_i, x_j-projection of a

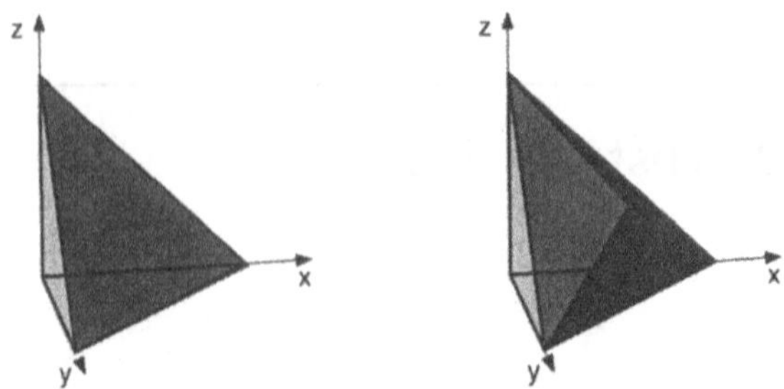

Fig. 8.1. Approximating a three-dimensional polyhedron with TVPI inequalities.

closed TVPI system, which is equivalent to applying the domain operations of general polyhedra on the whole TVPI system, albeit more efficient.

The next section proves this equivalence by reducing operations on TVPI polyhedra to planar polyhedra. Section 8.2 then comments on issues arising in an actual implementation, such as how general inequalities are approximated.

8.1 Principles of the TVPI Domain

For the sake of this section, let $var : Ineq \to \mathcal{P}(\mathcal{X})$ extract the variables that occur in an inequality; that is, $var(\mathbf{a} \cdot \mathbf{x} \leq c) = \{x_i \mid a_i \neq 0\}$. Let $Ineq^2 \subset Ineq$ denote all TVPI inequalities; that is, $Ineq^2 = \{\iota \in Ineq \mid |var(\iota)| \leq 2\}$. The set of all finite TVPI systems is therefore defined as $Two = \{I \subseteq Ineq^2 \mid |I| \in \mathbb{N}\}$.

In contrast to $Poly \subseteq \mathbb{Q}^n$, elements of Two are inequalities and we use the notation $[\![I]\!]$ to denote the set of points that is entailed by the TVPI inequality set $I \in Two$. This syntactic form is required to distinguish between closed and non-closed TVPI systems. To this end, define a family of syntactic projection operators $\pi_X(I) = \{\iota \in I \mid var(\iota) \subseteq X\}$ for all $X \subseteq \mathcal{X}$. The set of closed TVPI systems can now be defined as $Two^{cl} = \{I \subseteq Two \mid \forall \iota \in Ineq^2 \,.\, [\![I]\!] \sqsubseteq_P [\![\iota]\!] \Rightarrow [\![\pi_{var(\iota)}(I)]\!] \sqsubseteq_P [\![\iota]\!]\}$; that is, a TVPI system is closed if any TVPI inequality ι that is valid in the whole system I is also valid when considering only those inequalities of I that contain variables of ι. Intuitively, this definition implies that all information about a pair of variables $x_i, x_j \in \mathcal{X}$ is expressed as inequalities over these two variables. In particular, combining inequalities such as $ax_i - x_k \leq c_1$ and $x_k + bx_j \leq c_2$ with $ax_i + bx_j \leq c_1 + c_2$ does not add any new information about x_i and x_j. For instance, the system $I = \{x \leq y, y \leq z\}$ is not closed since $[\![\pi_{\{x,z\}}(I)]\!] = [\![\emptyset]\!] = \mathbb{Q}^2$, although the inequality $x \leq y$ fulfills $[\![I]\!] \sqsubseteq_P [\![x \leq z]\!]$ and $[\![\emptyset]\!] \not\sqsubseteq_P [\![x \leq y]\!]$.

A closed form always exists within Two, as stated by the following proposition.

Proposition 5. *For any $I \in Two$, there exists $I' \in Two^{cl}$ such that $I \subseteq I'$ and $[\![I]\!] = [\![I']\!]$.*

Proof. Define $[\![I]\!]_{x,y} = \exists_{\mathcal{X}\setminus\{x,y\}}([\![I]\!])$ for all $x, y \in \mathcal{X}$, and set $I_{x,y} \subseteq \mathit{Ineq}^2$ such that $[\![I_{x,y}]\!] = [\![I]\!]_{x,y}$. Let $I' = I \cup \bigcup_{x,y\in\mathcal{X}} I_{x,y}$. Since each $I_{x,y}$ is finite, I' is finite and hence $I' \in \mathit{Two}^{cl}$. By the definition of $\exists$, $[\![I]\!] \sqsubseteq_P [\![I_{x,y}]\!]$, and thus $[\![I \cup I_{x,y}]\!] = [\![I]\!]$ for all $x, y \in \mathcal{X}$ and $[\![I]\!] = [\![I']\!]$ follows. Moreover, $I \subseteq I'$ by construction.

In fact, any given TVPI system $I \in \mathit{Two}$ can be closed to obtain $I' \in \mathit{Two}^{cl}$ by calculating so-called resultants of I using $\mathit{result} : \mathit{Two} \to \mathit{Two}$, defined as follows:

$$\mathit{result}(I) = \left\{ ae\ z - db\ y \leq af - dc \;\middle|\; \begin{array}{l} \iota_1, \iota_2 \in I \quad\quad\;\; \wedge \\ \iota_1 \equiv ax + by \leq c \wedge \\ \iota_2 \equiv dx + ez \leq f \wedge \\ a > 0 \wedge d < 0 \end{array} \right\}$$

The purpose of the *result* function is to combine inequalities over the variables x, y and y, z with new inequalities over x, z, thereby making information on x, z explicit that was only implicitly available before. In particular, the combination above resembles Fourier-Motzkin variable elimination, where all information on x, z is made explicit in order to remove the variable y from the system. Nelson observed that merely adding inequalities (rather than removing those containing y) eventually leads to a closed system [138]. For instance, consider applying the *result* function to the following system of inequalities:

$$I_0 = \{x_0 \leq x_1, x_1 \leq x_2, x_2 \leq x_3, x_3 \leq x_4\}$$

We calculate $I_1 = \mathit{result}(I_0)$ and $I_2 = \mathit{result}(I_0 \cup I_1)$, resulting in the following sets:

$$\begin{aligned} \mathit{result}(I_0) &= \{x_0 \leq x_2, x_1 \leq x_3, x_2 \leq x_4\} \\ \mathit{result}(I_0 \cup I_1) &= I_1 \cup \{x_0 \leq x_3, x_0 \leq x_4, x_1 \leq x_4\} \end{aligned}$$

Here, $I_3 = \mathit{result}(\bigcup_{i=0}^{2} I_i)$ is a fixpoint in that $\mathit{result}(I_3) \subseteq I_3$. An important property of $I \cup \mathit{result}(I)$ is the way it halves the number of variables required to entail a given inequality $\iota \in \mathit{Two}$. Suppose $[\![I]\!] \sqsubseteq_P [\![\iota]\!]$. Then there exists $I' \subseteq I \cup \mathit{result}(I)$ such that $[\![I']\!] \sqsubseteq_P [\![\iota]\!]$ and I' contains no more than half the variables of I. Lemma 2 formalises this and is a reformulation of Lemma 1b of [138].

Lemma 2. *Let $I \in \mathit{Two}$ and $\iota \in \mathit{Ineq}^2$ such that $[\![I]\!] \sqsubseteq_P [\![\iota]\!]$. Then there exists $Y \subseteq \mathcal{X}$ such that $|Y| \leq \lfloor |\mathit{var}(I)|/2 \rfloor + 1$ and $[\![\pi_Y(I \cup \mathit{result}(I))]\!] \sqsubseteq_P \iota$.*

Lemma 2 suggests a way to obtain a closed system by applying the *result* function approximately $\log_2(|\mathit{var}(I)|)$ times to any system of inequalities $I \in \mathit{Two}$. More precisely, a TVPI system can be closed in $O(k^2 d^3 \log(d)(\log(k) +$

$\log(d)))$ steps, where d is the number of variables and k the maximum number of inequalities for any pair of variables. Empirical evidence suggests that k is bounded by a small constant in practice such that the bound collapses to $O(d^3(\log(d)^2))$ [172].

Rather than presenting such an algorithm, we detail how an initially empty inequality system can be incrementally closed each time a new inequality is added. An incremental closure is more amenable to abstract interpretation, where the meet operation is mostly used to add a few inequalities to a system before the inequality system is used in entailment checks, join calculations, and projection operations.

Interestingly, applying the planar algorithms from the last chapter to each syntactic projection $\pi_{\{x_i,x_j\}}(I)$ of a TVPI system $I \in \mathit{Two}^{cl}$ in most cases results in a closed system. Besides this practical property, the next sections also attest to the correctness of lifting the planar entailment check and the join and projection algorithms to the TVPI domain.

8.1.1 Entailment Check

We show that checking entailment between two closed TVPI systems can be reduced to checking entailment on each two-dimensional projection.

Proposition 6. *Let $I' \in \mathit{Two}^{cl}$ and $I \in \mathit{Two}$. Then $I' \sqsubseteq_P I$ iff $[\![\pi_Y(I')]\!] \sqsubseteq_P [\![\pi_Y(I)]\!]$ for all $Y = \{x, y\} \subseteq \mathcal{X}$.*

Proof. Suppose $[\![I']\!] \sqsubseteq_P [\![I]\!]$. Let $\iota \in \pi_Y(I)$. Then $[\![I']\!] \sqsubseteq_P [\![I]\!] \sqsubseteq_P [\![\iota]\!]$. Hence $[\![\pi_{var(\iota)}(I')]\!] \sqsubseteq_P [\![\iota]\!]$. Since $var(\iota) \subseteq Y$, $[\![\pi_Y(I')]\!] \sqsubseteq_P [\![\iota]\!]$ and therefore $[\![\pi_Y(I')]\!] \sqsubseteq_P [\![\pi_Y(I)]\!]$. Now suppose $[\![\pi_Y(I')]\!] \sqsubseteq_P [\![\pi_Y(I)]\!]$ for all $Y = \{x, y\} \subseteq \mathcal{X}$. Let $\iota \in I$. Then $\iota \in \pi_{var(\iota)}(I)$ and hence $[\![I']\!] \sqsubseteq_P [\![\pi_{var(\iota)}(I')]\!] \sqsubseteq_P [\![\pi_{var(\iota)}(I)]\!] \sqsubseteq_P [\![\iota]\!]$.

Note that the proposition does not require that both inequality systems be closed. This observation is interesting when applying operations to individual projections that can lead to a non-closed system.

As a consequence of Prop. 6, it suffices to check that entailment holds for all planar projections, which can be checked with the *entails* test presented as Alg. 2 in Sect. 7.2.1 of the previous chapter.

8.1.2 Convex Hull

In order to show that calculating the join of two TVPI polyhedra can be reduced to calculating the convex hull of each planar projection, we define the operation $\curlyvee^S$ as follows.

Definition 2. *The piece-wise convex hull $\curlyvee^S : \mathit{Two} \times \mathit{Two} \to \mathit{Two}$ is defined as $I_1 \curlyvee^S I_2 = \cup\{I_{x,y} \mid x, y \in \mathcal{X}\}$, where $[\![I_{x,y}]\!] = [\![\pi_{x,y}(I_1)]\!] \curlyvee [\![\pi_{x,y}(I_2)]\!]$.*

The following proposition states that calculating the convex hull on each planar projection results in a TVPI system that is closed if the two input TVPI systems were closed. The value of this observation lies in the fact that it is not necessary to calculate the $O(d^3(\log(d)^2))$ complete closure after each convex hull.

Proposition 7. $I'_1 \curlyvee^S I'_2 \in Two^{cl}$ *if* $I'_1, I'_2 \in Two^{cl}$.

Proof. Let $\iota \in Two$ such that $[\![I'_1 \curlyvee^S I'_2]\!] \sqsubseteq_P [\![\iota]\!]$. Let $x, y \in X$ and let $[\![I_{x,y}]\!] = [\![\pi_{\{x,y\}}(I'_1)]\!] \curlyvee [\![\pi_{\{x,y\}}(I'_2)]\!]$. By the definition of I'_1, $[\![\pi_{\{x,y\}}(I'_1)]\!] \sqsubseteq_P [\![I_{x,y}]\!]$, and therefore $[\![I'_1]\!] \sqsubseteq_P [\![I'_1 \curlyvee^S I'_2]\!]$. Likewise $[\![I'_2]\!] \sqsubseteq_P [\![I'_1 \curlyvee^S I'_2]\!]$; hence it follows that $[\![I'_1]\!] \sqsubseteq_P [\![\iota]\!]$ and $[\![I'_2]\!] \sqsubseteq_P [\![\iota]\!]$. Since $I'_1, I'_2 \in Two^{cl}$, $[\![\pi_{var(\iota)}(I'_1)]\!] \sqsubseteq_P [\![\iota]\!]$ and $[\![\pi_{var(\iota)}(I'_2)]\!] \sqsubseteq_P [\![\iota]\!]$; thus $[\![\pi_{var(\iota)}(I'_1)]\!] \subseteq [\![\iota]\!]$ and $[\![\pi_{var(\iota)}(I'_2)]\!] \subseteq [\![\iota]\!]$ and hence $[\![\pi_{var(\iota)}(I'_2)]\!] \cup [\![\pi_{var(\iota)}(I'_2)]\!] \subseteq [\![\iota]\!]$. Therefore $[\![\pi_{var(\iota)}(I'_1 \curlyvee^S I'_2)]\!] = [\![\pi_{var(\iota)}(I'_1)]\!] \curlyvee [\![\pi_{var(\iota)}(I'_2)]\!] \subseteq [\![\iota]\!]$ and hence $[\![\pi_{var(\iota)}(I'_1 \curlyvee^S I'_2)]\!] \sqsubseteq_P [\![\iota]\!]$ as required.

The following proposition states the correctness of reducing the convex hull on TVPI systems to planar convex hull operations on each projection $\pi_{\{x_i,x_j\}}(I)$.

Proposition 8. $[\![I'_1 \curlyvee^S I'_2]\!] = [\![I'_1]\!] \curlyvee [\![I'_2]\!]$ *if* $I'_1, I'_2 \in Two^{cl}$.

Proof. Since $[\![I'_1]\!] \sqsubseteq_P [\![I'_1 \curlyvee^S I'_2]\!]$ and $[\![I'_2]\!] \sqsubseteq_P [\![I'_1 \curlyvee^S I'_2]\!]$, and as $[\![I'_1 \curlyvee^S I'_2]\!]$ is convex, it follows that $[\![I'_1]\!] \curlyvee [\![I'_2]\!] \subseteq [\![I'_1 \curlyvee^S I'_2]\!]$. Suppose there exists $\langle c_1, \ldots, c_n\rangle \in [\![I'_1 \curlyvee^S I'_2]\!]$ with $\langle c_1, \ldots, c_n\rangle \notin [\![I']\!]$, where $[\![I']\!] = [\![I'_1]\!] \curlyvee [\![I'_2]\!]$. Thus $[\![\bigcup_{i=1}^n \{x_i \leq c_i, c_i \leq x_i\}]\!] \not\sqsubseteq_P [\![I']\!]$ and there exists $ax_j + bx_k \leq c \equiv \iota \in I'$ with $[\![\bigcup_{i=1}^n \{x_i \leq c_i, c_i \leq x_i\}]\!] \not\sqsubseteq_P [\![ax_j + bx_k \leq c]\!]$. But $[\![I'_1]\!] \sqsubseteq_P [\![I']\!] \sqsubseteq_P [\![\iota]\!]$ and $[\![I'_2]\!] \sqsubseteq_P [\![I']\!] \sqsubseteq_P [\![\iota]\!]$. Since $I'_1 \in Two^{cl}$ and $I'_2 \in Two^{cl}$, it follows that $[\![\pi_{\{x_j,x_k\}}(I'_1)]\!] \sqsubseteq_P [\![\iota]\!]$ and $[\![\pi_{\{x_j,x_k\}}(I'_2)]\!] \sqsubseteq_P [\![\iota]\!]$. Hence $[\![I'_1 \curlyvee^S I'_2]\!] \sqsubseteq_P [\![\iota]\!]$, and thus $[\![\bigcup_{i=1}^n \{x_i \leq c_i, c_i \leq x_i\}]\!] \sqsubseteq_P [\![I'_1 \curlyvee^S I'_2]\!]$ but $\langle c_1, \ldots, c_n\rangle \notin [\![I'_1 \curlyvee^S I'_2]\!]$, which is a contradiction. Thus, $[\![I'_1 \curlyvee^S I'_2]\!] = [\![I'_1]\!] \curlyvee [\![I'_2]\!]$.

Thus, an efficient way to calculate the convex hull of two closed TVPI systems I_1 and I_2 is to apply the planar convex hull algorithm in Sect. 7.2.3 to each pair $x_i, x_j \in \mathcal{X}$ of the syntactic projections $\pi_{\{x_i,x_j\}}(I_1)$ and $\pi_{\{x_i,x_j\}}(I_2)$.

8.1.3 Projection

Projection returns the most precise system without a given variable. An algorithmic definition of projection is easily possible for closed systems. Proposition 9 states that projection coincides with the definition of syntactic projection π; that is, projection can be implemented by removing all inequalities that contain variables that are to be eliminated. Furthermore, we prove that this operation preserves closure. We commence by defining syntactic projection.

Definition 3. *The syntactic projection operator* $\exists_x^S : \mathit{Two}^{cl} \to \mathit{Two}^{cl}$ *is defined as* $\exists_x^S(I) = \cup\{\pi_Y(I) \mid Y \subseteq X \setminus \{x\} \wedge |Y| = 2\}$.

Note that the projection function $\exists_x^S$ above operates on the inequality representation of a polyhedron rather than a set of points. The following proposition states that the syntactic projection defined above and the projection operator on polyhedra, which operates on sets of points, coincide.

Proposition 9. $\exists_x^S([\![I']\!]) = [\![\exists_x(I')]\!]$ *and* $\exists_x^S(I') \in \mathit{Two}^{cl}$ *for all* $I' \in \mathit{Two}^{cl}$.

Proof. By Fourier-Motzkin variable elimination, $\exists_x([\![I']\!]) = [\![I]\!]$, where $I = \{\iota \in I' \mid x \notin \mathit{var}(\iota)\}$. Observe that $[\![I]\!] \sqsubseteq_P [\![\exists_x^S(I')]\!]$. Now suppose $\iota \in I'$ such that $x \notin \mathit{var}(\iota)$. Since I' is closed, $[\![I']\!] \sqsubseteq_P [\![\iota]\!]$. Hence $[\![\pi_{\mathit{var}(\iota)}(I')]\!] \sqsubseteq_P [\![\iota]\!]$ and therefore $[\![\exists_x^S(I')]\!] \sqsubseteq_P [\![\iota]\!]$, and thus $[\![\exists_x^S(I')]\!] \sqsubseteq_P [\![I]\!]$ and hence $[\![\exists_x^S(I')]\!] = [\![I]\!]$ as required.

Now let $\iota \in \mathit{Ineq}^2$ such that $[\![\exists_x^S(I')]\!] \sqsubseteq_P [\![\iota]\!]$. Moreover, $[\![I']\!] \sqsubseteq_P [\![\exists_x^S(I')]\!] \sqsubseteq_P [\![\iota]\!]$ and hence $[\![\pi_{\mathit{var}(\iota)}(I')]\!] \sqsubseteq_P [\![\iota]\!]$. Since $x \notin \mathit{var}(\iota)$, $[\![\pi_{\mathit{var}(\iota)}(\exists_x^S(I'))]\!] \sqsubseteq_P [\![\iota]\!]$ as required.

As an example, consider again the system consisting of $x \le y$ and $y \le z$. Closure will introduce $x \le z$. Projecting out y will only preserve $x \le z$, which coincides with projection in that $\exists_y([\![\{x \le y, y \le z\}]\!]) = [\![x \le z]\!]$.

8.2 Reduced Product between Bounds and Inequalities

Given that most operations on a closed TVPI system are executed as operations on planar polyhedra over one specific pair of variables, it is prudent to use a data structure for storing TVPI systems that groups the inequalities by the pair of variables occurring in them. Given a domain of $n = |\mathcal{X}|$ variables, $n(n-1)/2$ unique combinations of variables exist that can be stored in a triangular matrix as shown in Fig. 8.2 for $n = 3$ and $n = 4$. The rows of the triangular matrix are mapped to a one-dimensional array that dynamically resizes. While indexing into this array is more complicated, it enables the implementation to keep the number of variables in the polyhedron to a minimum. This is important since the size of the matrix grows quadratically with n. Since not all variables $\mathcal{X}$ are present in the polyhedron at all times, it is possible to remove rows that correspond to variables that are projected out and add a new row whenever a variable is mentioned that is not currently represented in the matrix. If the underlying array is large enough, a new row with the index $n+1$ can be added by merely appending n new planar polyhedra to the end of the array, otherwise the matrix has to be copied to a larger array. In the likely case that some of the variables with indices $0 \ldots n$ are bounded from above or below, these bounds have to be inserted into the

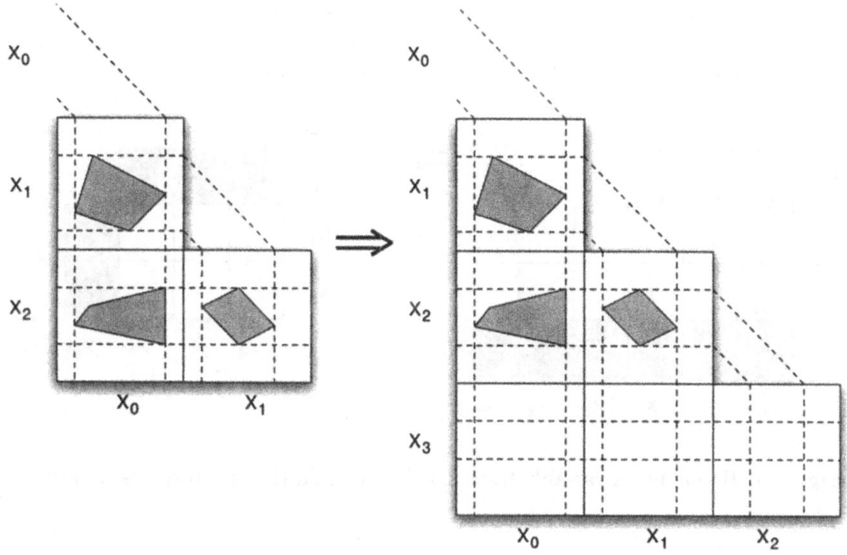

Fig. 8.2. Adding a variable to a TVPI domain adds a row to the triangular matrix.

newly added polyhedra. Rather than duplicating upper and lower bounds of each variable in each planar projection, we chose to implement the TVPI domain as a reduced product [50] between intervals and planar polyhedra that contain TVPI inequalities of the form $ax + by \leq c$, where both coefficients, a and b, are non-zero. The principle is shown in the left schematic drawing of Fig. 8.2. Here, the dashed lines depict the upper and lower bounds of each variable, which are stored separately from the TVPI inequalities. The grey polyhedra over $\langle x_0, x_1 \rangle$, $\langle x_2, x_0 \rangle$, and $\langle x_2, x_1 \rangle$ are defined by these bounds and the sets of TVPI inequalities that are specific to the given projection. In this representation, a fourth variable is introduced by merely adding a new range and three projections without any inequalities, one for each variable pair $\langle x_3, x_0 \rangle$, $\langle x_3, x_1 \rangle$, and $\langle x_3, x_2 \rangle$, as shown on the right of Fig. 8.2.

While the triangular matrix can always be extended with a new variable by adding a new row to the matrix, the removal of a variable might require the removal of a row that resides within the triangular matrix. For example, consider the removal of x_3 in the system depicted in Fig. 8.3. In order to avoid holes in the matrix, the projections of x_3 are replaced with those of the last row in the system, namely x_6. Specifically, the first three projections in the last row $\langle x_6, x_i \rangle$, $i = 1, \ldots 3$ simply replace the planar polyhedra $\langle x_3, x_i \rangle$ of the row that is to be removed. While the projection $\langle x_6, x_3 \rangle$ is merely deleted, the remaining two projections $\langle x_6, x_4 \rangle$ and $\langle x_6, x_5 \rangle$ cannot overwrite the projections $\langle x_4, x_3 \rangle$ and $\langle x_5, x_3 \rangle$ since the x and y axes of the projections do not

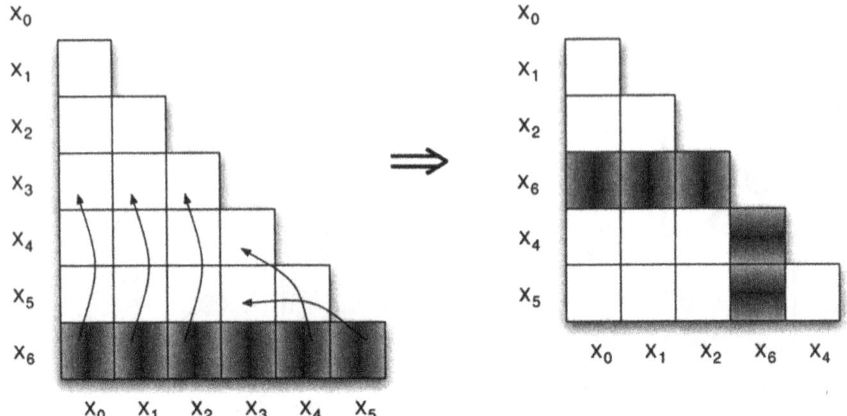

Fig. 8.3. Removing a variable from a TVPI domain that is stored as a matrix.

correspond. Thus, prior to replacing these target projections, the variables of the two planar polyhedra have to be swapped, which geometrically amounts to a mirroring along the $x = y$ line, which is a linear-time operation. Specifically, a polyhedron that is stored as an array of inequalities that itself is sorted by angle is mirrored by reversing the order of inequalities with an angle in $[0, \pi/2)$ and those in $[\pi/2, 2\pi)$. Afterwards, the x- and y-coefficients in each inequality are swapped.

Note that adding and removing rows provides an efficient framework to implement in situ updates of variables. Consider an assignment of the form `x=e` carried out on the state P, where `e` is an expression. In the special case where `e` is of the form $ax + b$, the projections over x can be updated by performing an affine transformation [62]. If e contains x in a more complex (e.g., non-linear) calculation, the result is approximated using several inequalities relating x to a fresh temporary variable t holding the result. In practice, a new row for the temporary variable t is added that contains the result of calculating the expression e. In order to set x to the value of t, the variable x is projected out. Since t is stored in the last row of the triangular matrix, the projections containing x are overwritten similarly to the removal in Fig. 8.3. Thus, updating a variable can be implemented by adding a new row, calculating the result within this new row, and replacing the target variable with the last row. If e does not mention x, the row of x can be emptied and updated with e. Thus, an update only requires minimal changes to the matrix.

Next to updates using the meet operation, the performance of the domain hinges on the join and entailment operations. The latter are mostly affected by the way memory is managed. In our implementation, a TVPI domain is merely a matrix in which each projection is a pointer to a planar polyhedron that can be shared among several TVPI domains. Thus, creating a new copy

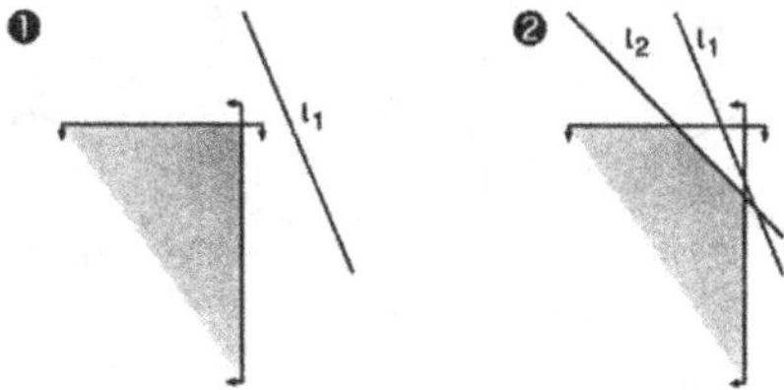

Fig. 8.4. Variations of entailment check between an inequality and interval bounds.

of a domain merely requires copying the pointers in the matrix. Furthermore, performing minor changes to the copy and calculating the join with the original is cheap: If only a few projections are changed, most projections will still refer to the same shared polyhedron. In this case, calculating the convex hull can be avoided since the result is the same shared polyhedron. Similarly, an entailment check holds trivially if the pointers to the projections are identical.

8.2.1 Redundancy Removal in the Reduced Product

The reduced product representation using interval bounds and TVPI inequalities requires some substantial changes to the redundancy removal algorithm. While the basic round-trip algorithm of Sect. 7.2.2 remains intact, special care has to be taken with respect to interval bounds. In particular, they cannot be converted to inequalities since they are often redundant with respect to other TVPI inequalities and would thus be removed. Thus, the redundancy removal algorithm only tests TVPI inequalities for redundancy and has to consult inequalities and interval bounds for entailment checks. In particular, special entailment tests are needed whenever inequalities lie in different quadrants; that is, if a bound lies angle-wise between two adjacent inequalities. Figure 8.4 shows the two necessary entailment tests between inequalities and interval bounds. The first schematic drawing depicts an inequality with no adjacent inequalities in its quadrant. In this case, it is necessary to test the inequality against the two nearest interval bounds. The second drawing shows a redundant inequality ι_1 that is the first in that quadrant. This inequality needs to be tested with respect to the two nearest bounds and the next inequality ι_2 in that quadrant. Similarly, the last inequality in each quadrant needs to be tested against the two nearest interval bounds and the previous inequality. In the case where an inequality has a neighbour on either side and within the same quadrant, the normal entailment check can be applied.

Besides testing inequalities for redundancy with respect to the interval bounds, the interval bounds may be too wide with respect to the inequalities. Since interval bounds are never removed, they need to be tightened. Interval

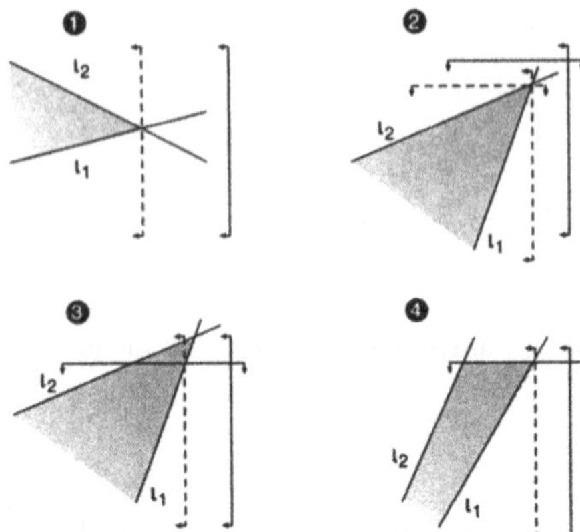

Fig. 8.5. Tightening interval bounds given two adjacent inequalities ι_1, ι_2. Bounds shown with solid lines are tightened to the bounds shown with dashed lines. Case one applies if $\iota_1 \measuredangle \iota_2 < \pi$ and one bound lies between them, case two applies if two bounds lie between ι_1 and ι_2, case three applies to those bounds that case two could not tighten, and case four applies if $\iota_1 \measuredangle \iota_2 > \pi$.

bounds are tightened in four principal ways, as shown in Fig. 8.5. The first case applies if the angle between the last inequality of a quadrant and the first inequality in the next quadrant is less than π, in which case the inequalities intersect in a point that might imply a tighter interval bound. The second case applies if two inequalities ι_1, ι_2 obey $\iota_1 \measuredangle \iota_2 < \pi$ and have an empty quadrant (i.e., two bounds) between them. In this case, their intersection point might tighten two bounds at once. If case two has not updated both bounds, one or both bounds might be tighter than what the intersection point suggests. In this case, the third schematic shows how one bound is tightened with respect to the intersection point of one of the inequalities and the other bound. The last graph in Fig. 8.5 depicts the fourth case, where the angle between two inequalities is greater than or equal to π. Here a single inequality might tighten the adjacent bound with the bound next to the adjacent bound. This completes the suite of entailment checks and tightenings for interval bounds.

The above entailment checks and tightenings have to be adapted to all four cardinal directions, which complicates the implementation of the redundancy removal algorithm considerably. As a consequence, the implementation of the actual algorithm is very technical and is therefore omitted. More interesting is the implementation of the closure, which implicitly uses the redundancy removal algorithm when adding inequalities to a projection.

8.2.2 Incremental Closure

Calculating the closure of a TVPI system as discussed in Sect. 8.1 can be implemented by a variant of the Floyd-Warshall algorithm [54], which infers

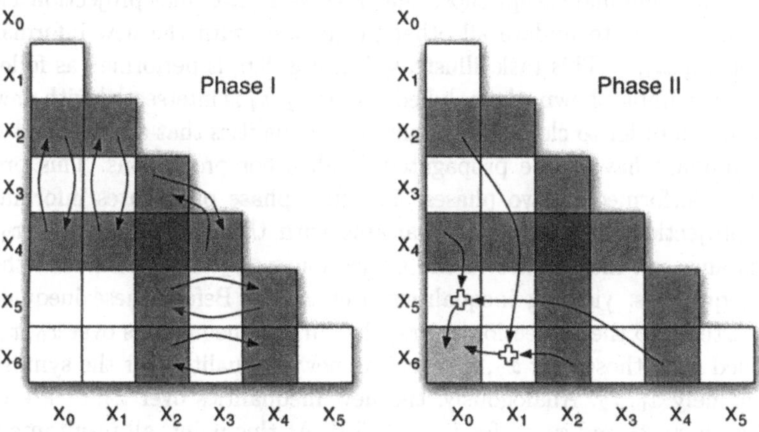

Fig. 8.6. Incremental closure after changing the projection x_2, x_4. New inequalities in x_3, x_5 serve to calculate distance-one projections (Phase I). For each set of these new inequalities, distance-two projections are calculated (Phase II).

the shortest paths between any pair of nodes in a graph. Specifically, the cubic Floyd-Warshall algorithm on n variables creates n different $n \times n$ matrices, where an element at $\langle x_i, x_j \rangle$ describes the cost of traversing the graph from node x_i to x_j. Note that the cost of travelling from x_i to x_j might be different from the cost of travelling from x_j to x_i. In the context of the triangular matrix of a TVPI domain, the planar polyhedron at index $\langle x_i, x_j \rangle$ represents both directions of travel due to the fact that inequalities of that polyhedron may have positive as well as negative coefficients. Apart from this oddity, the Floyd-Warshall algorithm is a closure operation on a TVPI system when adding the cost of two edges between x_i, x_j and x_j, x_k is replaced by calculating the resultants of the inequalities of the planar polyhedra over x_i, x_j and x_j, x_k. Updating the edge x_i, x_k with the smaller of current cost and the cost via x_j corresponds to inserting these resultants into the target polyhedron x_i, x_k.

Unfortunately, calculating a complete closure of a TVPI system is at odds with the needs of program analysis. Here, inequalities are usually added one by one through conditionals or assignments. Furthermore, adding inequalities to the domain is interleaved with variable removal and the calculation of joins. Thus, an incremental closure is required that takes a closed system and a set of inequalities that are to be intersected with a given projection x_i, x_j and returns a new, closed system in which the inequalities are incorporated. The incremental closure uses operations similar to those used in the Floyd-Warshall algorithm, albeit with a different strategy. Specifically, the Floyd-Warshall algorithm calculates a sequence of n matrices $m_1, \ldots m_n$ such that the distance between nodes in matrix m_i that are no more than i edges apart is minimal. In contrast, the incremental closure operates on a system in which all distances are minimal except those that involve a particular projection x_i, x_j. Thus, the task is to update all other projections with the new information available on x_i, x_j. This task, illustrated in Fig. 8.6, is performed as follows.

In the example shown, the polyhedron for x_2, x_4 is intersected with new inequalities. In order to close the system, all inequalities that are new and were non-redundant have to be propagated to all other projections. This propagation is performed in two phases. The first phase propagates information to all projections that share one variable with the inequalities over x_2, x_4. For instance, all inequalities in the projection x_0, x_2 are combined with the new inequalities, yielding inequalities over x_0, x_4. Before these inequalities are inserted into the projection x_0, x_4, the current inequalities over x_0, x_4 are combined with those over x_2, x_4, yielding new inequalities for the symmetric case, namely x_0, x_2. Analogously, the new inequalities over x_2, x_4 are combined with x_i, x_2 and x_i, x_4 for $i = 1, 3, 5, 6$. At this point, all resultants that share one variable with the projection x_2, x_4 are up-to-date. In terms of a graph, all nodes (variables) that are immediate neighbours of the nodes x_2 and x_4 are up-to-date; the polyhedra on these edges are called distance-one results. The second phase uses the distance-one results to update the remaining projections that have no variables in common with the x_2, x_4-projection. Specifically, each projection x_i, x_j, where $i, j \in \{0, 1, 3, 5, 6\}, i \neq j$, is updated by calculating the resultants of the projections x_i, x_2 and x_2, x_j in addition to the resultants of the projections x_i, x_4 and x_4, x_j. For instance, as shown on the right of Fig. 8.6, the projection x_0, x_6 is updated by first calculating the resultants of x_0, x_2 and x_2, x_6 followed by x_0, x_4 and x_4, x_6. The projections x_i, x_j are called distance-two results because, in the context of the graph interpretation, these edges are two nodes away from the nodes x_2 and x_4 for which closure is run. After all distance-two projections have been updated, the TVPI system is closed.

Algorithm 7 sketches the structure of the incremental closure algorithm. The algorithm shown assumes that the domain is not implemented as a reduced product between TVPI inequalities and intervals, thereby simplifying the presentation significantly. Furthermore, a triangular matrix stored in a dynamically resizable array requires that each projection x_i, x_j be stored at the index $j(j-1)/2+i$ if $i < j$; see [112]. This implies that the indices i and j have to be swapped whenever $i > j$, and the calculation of the resultants requires extra arguments that determine which variable to eliminate. Algorithm 7 simply defines inequality sets $I_{\{i,j\}}$ for every variable set $\{x_i, x_j\}$. As the index is a set, $I_{\{i,j\}} = I_{\{j,i\}}$ follows. Furthermore, the sets are not a partitioning of I since inequalities ι with $var(\iota) = \{x_i\}$ appear in all sets $I_{\{x_i,x_j\}}$ with $j \neq i$. The correctness of this simplified algorithm is stated as follows.

Proposition 10. *Suppose* $I_{new} \in Two$ *is given with* $|var(I_{new})| = 2$. *Let* $I \in Two^{cl}$ *and let* $I' = intersect(I, I_{new})$. *Then* $[\![I \cup I_{new}]\!] = [\![I']\!]$ *and* $I' \in Two^{cl}$.

Algorithm 7 Intersecting with inequalities over x_j, x_k and performing closure

procedure *intersect*(I, I_{new})
1: $\{x_j, x_k\} \leftarrow var(I_{new})$
2: $I_{\{i,j\}} \leftarrow \{\iota \in I \mid var(\iota) \subseteq \{x_i, x_j\}\}$
3: $I'_{\{j,k\}} \leftarrow nonRedundant(I_{\{j,k\}} \cup I_{new})$
4: **if** $[\![I'_{\{j,k\}}]\!] = \emptyset$ **then**
5:     **return** $\{0 \leq -1\}$
6: **end if**
7: $n \leftarrow |var(I)|$
8: **for** $i \in [0, n-1] \setminus \{j,k\}$ **do**
9:     $I'_{\{j,i\}} \leftarrow nonRedundant(I_{\{j,i\}} \cup result(I'_{\{j,k\}} \cup I_{\{k,i\}}))$
10:     $I'_{\{k,i\}} \leftarrow nonRedundant(I_{\{k,i\}} \cup result(I'_{\{k,j\}} \cup I_{\{j,i\}}))$
11:     **if** $[\![I'_{\{j,i\}}]\!] = \emptyset \vee [\![I_{\{k,i\}'}]\!] = \emptyset$ **then**
12:         **return** $\{0 \leq -1\}$
13:     **end if**
14: **end for**
15: **for** $x \in [0, n-1] \setminus \{j,k\}$ **do**
16:     **for** $y \in [x+1, n-1] \setminus \{j,k\}$ **do**
17:         $I'_{\{x,y\}} \leftarrow nonRedundant(I_{\{x,y\}} \cup result(I'_{\{x,j\}} \cup I'_{\{j,y\}}) \cup result(I'_{\{x,k\}} \cup I'_{\{k,y\}}))$
18:         **if** $[\![I'_{\{x,y\}}]\!] = \emptyset$ **then**
19:             **return** $\{0 \leq -1\}$
20:         **end if**
21:     **end for**
22: **end for**
23: **return** $\{I'_{x,y} \mid 0 \leq x < y < n\}$

Proof. For convenience, let $I'_{i,j} = \{\iota \in I' \mid var(\iota) \subseteq \{x_i, x_j\}\}$. Show that $[\![I \cup I_{new}]\!] = [\![I']\!]$. Note that, for all $x, y \in [0, n-1]$, lines 3, 9, 10, and 17 ensure that $I'_{\{x,y\}} = nonRedundant(I_{\{x,y\}} \cup \bar{I})$ for $0 \leq x < y \leq n-1$ and some $\bar{I} \subseteq Two$. Due to line 3, $I'_{\{j,k\}} = nonRedundant(I_{\{j,k\}} \cup I_{new})$ and hence $[\![I'_{\{j,k\}}]\!] \sqsubseteq_P [\![I_{\{j,k\}} \cup I_{new}]\!]$, and thus, by line 23, $[\![I']\!] \sqsubseteq_P [\![I \cup I_{new}]\!]$. Since $\bar{I}$ are resultants of inequalities from I, $[\![I \cup \bar{I}]\!] = [\![I]\!]$ and hence $[\![I \cup I_{new}]\!] \sqsubseteq_P [\![I']\!]$, and thus $[\![I \cup I_{new}]\!] = [\![I']\!]$.

Now show $I' \in Two^{cl}$. Suppose $I' \notin Two^{cl}$ and choose $\bar{I} \subseteq Ineq^2$ minimal with $I' \cup \bar{I} \in Two^{cl}$. Let $\iota \in \bar{I}$. We show that $\iota \in I'$ and hence $\bar{I} = \emptyset$. Since $I \in Two^{cl}$, $\iota \in result(\ldots result(I \cup I_{new}) \ldots)$ and specifically $\iota \in result(\ldots result(I \cup I'_{\{j,k\}}) \ldots)$. Let $\iota = c_a x_a + c_b x_b \leq c$. Note that $\{a, b\} \neq \{j, k\}$ since otherwise $\iota \in I'_{\{j,k\}}$ and hence $\iota \in I'$ by line 23. Thus, suppose $a \in \{j,k\}$ and $b \notin \{j,k\}$. If $a = j$, then line 9 requires that $\iota \in I'_{\{j,b\}}$. Since $\iota \notin I_{\{j,b\}}$ as $I \in Two^{cl}$, it follows that $\iota \in result(I'_{\{j,k\}} \cup I_{\{k,b\}})$ and hence $\iota \in I'$. Similarly for $a = k$ and line 10.

Now suppose $\{a,b\} \cap \{j,k\} = \emptyset$ and $\iota \in I'_{\{a,b\}}$ as defined on line 17. Since $\iota \notin I_{\{a,b\}}$ as $I \in Two^{cl}$, either $\iota \in result(I'_{\{a,j\}} \cup I'_{\{j,b\}})$ or $\iota \in result(I'_{\{a,k\}} \cup I'_{\{k,b\}})$.

In the first case, $\iota \in \mathit{result}(\mathit{result}(I_{\{a,j\}}, I'_{\{j,k\}}), \mathit{result}(I'_{\{k,j\}}, I_{\{j,b\}}))$, and thus $\iota \in I'$. Analogously for the second case.

Note that the algorithm above only handles inequalities with exactly two variables. Inserting inequalities over a single variable can be implemented by a simpler algorithm that merely updates the upper and lower bounds of a single row or column. Adding inequalities with more than two variables is considered in the next section.

8.2.3 Approximating General Inequalities

Figure 8.1 at the beginning of this chapter depicts the problem of adding the inequality $x+y+z \leq 1$ over three variables to the TVPI domain. The resulting domain is necessarily an approximation, as only inequalities with at most two variables can be represented. In general, calculating the intersection of $I \in \mathit{Two}$ with an inequality of the form $a_1x_1 + \ldots a_nx_n \leq c$ can be approximated by inserting the set of inequalities $a_jx_j + a_kx_k \leq c - c_{j,k}$ into I, where $1 \leq j < k \leq n$ and $c_{j,k} = \mathit{minExp}(\sum_{i \in [1,n] \setminus \{j,k\}} a_ix_i, I)$. If the bound $c_{j,k}$ is infinite, no approximation is possible. The number of inequalities that are generated this way may be as large as $\binom{n}{2}$; that is, the number is quadratic in the number of non-zero coefficients n. In practice, Core C programs rarely give rise to inequalities with more than three variables, so the number of TVPI inequalities needed to approximate a single inequality is not a bottleneck.

Note that the approximation of inequalities presented is not always optimal. Consider the task of adding $x - 2y + z \leq 0$ to the closed TVPI system $I = \{x - y = 0\}$. Since neither x, y nor z have an upper bound or the term $x - 2y$, the approximation above would fail to deduce any information. However, given the inequality $x - y = 0 \in I$, the new inequality rewritten to $x - y - y + z \leq 0$ can be approximated with $-y + z \leq 0$. While a better approximation algorithm is certainly desirable, it is not too crucial in the context of our analysis, as inequalities with more than two variables tend to have coefficients of 1 or -1. Furthermore, variables are usually bounded.

Thus, an efficient implementation of the strategy above to approximate an n-dimensional inequality with TVPI inequalities hinges on an efficient linear programming algorithm to implement *minExp*, which is discussed next.

8.2.4 Linear Programming in the TVPI Domain

Inferring the minimum of a linear expression in the context of a TVPI system can be solved straightforwardly using general linear programming techniques such as Danzig's simplex method. Interestingly, a linear programming algorithm that exploits the special structure of a TVPI system has only been found recently [186]. This algorithm runs in $O(m^3n^2 \log B)$, where m is the number of inequalities, n the number of variables, and B the upper bound on the constants. Observe that, for a closed TVPI system, Prop. 9 states that the algorithm may be run on just those variables that appear in the expression to be minimised. While this also reduces the number of inequalities m that the algorithm runs on, a closed system contains many redundant inequalities that increase m unnecessarily.

In the context of analysing programs, the minimum of an expression is usually queried on a few variables, namely to approximate inequalities as described in the last section or to infer bounds of pointer offsets. Hence, rather than implementing an algorithm for arbitrary dimensions that is cubic in the number of inequalities, we chose to implement a simple but sufficient technique to infer bounds on expressions with up to four variables as follows. To this end, note that the bound of a single variable is readily available and that an expression over two variables can be bounded by using the planar linear programming algorithm presented in the last chapter.

For expressions over more than two non-zero variables, the problem can be decomposed into several planar linear programming problems. For instance, the minimum of $c_a x_a + c_b x_b + c_c x_c \in \mathit{Lin}$ in the TVPI domain I is the minimum of the following expressions, where x_i^l denotes the lower bound of x_i:

$$\begin{aligned}
&\mathit{minExp}(c_a x_a + c_b x_b, I) + c_c x_c^l \\
&\mathit{minExp}(c_a x_a + c_c x_c, I) + c_b x_b^l \\
&\mathit{minExp}(c_b x_b + c_c x_c, I) + c_a x_a^l
\end{aligned}$$

Similarly, the minimum value of an expression $c_a x_a + c_b x_b + c_c x_c + c_d x_d \in \mathit{Lin}$ is the minimum of the following expressions:

$$\begin{aligned}
&\mathit{minExp}(c_a x_a + c_b x_b, I) + \mathit{minExp}(c_c x_c + c_d x_d, I) \\
&\mathit{minExp}(c_a x_a + c_c x_c, I) + \mathit{minExp}(c_b x_b + c_d x_d, I) \\
&\mathit{minExp}(c_a x_a + c_d x_d, I) + \mathit{minExp}(c_b x_b + c_c x_c, I)
\end{aligned}$$

With these definitions, it is possible to calculate $\mathit{minExp}(e, I)$ for all $e \in \mathit{Lin}$ with $\mathit{var}(e) \leq 4$ and thus to approximate inequalities with up to six variables. While this approach might not be satisfactory in the general case, it is sufficient for a static analysis, as more complex expressions in C programs are always simplified by the compiler using assignments to temporary variables. This simplification happens as part of the translation to three-address code and is therefore present in the Core C code produced.

The last operation that exhibits an interesting behaviour when lifted from planar polyhedra to TVPI polyhedra is widening, which is the topic of the last section in this chapter.

8.2.5 Widening of TVPI Polyhedra

As pointed out in Sect. 7.2.5, widening the individual planar projections of the TVPI domain is simpler than for general planar polyhedra, as the TVPI

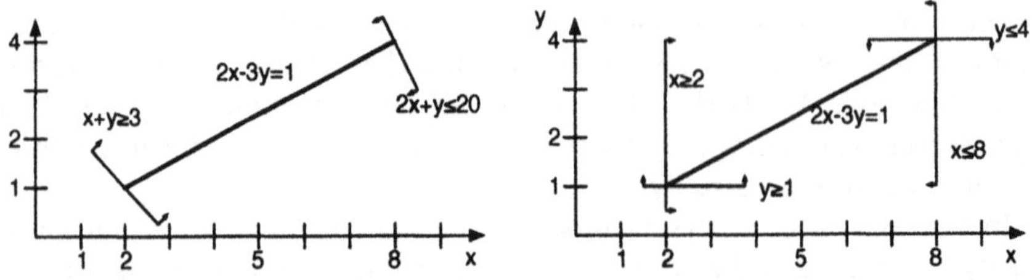

Fig. 8.7. Polyhedra over n variables that extend in fewer than n dimensions (degrees of freedom) have several inequality sets that represent them.

domain is implemented as the product of interval bounds and TVPI inequalities. For instance, Figure 8.7 shows two inequality sets that describe the same polyhedron. While the left graph shows a non-redundant set of TVPI inequalities, two of the interval bounds in the right graph are redundant. However, the representation using bounds and TVPI inequalities is unique in that no other set of bounds and inequalities defines the same polyhedron. As a consequence, widening, which removes facets of a polyhedron that are unstable, reduces to a simple set-difference operation. The details of an actual implementation of widening are omitted at this point because Chap. 12 will present a function that extrapolates the change between two consecutive iterates. Widening is a special case of this function in the sense that the change between two consecutive iterates is extrapolated by an infinite amount.

Note that widening each planar projection results in a TVPI system that is not closed in general. For instance, let $I_1 = \{x \leq y+1, y \leq z+1, x \leq z+1\}$ and $I_2 = \{x \leq y+1, y \leq z+1, x \leq z+2\}$ represent two consecutive loop iterates. The result of $I = I_1 \nabla I_2$ discards the inequality $x \leq z+1$, as it has changed to $x \leq z+2$. However, $\mathit{result}(I) = \{x \leq z+2\} \notin I$, and thus the widened TVPI system I is not closed. In fact, $\mathit{result}(I) \cup I = I_2$, which hints at the fact that closure might interfere with widening in that it re-introduces inequalities that were widened away. Section 10.3 in the chapter on interfacing the TVPI domain with the analysis will illustrate how widening and closure have to be applied in order to guarantee the convergence of fixpoint calculations.

While widening is a prerequisite to ensure that a fixpoint calculation using the TVPI domain will terminate, it may not be sufficient. Infinite chains manifest themselves in a continuously growing number of inequalities and infinitely increasing coefficients. While widening tackles both aspects, it is not always sufficient to ensure that coefficients in inequalities stay tractable. The next chapter therefore presents techniques to reduce the size of coefficients by tightening the inequalities around the set of integral points that they enclose.

8.3 Related Work

Using n-dimensional polyhedra as an abstract domain for program analysis is expressive but expensive [62]. Recent proposals have been made to only infer certain inequalities that are deemed to be important to prove a property [153], only use a special geometric shape of polyhedron [35], impose a fixed dependency between variables [154], or simply approximate the exponential operations when the size of the system becomes too large [169]. In contrast, the TVPI domain limits the precision of the inferred polyhedra up front. TVPI polyhedra form a so-called weakly relational domain and thereby constitute a proper sub-class of general polyhedra. Other sub-classes include difference bounds matrices (DBMs for short) [7,127,161], the Octagon domain [128,130], and the Octahedron domain [47]. The abstract domain of DBMs represents inequalities of the form $x_i - x_j \leq c_{ij}$, $x_i, x_j \in \mathcal{X}$ by storing c_{ij} in an $n \times n$ matrix such that the entry at position i, j is c_{ij}. A special value ∞ is stored at this position if $x_i - x_j$ is not constrained. Closure is computed with an all-pairs Floyd-Warshall shortest-path algorithm that is $O(n^3)$ and echoes ideas in the early work of Pratt [145]. The Octagon domain [128] represents inequalities of the form $ax_i + bx_j \leq c$, where $a, b \in \{1, 0, -1\}$ and $x_i, x_j \in \mathcal{X}$. The key idea of [128] is to simultaneously work with a set of positive variables x_i^+ and negative variables x_i^- and consider a DBM over $\{x_1^+, x_1^-, \ldots, x_n^+, x_n^-\}$, where $n = |\mathcal{X}|$. Then $x_i - x_j \leq c$, $x_i + x_j \leq c$, and $x_i \leq c$ can be encoded respectively as $x_i^+ - x_j^+ \leq c$, $x_i^+ - x_j^- \leq c$ and $x_i^+ - x_i^- \leq 2c$. Thus a $2n \times 2n$ square DBM matrix is used to store this domain. The Octagon domain has been successfully applied to verify large-scale embedded software [30, 31]. While the matrix representation makes adding and removing variables cumbersome, matrix elements can be simple integers or floating-point variables rather than arbitrary-precision integers as required for the TVPI domain. In fact, the operations of the Octagon abstract are so simple that they can be implemented efficiently on high-end graphics hardware [21]. The Octagon domain was generalised into the Octahedron domain [47], allowing more than two variables with zero or unary coefficients while maintaining a hull operation that is polynomial in the number of variables. Miné has generalised DBMs [129] to a class of domains that represent invariants of the form $x - y \in C$, where C is a non-relational domain that represents, for example, a congruence class [86]. This work is also formulated in terms of shortest-path closure and illustrates the widespread applicability of the closure concept. In particular, closure has been applied to check for satisfiability of TVPI systems. Some of these approaches are discussed in the Related Work section of the next chapter.

9 The Integral TVPI Domain

All properties that the analysis presented infers can be expressed with integral numbers. Hence, it is possible to restrict the inferred polyhedra to the contained integral points. In fact, shrinking a polyhedron around the contained integral points is highly desirable for precision as well as for performance, as discussed in the next section. However, ensuring that all vertices of a polyhedron are integral is a computationally hard task. A first step towards an integral TVPI domain is to tighten each individual inequality $ax + by \leq c$, $a, b, c \in \mathbb{Z}$ by replacing it with the inequality $(a/d)x + (b/d)y \leq \lfloor c/d \rfloor$, where $d = \gcd(a, b)$. Note that every integral point $\langle x', y' \rangle$ with $ax' + by' \leq c$ satisfies the tightened inequality since $ax' + by'$ and $(ax' + by')/d$ are integral and thus $(ax' + by')/d \leq \lfloor c/d \rfloor$ [148]. However, tightening individual inequalities is not enough to ensure that the vertices of the polyhedron are integral, and additional inequalities need to be added. One way to add these extra inequalities is Gomory's famous cutting plane method [157, Chap. 23]. This method systematically infers inequalities $ax + by \leq c$ for a given polyhedron $I \in Two$ such that $[\![ax + by \leq c]\!] \subseteq [\![I]\!]$. The tightened inequality $(a/d)x + (b/d)y \leq \lfloor c/d \rfloor$ where $d = \gcd(a, b)$ is then added to the representation of P, thereby cutting off space of I that contains no integral points. This process is repeated until no more inequalities $ax + by \leq c$ can be inferred in which $\lfloor c/d \rfloor < c/d$, at which point the polyhedron is integral. The method terminates after generating a finite number of cutting planes; however, the number of cutting planes may be exponential in the width of the polyhedron. This is illustrated by an example presented in [157, p. 344]. Consider the rational polyhedron shown in Fig. 9.1 that is defined by the vertices $\langle 0, 0 \rangle$, $\langle 0, 1 \rangle$, and $\langle k, \frac{1}{2} \rangle$. One step of Gomory's algorithm infers new inequalities such that $\langle k-1, \frac{1}{2} \rangle$ is a new vertex. By induction, $k - 1$ further steps are necessary to derive the $\mathbb{Z}$-polyhedron containing only the vertices $\langle 0, 0 \rangle$ and $\langle 0, 1 \rangle$.

While all but the last planes in this example were redundant, even the number of non-redundant inequalities that need to be added to define an integral polyhedron can be exponential in the number of inequalities that describe the

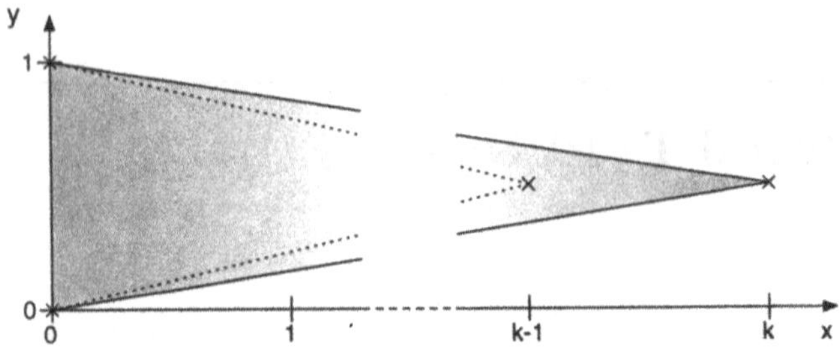

Fig. 9.1. Calculating a new set of cutting planes will refine the rational vertex $\langle k, \frac{1}{2}\rangle$ to $\langle k-1, \frac{1}{2}\rangle$. Thus $k-1$ more steps are necessary to obtain a $\mathbb{Z}$-polyhedron.

rational input polyhedron. Thus, it is not possible to implement an efficient static analysis using $\mathbb{Z}$-polyhedra as the abstract domain. For the special case of planar polyhedra, Harvey proposed an efficient algorithm to shrink a rational polyhedron around the contained integral points [96]. Section 9.2 presents Harvey's algorithm and its implementation in the context of the TVPI domain when it is realised as a reduced product between interval bounds and TVPI inequalities. Not surprisingly, shrinking each planar projection is not sufficient to obtain an integral n-dimensional TVPI system. In fact, testing whether a TVPI polyhedron has an integral solution is NP-complete [114]. Hence, Sect. 9.3 discusses how Harvey's algorithm can be combined with the TVPI closure presented in the previous chapter to approximate an integral TVPI domain. The chapter concludes with an overview of related work.

9.1 The Merit of $\mathbb{Z}$-Polyhedra

This section motivates the use of $\mathbb{Z}$-polyhedra in an analysis. On the one hand, the precision of an analysis can be improved by removing non-integral state space. On the other hand, shrinking the state space around the contained integral points avoids excessive growth of coefficients in the inequalities that describe the polyhedron. The following sections discuss each aspect in turn.

9.1.1 Improving Precision

Restricting the solution set of a polyhedron to integral points can improve the precision of an analysis to the extent that certain properties can be verified that are too coarsely approximated when using rational polyhedra. Spurious state space that contains no integral points may be transformed by scaling (that is, evaluating multiplication operations) to a state that contains spurious integral points. For instance, consider Fig. 9.2, which shows the state space after executing the first assignment in the following C function:

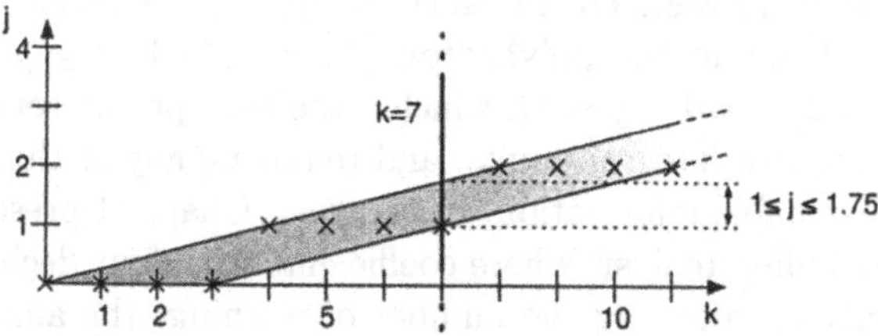

Fig. 9.2. The state space after executing `j=k/4`, where `j` and `k` are integers. The crosses mark possible variable valuations. The dashed lines denote the admissible solutions for j after intersecting the polyhedron with $k = 7$.

```
void f(unsigned int k) {
  int i,j;
  j = k/4;
  i = j*2;
  if (k==7) { assert(i<3); }
};
```

Integer division in C rounds towards zero and, when assuming that $0 \leq k \leq 2^{32} - 1$, the smallest polyhedron that contains all solutions of the division is $[\![\{0 \leq k \leq 2^{32} - 1, 4j \leq k \leq 4j + 3\}]\!]$, which is shown in grey. The multiplication `i=j*2` adds $i = 2j$ to the description, yielding $2i \leq k \leq 2i + 3$ as the relationship between k and i. The assertion in the branch of the conditional therefore does not seem to hold since with $k = 7$ it only follows that $2i \leq 7 \leq 2i + 3$; i.e., $i \in [2 \ldots 3.5]$. However, when $i = 3$, then $j = 1.5$, which is not a possible state in the actual program. In fact, the largest value of j for $k = 7$ is 1, and hence the maximal value for i is 2. The necessary precision to verify the assertion can be attained by shrinking the polyhedron around the containing integral points after the intersection with $k = 7$. While the possible rational values for j are $[1, 1.75]$, the only integral point in this polyhedron is $j = 1, k = 7$, which indeed implies $i = 2$ and thus verifies the assertion.

9.1.2 Limiting the Growth of Coefficients

While improved precision is important in some circumstances, a more pressing reason to perform tightening around the integral grid is the growth of coefficients that can occur otherwise. Specifically, repeated application of the join operator during a fixpoint calculation can lead to coefficients that are excessively large [169]. In principle, widening can be applied to remove inequalities with excessive coefficients. In practice, coefficients may grow drastically before widening is applied. Thus, analyses that do not restrict the size of coefficients may grind to a halt due to expensive arithmetic on very large numbers. Even if widening is applied regularly, the question of whether an inequality contains an excessively large coefficient and should therefore be discarded has no straightforward answer. For instance, wrapping the variable x to an unsigned 32-bit integer in the polyhedron $[\![\{x = y - 1, 0 \leq y \leq 1\}]\!]$ yields $\{x+(2^{32}-1)y = 2^{32}-1, 0 \leq y \leq 1\}$, which is the most precise set of inequalities that contains the two integral points, and removing any of these inequalities would discard valuable information. In contrast, Chap. 11 presents an example from string buffer analysis where coefficients with four decimal digits can be removed without affecting the number of warnings the analysis emits. A more principled way to prevent coefficients from growing excessively is, again, to shrink the polyhedron around its integral points. For instance, adding the inequality $x \leq 7$ to the above system $\{x + (2^{32} - 1)y = 2^{32} - 1, 0 \leq y \leq 1\}$ results in a polyhedron that only contains the integral point $\langle x, y\rangle = \langle 0, 1\rangle$. Tightening the rational polyhedron around this integral point results in the inequality set $\{0 \leq x \leq 0, 1 \leq y \leq 1\}$, which contains none of the large coefficients of the rational system. In general, the coefficients of inequalities in a $\mathbb{Z}$-polyhedron are bound by the admissible range of the variables in the inequality. For instance, the wrapped system above constitutes an inequality set that spans the whole 32-bit range of x and whose coefficients are bound by 2^{32}. In fact, this system depicts the worst-case scenario, as any other system of inequalities that contains the values 0 and $2^{32}-1$ for x has the same or smaller coefficients for x. Thus, tightening combined with wrapping guarantees upper bounds on the coefficient sizes.

The next section details this tightening process and its implications.

9.2 Harvey's Integral Hull Algorithm

The integral TVPI domain used in our analysis is based on Harvey's integral hull algorithm, which tightens a planar polyhedron around the contained integral points. A general $\mathbb{Z}$-polyhedron is characterised by the fact that all vertices have coordinates in $\mathbb{Z}^{|\mathcal{X}|}$. Thus, testing if a set of inequalities defines an integral polyhedron by calculating the vertices is exponential, as the number of vertices grows exponentially with the number of inequalities. In contrast, a planar polyhedron is integral if all inequalities that are adjacent with respect to their angle intersect in an integral point. Thus, testing if a planar polyhedron is integral is a linear-time operation. In fact, the idea of the integral hull algorithm is to calculate cutting planes between adjacent inequalities. The number of new cutting planes is bounded logarithmically by the size of the (coefficients of the) polyhedron; hence, Harvey's algorithm runs in $O(n \log A)$, where A represents the maximum coefficient in the inequality set. Calculating cutting planes between two adjacent inequalities is presented next. Section 9.2.2 describes the integral hull algorithm for the reduced product between intervals and TVPI inequalities with two non-zero coefficients, thereby yielding a simpler implementation than Harvey's original proposal.

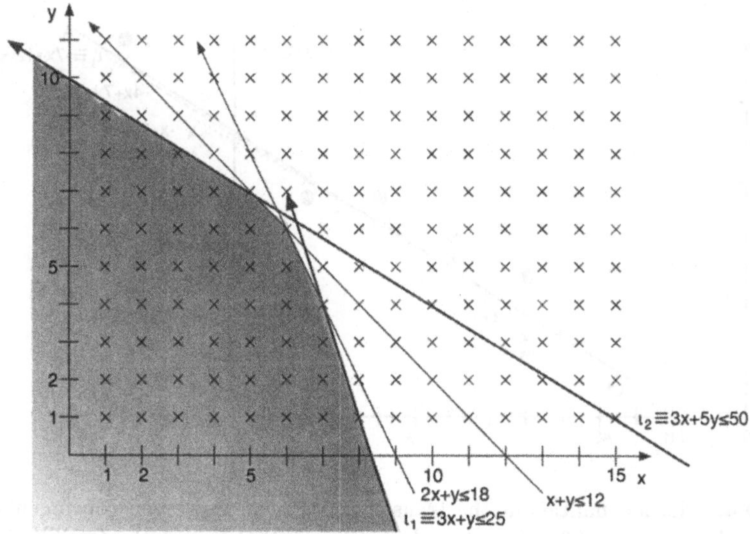

Fig. 9.3. Given two neighbouring inequalities, Harvey's algorithm calculates so-called cuts that tighten these inequalities around the contained integral points.

9.2.1 Calculating Cuts between Two Inequalities

The building block of Harvey's algorithm calculates cuts between two adjacent inequalities that have a rational intersection point. These cuts correspond to Gomory's cutting planes, except that they are always non-redundant. Suppose the following inequalities are adjacent in the input polyhedron:

$$\iota_1 \equiv 3x + y \leq 25$$
$$\iota_2 \equiv 3x + 5y \leq 50$$

Figure 9.3 shows that the intersection point of these inequalities is not integral. In order to calculate the cuts $2x + y \leq 18$ and $x + y \leq 12$ shown, the initial inequalities are mapped to a different coordinate system by applying a linear transformation $T \in \mathbb{Z}^{2\times 2}$ to the coefficients such that $\det(T) \in \{1, -1\}$ and

$$\begin{pmatrix} 3 & 1 \\ 3 & 5 \end{pmatrix} T = \begin{pmatrix} t & u \\ 1 & 0 \end{pmatrix};$$

that is, the inequality $\iota_1 \equiv 3x + y \leq 25$ is mapped to $\iota_1' \equiv tx + uy \leq 25$ and $\iota_2 \equiv 3x + 5y \leq 50$ to $\iota_2' \equiv x \leq 50$. In addition, the transformation matrix is unimodular (that is, $\det(T) \in \{1, -1\}$), which implies that the transformation maps every integral point in the original system to an integral point in the new coordinate system [67]. In the example above, a suitable matrix is

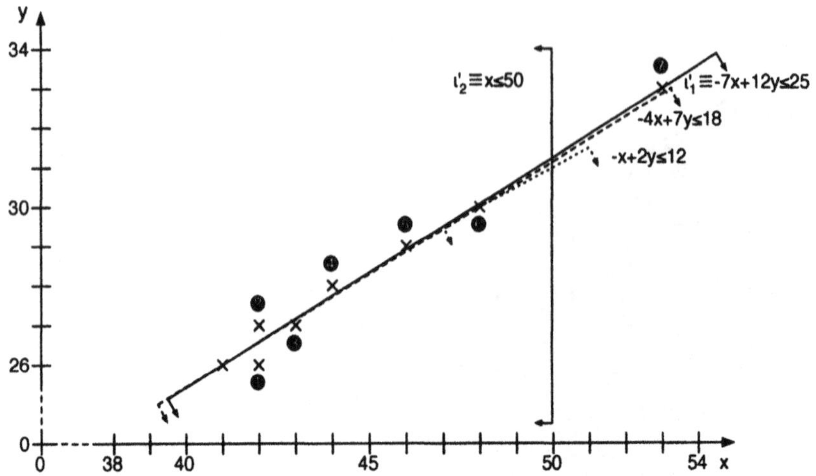

Fig. 9.4. The inequalities in the transformed space. The seven convergents are shown relative to $\langle 41, 26\rangle$, the integral point on ι_1' that lies on the feasible side of $\iota_2' = x \leq 50$.

$$T = \begin{pmatrix} -3 & -5 \\ 2 & -3 \end{pmatrix}.$$

Applying this transformation matrix to the inequality ι_1 yields $\iota_1' \equiv -7x + 12y \leq 25$, which is shown together with ι_2' in Fig. 9.4. As before, ι_1' has a non-integral intersection with $\iota_2' \equiv x \leq 50$. However, observe that the first feasible integral point that lies on the boundary of ι_1' is at $\langle 41, 26\rangle$, such that the problem of calculating cuts reduces to finding integral points with $41 \leq x \leq 50$. The idea is to consider the slope of ι_1' as a fraction $12/7$ and to calculate approximations to $12/7$ using fractions made up of smaller numbers. These approximations provide potential slopes for a cut that originates in $\langle 41, 26\rangle$.

To this end, observe that every rational number can be represented as a finite continued fraction that takes on the following form:

$$a_1 + \cfrac{1}{a_2 + \frac{1}{a_3 + \cdots \frac{1}{a_n}}}$$

The coefficients a_i can be inferred by observing the intermediate results of Euclid's greatest common divisor algorithm when applied to 12 and 7 [67]:

$$\begin{aligned} 12 &= 1 \times 7 + 5 \\ 7 &= 1 \times 5 + 2 \\ 5 &= 2 \times 2 + 1 \\ 2 &= 2 \times 1 + 0 \end{aligned}$$

The continued fraction representation of 12/7 can be derived by dividing the equalities by $7, 5, and$ 2, respectively, and by substituting the reciprocal right-hand side of each equation into the previous equality. The coefficients are thus $a_1 = 1, a_2 = 1, a_3 = 2, a_4 = 2$. Approximations to 12/7 can now be derived by calculating the continued fraction of a prefix of these coefficients. Rather than operating on rational numbers, the following recurrence equations provide a way to calculate these approximations using integral numbers only:

$$A_0 = 1 \qquad A_1 = a_1 \qquad A_m = a_m A_{m-1} + A_{m-2}$$
$$B_0 = 0 \qquad B_1 = 1 \qquad B_m = a_m B_{m-1} + B_{m-2}$$

The following table shows the values of A_i and B_i. For all coefficients $a_i > 1$, we calculate the values of $1, \ldots a_i - 1$ first, as they provide additional slopes for potential cuts. The index i for this case is written $i.j$, where $j = 1, \ldots a_i$.

#	1	2	3	4	5	6	7
i	0	1	2	3.1	3.2	4.1	4.2
A_i	1	1	2	3	5	7	12
B_i	0	1	1	2	3	4	7

These seven slopes are shown in Fig. 9.4 as a displacement to the integral point $\langle 41, 26\rangle$. Note that slopes with odd indices i are not feasible with respect to ι'_1 and can therefore be discarded as an endpoint for a cut. In particular, the coefficients for the first cut are taken from the largest even index i (or sub-index $i.j$ with i even) that yields a point that is still satisfied by $\iota'_2 \equiv x \leq 50$. The first cut in the transformed space is therefore $-4x + 7y \leq 18$ using the sixth fraction. The next cut originates in the endpoint of the first cut, which is $\langle 48, 30\rangle$. Calculating the next cut is a matter of approximating the slope 7/4 of the first cut. Since the continued fraction coefficients of 7/4 form a suffix of those of 12/7, we can reuse the table above to find a suitable slope. The slope $\langle 2, 1\rangle$ gives a displacement that reaches $\langle 50, 31\rangle$, which lies on the boundary of ι'_2. Thus the corresponding inequality $-x + 2y \leq 12$ is the final cut with respect to the two input inequalities. The two cuts are translated to the original coordinate system by multiplying the coefficients with

$$T^{-1} = \begin{pmatrix} 3 & 5 \\ 2 & 3 \end{pmatrix},$$

yielding $2x + y \leq 18$ for the first cut and $x + y \leq 12$ for the second, as shown in Fig. 9.3. Given that Euclid's algorithm requires $\log(A)$ steps, where A is the larger of the two input coefficients, and the fact that the fraction in each step may give rise to at most one cut, no more than $\log(A)$ new inequalities are generated. However, some of these new cuts may be redundant with respect to adjacent inequalities of the rational polyhedron. The next section discusses the challenges of implementing an algorithm that combines tightening of two inequalities with redundancy removal and thereby provides a practical implementation of Harvey's algorithm in $O(n \log A)$ time if all the new inequalities are sorted. Here n is the sum of new and existing inequalities.

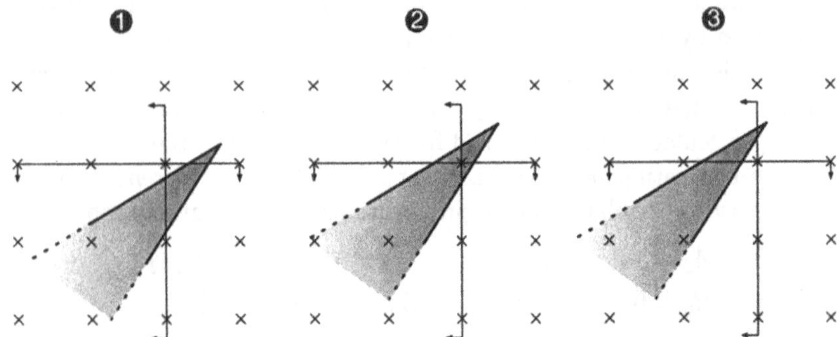

Fig. 9.5. Determining the bounds with which to calculate integral cuts.

9.2.2 Applying the Integer Hull in the Reduced Product Domain

Based on the method of calculating cuts between two inequalities, Harvey suggested an incremental algorithm that tightens a rational input polyhedron by adding its inequalities one by one to an initially empty tree of inequalities that constitutes the output. The complexity of $O(n \log A)$ is based on a level-linked finger tree [124] that is implemented in a circular fashion. In this section, we present a way to calculate the integral hull in $O(n \log A)$ time, which is likely to be faster for small n that occur in program analysis and more in tune with the reduced product representation of the TVPI domain where interval bounds are represented separately from the relational information and the latter is stored as an ordered, non-redundant array of inequalities.

Calculating the integral hull of a polyhedron has to be performed whenever new inequalities are added through the meet operation since this may create non-integral intersection points. On the one hand, any redundant inequality that arises when adding new inequalities needs to be removed, as cuts must be calculated on pairs of inequalities that are themselves non-redundant. On the other hand, any new cut may make other inequalities redundant such that the redundant inequalities must be removed while calculating cuts. Combining both algorithms into one is difficult, as the redundancy removal algorithm in Sect. 7.2.2 on p. 132 reduces the number of inequalities until a fixpoint is reached, while calculating cuts creates new inequalities. Thus, we present a strategy that separates these concerns by exploiting the fact that the TVPI domain is implemented as a reduced product between TVPI inequalities and interval bounds. In particular, by observing that interval bounds are tightened explicitly during redundancy removal, as shown in Fig. 8.5 on p. 156, we propose to tighten the interval bounds further, namely to the values that they will take on in the final $\mathbb{Z}$-polyhedron. Given these tightened interval bounds, cuts can be calculated separately within each quadrant without

Algorithm 8 Test for a feasible integral point on the upper bounds.

procedure *hasZPoint4th*$(\iota_1, \iota_2, x_u, y_u)$ **where** $x_u, y_u \in \mathbb{Z} \cup \{\infty\}$ **and** $\iota_1, \iota_2 \in Ineq$ **with** $\frac{3}{2}\pi < \iota_1 < 2\pi \wedge \iota_1 \measuredangle \iota_2 \leq 2\pi \wedge \theta(\iota_2) \neq 0 \wedge \theta(\iota_2) \neq \frac{\pi}{2}$

1: $a_1 x + b_1 y \leq c_1 \leftarrow \iota_1$
2: $a_2 x + b_2 y \leq c_2 \leftarrow \iota_2$
3: **if** $x_u = \infty$ **then**
4:    **return** *true*
5: **end if**
6: $\{\langle \_, lower \rangle\} \leftarrow [\![\{a_1 x + b_1 y = c_1, x = x_u\}]\!]$
7: $\{\langle \_, upper \rangle\} \leftarrow [\![\{a_2 x + b_2 y = c_2, x = x_u\}]\!]$
8: **if** $y_u < \infty \wedge upper > y_u$ **then**
9:    **if** $lower \leq y_u$ **then**
10:       **return** *true*
11:    **end if**
12:    $\{\langle upper, \_ \rangle\} \leftarrow [\![\{a_1 x + b_1 y = c_1, y = y_u\}]\!]$
13:    $\{\langle lower, \_ \rangle\} \leftarrow [\![\{a_2 x + b_2 y = c_2, y = y_u\}]\!]$
14: **end if**
15: **return** $\lceil lower \rceil \leq \lfloor upper \rfloor$

requiring a fixpoint computation to remove inequalities that become redundant with respect to the calculated cuts.

As a first step, we describe how to tighten interval bounds to the bounds of the final $\mathbb{Z}$-polyhedron. Since a $\mathbb{Z}$-polyhedron is characterised by the fact that all vertices are integral, it follows that a $\mathbb{Z}$-polyhedron has at least one feasible integral point on each bound of its bounding box. Thus, in order to find this bounding box, the intervals of the rational polyhedron must be tightened until at least one integral point lies on each bound. Suppose that the bounds are rationally tightened such that the intersection of two adjacent bounds defines a feasible (but possibly rational) point corresponding to graphs 2–4 in Fig. 8.5. Rounding the bounds to the nearest feasible integral values may lead to one of the situations depicted in Fig. 9.5.

The interval bounds need no tightening if a point lies on them that is integral and feasible, such as in the second graph. Algorithm 8 implements this test for pairs of inequalities ι_1, ι_2 where the normal vector of ι_1 points towards the fourth quadrant, as is the case in Fig. 9.5. Only the upper interval bounds, namely x_u and y_u, are relevant for this test. The algorithm returns *false* if the feasible section of a bound contains no integral point. In particular, lines 6 and 7 calculate the lower and upper y-values of the intersection point of the inequalities with the x-bound x_u, which corresponds to the first graph in Fig. 9.5. Line 8 tests if these intersection points are feasible with respect to the upper bound on y, namely y_u. If these two bounds lie on either side of y_u, then an integral point has been found since $y_u \in \mathbb{Z}$ (lines 9–11 and graph two in the figure). Otherwise, the upper and lower x-values of the intersection between the inequalities and y_u are calculated (lines 12–13 and graph three in the

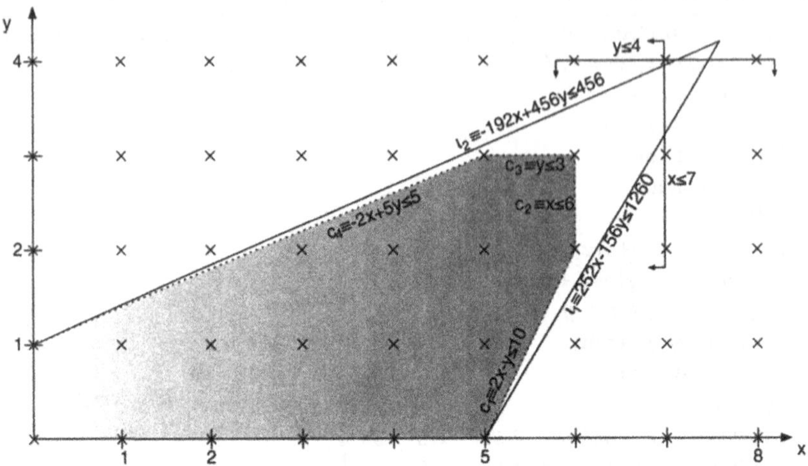

Fig. 9.6. Deriving new bounds by calculating intersection points between cuts.

figure). If rounding these values towards each other results in a non-empty interval, a feasible point on one of the bounds has been found and line 15 returns *true*. Two special cases may occur during the test. Firstly, $\iota_1 \measuredangle \iota_2 = \pi$; that is, the inequalities define an equality. In this case, the polyhedron may have no upper bounds and is thus automatically a valid $\mathbb{Z}$-polyhedron and lines 3–5 return prematurely. Secondly, if the inequalities ι_1 and ι_2 lie in adjacent quadrants (graph one in Fig. 8.5 on p. 156), y_u may be infinite and line 8 ensures that calculating the intersection with y_u in lines 9–13 is skipped.

Since Alg. 8 only tests whether an integral point exists on the upper x-bound, three more variants of this test are necessary for the other bounds. If these tests return *true*, shrinking the polyhedron around the contained integral points will not affect the corresponding bound. If *false* is returned, the corresponding bound must be tightened until it contains a feasible integral point. In order to find such an integral point, we calculate a sequence of cuts $c_0, c_1, \ldots$ from the two inequalities ι_1 and ι_2 as follows. Set $c_0 = \iota_1$ unless $\iota_1 \measuredangle \iota_2 = \pi$, in which case c_0 is set to the cut between ι_1 and the next bound. (In the example, this is the upper x-bound such that the cut is calculated with respect to $1x + 0y \leq x_u$). Let c_i denote the cut between the inequalities c_{i-1} and ι_2. Suppose there are n such cuts such that $c_0, \ldots c_n, c_{n+1}$ denotes a sequence of inequalities with integral intersection points where $c_{n+1} = \iota_2$. Furthermore, let i denote the smallest index such that $\mathit{class}(c_i) \neq \mathit{class}(\iota_1)$; that is, the cut c_i lies in the next quadrant. Then the intersection between $c_{i-1} and c_i$ is an integral vertex that represents the largest extent of the polyhedron towards that direction. Analogously, let j denote the largest index such that $\mathit{class}(c_j) \neq \mathit{class}(\iota_2)$, and use the intersection of c_j and c_{j+1} to refine the

Algorithm 9 Calculating all cuts between two non-redundant bounds.

```
procedure tightenFirstQuadrant(I, x_u, y_u), I sorted, x_u, y_u ∈ ℤ ∪ {∞}
if x_u < ∞ then
  I ← ⟨1x + 0y ≤ x_u⟩ · I
end if
if y_u < ∞ then
  I ← I · ⟨0x + 1y ≤ y_u⟩
end if
O ← ∅
while |I| > 0 do
  if |O| = 0 then
    ⟨ι_0, ... ι_n⟩ ← I
    I ← ⟨ι_1, ... ι_n⟩
    O ← ⟨ι_0⟩
  else
    ⟨o_1, ... o_m⟩ ← O
    ⟨ι_1, ... ι_n⟩ ← I
    if |I| > 1 ∧ {o_m, ι_2} ⊑ ι_1 then
      O ← ⟨o_1, ... o_{m-1}⟩
      I ← ⟨o_m, ι_2, ... ι_n⟩
    else if intersect(o_m, ι_1) ∈ ℤ² then
      O ← ⟨o_1, ... o_m, ι_1⟩
      I ← ⟨ι_2, ... ι_n⟩
    else
      O ← ⟨o_1, ... o_{m-1}⟩
      I ← ⟨o_m, calculateCut(o_m, ι_1), ι_1, ... ι_n⟩
    end if
  end if
end while
return O
```

next bound. For example, consider Fig. 9.6, where $\iota_1 = c_0$ and $\iota_2 = c_5$ define the first and the last cuts. Here, $i = 2$ and $j = 3$ such that the intersection point between c_1 and c_2 defines the upper x-bound and similarly c_3 and c_4 define the upper y-bound. After tightening the bounds, all cuts are discarded and the redundancy removal algorithm continues, possibly identifying ι_1 or ι_2 as redundant, in which case the bounds might need tightening again.

By applying the procedure above for all quadrants of the planar space, the redundancy removal algorithm will infer a polyhedron in which the bounds coincide with the bounds of the corresponding $\mathbb{Z}$-polyhedron. With cuts being calculated on-the-fly rather than being inserted into the sequence of inequalities, there is no need to alter the fixpoint calculation. The integral bounds can now serve to tighten each quadrant of the polyhedron separately, as implemented by Alg. 9 for inequalities $I = \{\iota_1, \ldots \iota_n\}$ with

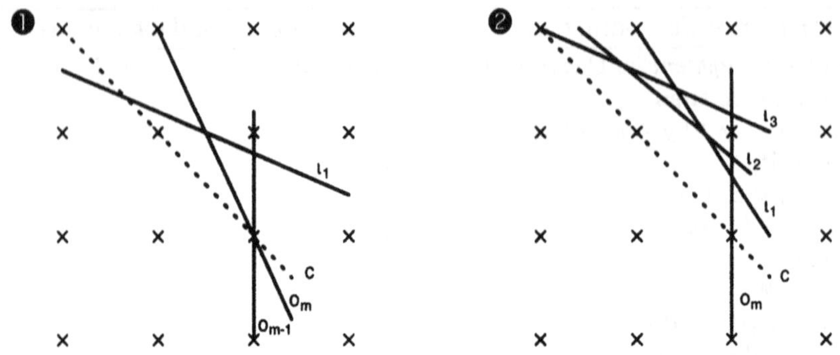

Fig. 9.7. Removing inequalities that were made redundant by a new cut.

$0 < \theta(\iota_1) < \ldots < \theta(\iota_n) < \frac{\pi}{2}$. The first three lines prepend the upper x bound as the inequality to I (using a dot to denote concatenation), thereby ensuring that the sequence begins with a non-redundant inequality that is known to be part of the output $\mathbb{Z}$-polyhedron. Analogously, lines 4–6 append the upper y-bound. Note that if a bound is infinite, the nearest inequality is in both cases a non-redundant ray that is satisfiable by an integral point and must therefore be part of the output $\mathbb{Z}$-polyhedron. While lines 1–6 are specific to the first quadrant, the code in lines 7–28 is applicable to all quadrants. The loop shown examines the polyhedron described by I and calculates its $\mathbb{Z}$-polyhedron in form of the initially empty sequence O. Specifically, lines 9–12 ensure that O contains at least the first element of I, which is known to be in the output $\mathbb{Z}$-polyhedron by the previous argument. The following lines ensure that O only contains inequalities that are non-redundant and that intersect in an integral point. Non-redundancy of ι_1 is ensured by the test in line 16, which holds if ι_1 is the last inequality in I or if it is not entailed by its neighbouring inequalities. Furthermore, integrality is ensured by the test in line 19, in which case lines 20–21 move the head ι_1 of I to the tail of O. If the intersection point is not integral, lines 23–24 calculate a cut that has an integral intersection with o_m. However, the new cut may make o_m redundant, which is illustrated in the first graph of Fig. 9.7. Since o_m was appended to O, it is non-redundant with respect to o_{m-1} and ι_1 and there exists an integral point $p = \mathit{intersect}(o_{m-1}, o_m)$. It follows that o_m can only be redundant if $p = \mathit{intersect}(o_m, c)$. In this case, the current o_m becomes ι_1 in the next loop iteration and is removed by lines 16–18. The loop iteration thereafter will find that o_{m-1} and the cut c (now the new ι_1) intersect in the same integral point p and therefore append the cut to O. Thus, no more than one element of O is ever taken out of O. On the contrary, inserting a new cut c in line 24 may render several of the following inequalities $\iota_1, \iota_2, \ldots$ redundant, as shown in the second graph of Fig. 9.7. These are consecutively removed by lines 16–18. Since the number of possible cuts between two inequalities is bounded and the three other branches of the loop (lines 10–12, 17–18, and 20–21) reduce the length of I, the loop will terminate eventually. In particular, since each element in O is put back into I at most once, the algorithm is linear in the size of the output set O.

With respect to assessing the complexity of the tightening methods above, observe that the linear redundancy removal is augmented with the calculation of cuts between interval bounds and the adjacent inequalities. Harvey observes that no more than $O(\log A)$ cuts can exist between any two inequalities whose coefficients are bound by A. Furthermore, even if inequalities that are adjacent to the bounds become redundant and each rational input inequality is removed, the whole redundancy removal will still terminate in $O(n \log A)$. Similarly, calculating cuts in each quadrant terminates after creating at most $O(\log A)$ cuts between each pair of adjacent inequalities, giving an overall running time of $O(n \log A)$. Note that Harvey's algorithm requires only $O(n \log A)$ steps even if the input inequalities are not sorted by angle. However, Alg. 9 can be implemented using a dynamically growing array for the output O. The simpler data structure is likely to make up for the requirement of sorting the input inequalities. This is particularly true when using the TVPI domain for the analysis presented, as the occurring planar polyhedra are very small, such that the overall running time is dominated by factors such as the cost of arithmetic on multiple precision integers rather than the complexity class.

The next section presents a closure algorithm that builds on the integral hull and the redundancy removal algorithm.

9.3 Planar $\mathbb{Z}$-Polyhedra and Closure

Given an efficient algorithm that shrinks a given planar polyhedron around the integral points that it contains, we now consider the problem of closing a system of planar integral polyhedra. Nelson originally proposed the calculation of the closure of a TVPI system as a way to check satisfiability [138]. However, checking if a TVPI system has an integral solution is NP-complete [114]. Indeed, the closure algorithm presented in the last chapter is incomplete when combined with the planar integral hull algorithm. In this section, we shall explore how this fact manifests itself in practice.

9.3.1 Possible Implementations of a $\mathbb{Z}$-TVPI Domain

The complexity of the problem could be circumvented by implementing a rational TVPI domain and merely tightening planar projections whenever the value of a variable is queried. However, this approach does not prevent the excessive growth of coefficients and does not fully exploit the precision improvement due to tightening. The latter is illustrated by Fig. 9.8, which depicts the closure of a TVPI system containing inequalities over x, z and z, y. Here, rational inequalities are shown as solid lines, whereas the contained

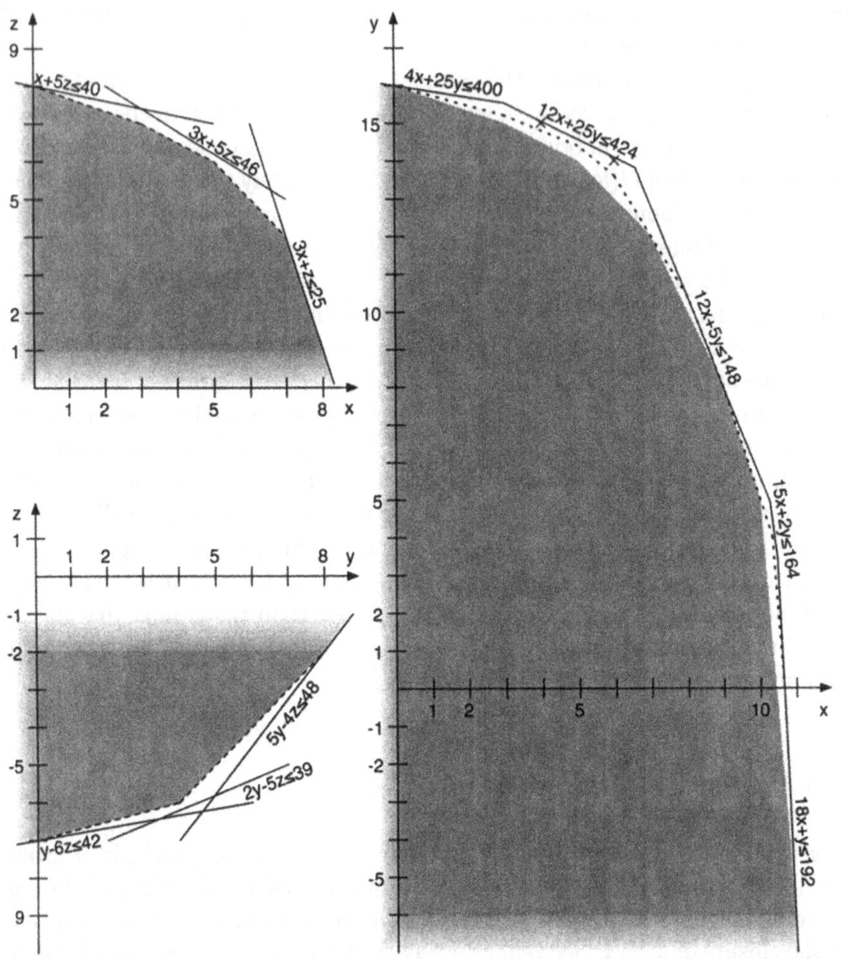

Fig. 9.8. Closing a system over x, z and y, z to yield x, y. Shrinking the initial system around the integral grid removes two integral points in the x, y system.

integral polyhedra are shown as the grey area. Calculating the resultants of the inequalities in the two graphs on the left yields five non-redundant inequalities, shown in the x, y projection. The dashed lines in the left two graphs denote the cuts that define the integral polyhedra in the x, z and z, y projections. Closing the system using these cuts results in the inequalities indicated by the dashed line in the x, y projection. Note that the two integral points $\langle 4, 15\rangle$ and $\langle 6, 14\rangle$ are no longer part of the feasible state space, which shows that tightening every planar projection leads to more precise relationships between other variables of the system. However, the x, y resultants calculated from the integral x, z and z, y cuts do not define a $\mathbb{Z}$-polyhedron, and tightening these resultants around the integral grid is necessary, which results in the grey polyhedron. The example shows that tightening each individual projection around the integral grid is more precise than inferring the best integral solution in a rational TVPI polyhedron. Hence, both growth of coefficients and precision demand that each projection be tightened around the contained integral points.

9.3.2 Tightening Bounds across Projections

In order to adapt the incremental closure algorithm to use Harvey's tightening algorithm, observe that closure adds a set of inequalities to a projection, removes redundant inequalities, and uses the non-redundant subset of the new inequalities to close the TVPI system. When performing integral tightening after running the redundancy removal algorithm, the set of new, non-redundant inequalities, is not a subset of the new inequalities as tightening may have added cuts that are necessary to describe the integral polyhedron. This behaviour requires that the closure algorithm store inequalities on the heap, which is less efficient than stack-allocated arrays, which are sufficient for the rational closure. Repeated allocation is avoided in our implementation by using two large heap-allocated arrays, one for all distance-one resultants and one for distance-two resultants.

Except for calculating the cuts, performing closure with tightening exhibits the same complexity as the rational closure. However, an integral closure algorithm would provide a way to test for integral satisfiability, which in turn is NP-complete. In fact, performing the closure with tightening does not generally lead to a closed system, and the resulting system might not be integral. For instance, consider the TVPI system over the three variables x, y, z shown in Fig. 9.9. Suppose the initial system consists of $\{x = 2z\}$ and that the inequalities $2x+3y \leq 27, -2x+3y \leq 3, -2x-3y \leq -15, 2x-3y \leq 9$ are then added as indicated by the solid lines in the upper left system. No integral tightening is necessary, as all intersection points of the inequalities are integral. Closing the system calculates the resultants of these inequalities and the empty y, z projection before combining the x, y projection with $\{x = 2z\} = \{x - 2z \leq 0, -x + 2z \leq 0\}$, which effectively scales the wider extent of the rhombus by one-half. During redundancy removal, the interval bounds of z are tightened to $2 \leq z \leq 4$. Now the scaled rhombus $4z + 3y \leq 27, -4z + 3y \leq 3, -4z - 3y \leq -15, 4z - 3y \leq 9$, depicted by the solid lines, has non-integral intersection points with the bounds of z. Thus, the inequalities are tightened around the integral grid, yielding $2z + y \leq 11, -2z + y \leq -1, -2z - y \leq -7, 2z - y \leq 5$, as indicated by the dashed lines. In order to ensure that the interval bounds are maximally tight, all projections are checked for inequalities that fulfil cases three and four of Fig. 8.5. In the example, the x, z projection thereby tightens the interval of x to $4 \leq x \leq 8$. The incremental closure stops at this point. Since the bound on z was updated after the x, y projection was tightened, it now

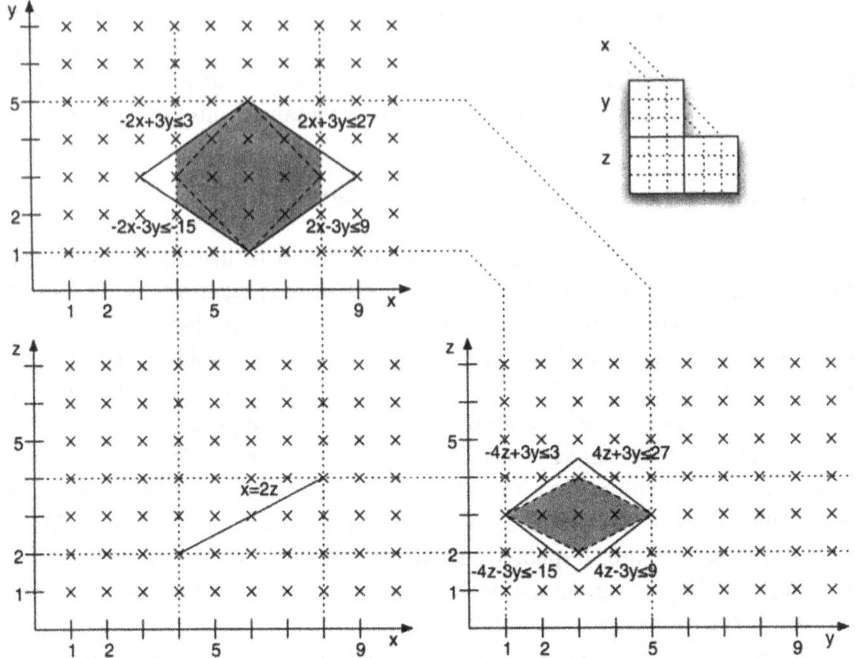

Fig. 9.9. Tightening inequalities around their integral grid can affect interval bounds, which requires a new tightening step in previously visited projections.

contains non-integral intersection points. The dashed rhombus shown can be obtained either by tightening the existing inequalities to the integral grid or by calculating new resultants from the rhombus in the z, y projection and the line segment in the z, y projection. In order to ensure that a TVPI system is closed, closure and tightening have to be applied repeatedly until a fixpoint is reached. To ensure efficiency, our implementation stops with the state space shown in grey and thereby admits non-integral intersection points with the interval bounds. This implies that vertex-based algorithms such as the planar convex hull may operate on non-integral vertices and may thus create inequalities that intersect in non-integral points. For instance, the join of the integer polyhedron in Fig. 3.6 on p. 60, after being intersected with integral interval bounds, creates inequalities that intersect in non-integral points. These rational intersection points are only removed if new inequalities are added and integrality of the vertices is tested again.

9.3.3 Discussion and Implementation

Note that, rather unexpectedly, repeated application of integral closure does not constitute a decision procedure for integral TVPI satisfiability. Consider

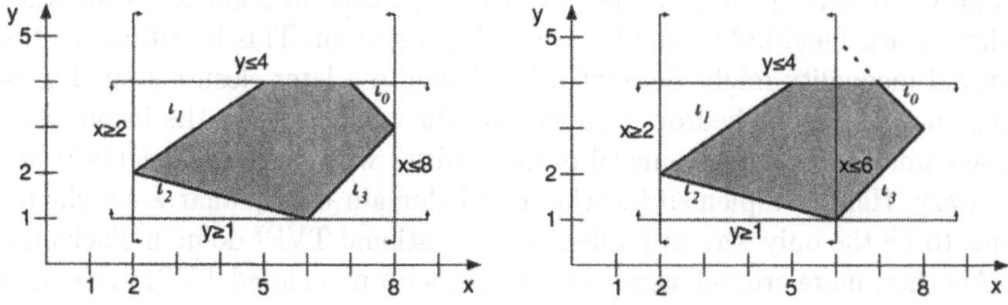

Fig. 9.10. Allowing interval bounds to tighten after calculating the integral hull can lead to intersection points outside the original polyhedron.

the system comprised of the inequalities $0 \leq 4y - 7z \leq 1$ for the y, z projection and $-6 \leq 4x - 7z \leq -7$ for the x, z projection. Closing this TVPI system adds $4x - 4y \leq -7, -4x + 4y \leq -5$ to the x, y projection. The latter can be tightened without losing integral solutions to $x - y \leq \lfloor\frac{-7}{4}\rfloor, -x + y \leq \lfloor\frac{-5}{4}\rfloor$, which is equivalent to $x - y = -2$. The resulting system is in fact closed, and all projections are integral. This is peculiar since the TVPI system is actually unsatisfiable in $\mathbb{Z}$, which becomes apparent when adding the inequality $z \geq 1$: applying closure will result in a non-integral intersection between the z-bound and alternately the y, z- and the x, z projections without ever stabilizing.

Another consequence of bounds being tightened after a projection has been shrunk around the integral grid is that inequalities might become redundant. Suppose the integral polyhedron in the left graph of Fig. 9.10 has just been tightened around the integral grid. If tightening in a different projection reduces the upper bound on x to $x \leq 6$, the polyhedron will contain two redundant inequalities, ι_0 and ι_3, as shown in the right graph. Applying the convex hull algorithm to this system of inequalities will calculate an intersection point between the upper bound on y and the boundary of ι_0, as these are angle-wise adjacent. However, the resulting point lies outside the polyhedron, and the convex hull algorithm calculates a result that is incorrect. Thus, inequalities that are redundant due to tightened bounds have to be removed before applying the convex hull or other planar algorithms that require a non-redundant input system. However, note that these excess inequalities can be removed on-the-fly by merely using case two of the tests in Fig. 8.4 on p. 155, rather than by applying the full redundancy removal algorithm.

Working on non-closed TVPI systems implies that the analysis is not as precise as possible. Worse, since the closure calculation has to be stopped at some point, the specific implementation of the TVPI domain determines the precision of the analysis and thus the number of warnings the analysis infers. While this is an argument against integer tightening, observe that the integral TVPI domain is always more precise than its rational counterpart. Furthermore, since coefficients in the inequalities of a rational TVPI system can grow excessively, inequalities have to be removed in order to ensure scalability, which inevitably leads to a non-closed system. This is critical, as the removed inequality might be reintroduced due to a later closure step. Tightening the planar polyhedron ensures that the coefficients in the inequalities remain small so that the removal of inequalities with large coefficients is unnecessary. Hence, implementing the TVPI domain over planar $\mathbb{Z}$-polyhedra seems to be the only way to implement the rational TVPI domain efficiently.

Another, more critical aspect of working with non-closed TVPI systems is that fixpoint computations might be jeopardised. Suppose a reference TVPI system is stored at the head of a loop. Propagating a copy of this system around the loop so that operations modify the domain in such a way that it contains inequalities whose resultants are not explicitly expressed may make the entailment check fail, even if the propagated state defines a smaller state than the state that the reference system describes. This situation is unlikely to arise, as the reference system will be created by evaluating the very same operations in the loop body. However, at the time of writing, we have neither a proof that no infinite chain can arise nor an example that exhibits an infinite chain. Note, though, that Prop. 6 shows that it is not necessary to close the reference system stored at the head of a loop.

We conclude this chapter by reviewing work on rational and integral satisfiability of the TVPI domain.

9.4 Related Work

Inequalities with at most two variables have given rise to much research in recent decades, not least due to the fact that general network-flow problems can be expressed using a TVPI system. Integer TVPI systems describe a special class of flow problems where flows consist of discrete units.

The closure operation presented in Sect. 8.2 stems from an idea of Nelson to check for satisfiability of a rational TVPI system [138]. It turns out that more efficient methods exist for this task [102]. However, Shostak used closure algorithms to check for satisfiability of integer TVPI problems [164], although his procedure is not guaranteed to either terminate or detect satisfiability. In the context of weaker TVPI classes, Jaffar et al. [106] show that satisfiability of two variables per inequality constraints with unit coefficients can be solved in polynomial time and that this domain supports efficient entailment checking and projection. More recently, Harvey and Stuckey [97] have shown how to reformulate this solver to formally argue completeness, which gave rise to the planar integer hull algorithm [96]. Su and Wagner [177] present an algorithm for calculating the least integer solution of a system of two variable inequalities. They claim that their algorithm is polynomial; however, it turns out that solving integer two variable per inequality constraints is NP-complete [114]. However, checking integral satisfiability of a TVPI system is polynomial if all inequalities are monotone [101] – that is, if all inequalities have the form $ax - by \leq c$, where $a, b \in \mathbb{N}$. A practical implementation of the integer satisfiability for general polyhedra is the Omega test [148], an extension of Fourier-Motzkin variable elimination that is complete over the integers but that might not terminate in general. Other classic integer decision procedures include the SUP-INF algorithm [163] and Cooper's algorithm [53]. These techniques are widely applied in verification but do not provide the operations necessary for domains used in abstract interpretation.

10

Interfacing Analysis and Numeric Domain

The principles of program analysis are often explained in the context of analysing toy languages that only require the standard lattice operations and possibly a widening operator. This chapter details additional operations and implementation techniques for the TVPI domain that are required to deal with advanced techniques such as populating the map of fields on demand and analysing programs with many variables and constants. In terms of efficiency, Section 10.1 comments on how to minimise the number of variables in a domain and, in particular, how to minimise the number of variables in the arguments to the quadratic TVPI domain operations. Another key problem that needs to be solved in any polyhedral domain is the addition of new variables to the domain. While this task is not difficult in itself, Sect. 10.2 introduces the concept of typed domain variables, which is essential in terms of precision when adding new domain variables within a loop. Section 10.3 concludes with a discussion on how the analysis has to apply widening in order to ensure that fixpoint computations terminate.

10.1 Separating Interval from Relational Information

The domain of general polyhedra has not yet found its way into large-scale program analysis due to its poor scalability. The TVPI domain promises strongly polynomial performance; however, even resource consumption that is quadratic in the number of variables is prohibitive, as a typical program may have a few hundred live variables at a given program point. Thus, for any real-world program, it is important to reduce the number of variables in the TVPI domain. An important observation is that the way the analysis interprets a program leads to many constant domain variables. For instance, most pointers in C have no offset, such that the variable representing the pointer offset is constant zero. Thus, the number of variables in the TVPI domain can be reduced by storing constant-valued domain variables separately.

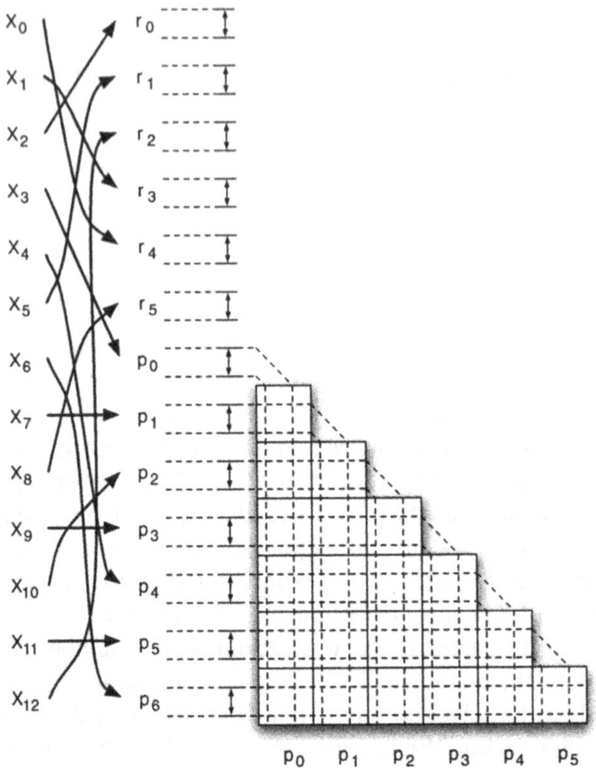

Fig. 10.1. Storing the information on domain variables x_i as a set of ranges r_i and a TVPI domain containing relational information on the variables p_i.

To this end, note that performing updates on the triangular matrix of the TVPI domain permutes the rows. If domain variables are mere indices into the triangular matrix, then variables in the analysis must be renamed, which amounts to modifying all data structures representing fields of memory regions, heap sizes, etc., which in turn is impractical. Hence, the domain keeps a permutation map between abstract variables $\mathcal{X}$ and so-called domain variables that represent indices into the triangular matrix. Figure 10.1 shows this permutation map for a domain containing 13 variables. As a side effect of the permutation map, it is straightforward to map some of the abstract variables to a range r_i as is done for $x_0, \ldots x_2, x_5$, and x_8. A range is simply an interval but is called a range to distinguish it from the intervals stored as part of the TVPI domain. Having both ranges and TVPI variables renders operations that take two domains as their argument more complicated. For instance, when checking the entailment $N_1 \sqsubseteq_N N_2$, it is necessary to promote any interval in N_1 to the TVPI domain if this variable is stored in the TVPI domain in N_2. On the contrary, if x_i is stored in the TVPI domain in N_1

but as a range in N_2, then entailment can be checked by comparing the two intervals. In the context of the join, a variable that is stored as a range in both arguments is promoted to a TVPI variable since the convex hull operation may infer a linear relationship between the arguments. The downside of this opportunistic promotion strategy is that applying domain operations may promote ranges into the TVPI domain even though they remain representable as ranges. Thus, after certain operations such as the join, an operation is run that tries to demote variables – that is, to remove variables from the TVPI domain that can be represented as ranges. A variable p_i can be represented as a range if its interval is a constant or all projections involving p_i are empty.

Besides improving performance by storing fewer variables in the TVPI domain, it is important to reduce the overall number of variables that are necessary to describe the program state. Temporary variables, for instance, should always be projected out as soon as they are not needed anymore. Furthermore, it is beneficial to perform a liveness analysis on the local variables of a function. The removal of all domain variables associated with dead local variables results in a speedup of 143% on our larger example program.

The next section discusses how variables are added to the domain and thereby addresses the question of what the value of a variable is that is not mapped to either a range or a TVPI variable.

10.2 Inferring Relevant Fields and Addresses

This section addresses a precision loss that may occur when adding new variables to a polyhedron. Specifically, we consider the problem of refining the way dynamically allocated memory regions are summarised. The analysis presented in Chap. 5 proposes to merely summarise dynamically allocated memory by the location of the call to `malloc`. Many programs, however, wrap `malloc` in a function that terminates the program if the allocation fails. Summarising the dynamic memory regions in the proposed way would, in this case, result in all dynamically allocated memory being summarised into a single abstract memory region. This results in a severe loss of precision, which can already be observed in the simpler setting of a points-to analysis [140]. In the actual implementation, dynamic memory regions are therefore summarised if their locations in the program and the current call stack coincide. When using this heuristic, a wrapper function around `malloc` has no effect on how allocation sites are merged. On the downside, the number of summarised heap regions is not known up-front since the call stacks by which regions are summarised only arise during the analysis. Thus, the map $H : \mathcal{D} \to \mathcal{X} \times \mathcal{X}$ of summarised memory regions has to be populated during the analysis.

In order to illustrate the problems with this approach, consider the control-flow graph in Fig. 10.2, which depicts a simplified **while**-loop that contains a call to `malloc`. The edges of the control-flow graph are decorated with the

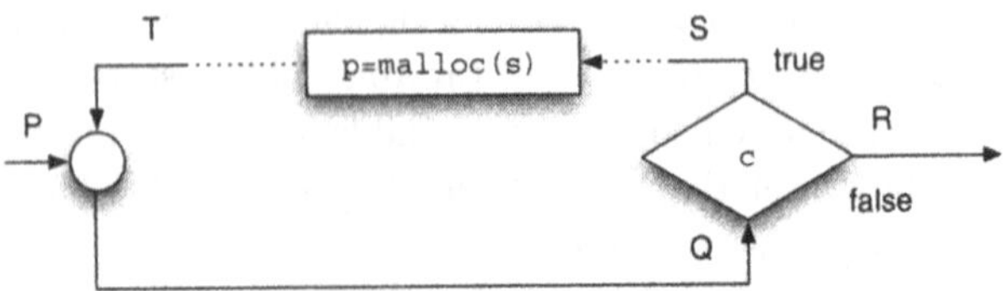

Fig. 10.2. Allocating a dynamic memory region $m \in \mathcal{D}$ in a loop requires a special join if the variables x_m and x_s, $\langle x_m, x_s \rangle = H(m)$, are initialised on-the-fly.

abstract states $P, Q, R, S, T \in Num$, where $Q = P \sqcup_N T$ and where S, R are defined in terms of Q and an unspecified condition `c`. In order to define T, suppose that the allocated memory is summarised by the abstract memory region $m \in \mathcal{D}$. For the sake of the example, consider only the two variables $\langle x_m, x_s \rangle = H(m)$, where x_m denotes the number of concrete memory regions m summarises and x_s denotes their size. Chapter 5 on the abstract semantics suggests settin both variables to zero at the start of the analysis.

Suppose now that the summarised memory region $m \in \mathcal{D}$ is created on-the-fly; that is, during the first evaluation of the `malloc` statement. Creating this new memory region m also creates new polyhedral variables $\langle x_m, x_s \rangle = H(m)$. To infer any information, these variables must be initialised before evaluating the abstract transfer function of `malloc`. Thus, on creation of the new abstract memory region m, the state space feeding into the `malloc`-statement is intersected with $x_m = 0$ and $x_s = 0$. The transfer function that was defined in Fig. 6.5 on p. 122 remains unchanged and thus increments x_m to one (indicating that m summarises one region) and sets x_s to the argument `s`.

Initialising x_m and x_s to zero on creation of the memory region $m \in \mathcal{D}$ is straightforward, such that the semantics of `malloc` in the first iteration of the loop can be summarised as $T = S \triangleright x_m := 1 \triangleright x_s := s$, where s is the value of the parameter to `malloc`. Now consider the result of calculating $Q = P \sqcup_N T$. Since both variables, x_m and x_s, have not been used before, the sets of values $P(x_m)$ and $P(x_s)$ are both unbounded. Thus, the join $P \sqcup_N T$ will remove all information on x_m and x_s. Any access to m will now trigger a warning, as $x_m > 0$ cannot be guaranteed, thereby indicating that the memory region might already be freed; furthermore, x_s is arbitrary, rendering every access out-of-bounds. Ideally, x_m and x_s would have been initialised at the start of the analysis such that $P(x_m) = 0$ and $P(x_s) = 0$, which would yield the best precision in our abstraction of `malloc`. Since this is not possible when adding new memory regions $m \in \mathcal{D}$ on demand, the state P has to be refined after m is created. One possibility is to add the equalities $x_m = 0$ and $x_s = 0$ to all non-empty polyhedra that are stored in the analyser, which corresponds to initialising the two variables at the start of the analysis. This strategy is less than ideal since it requires the modification of all polyhedra stored in the analyser, even if they represent stable states that will never be revisited again.

10.2.1 Typed Abstract Variables

A more lightweight method of avoiding the precision loss above is to modify the join operation such that it initialises the two variables while calculating the join; i.e., calculating $Q = (P \sqcap_N \{x_m = 0, x_s = 0\}) \sqcup_N T$. In this ad hoc method, each abstract variable $x \in \mathcal{X}$, where $x \notin \mathcal{X}^T$ has a type associated with it that defines a safe range for that variable. The range must be safe in the sense that it constitutes a sound assumption about the value of the variable at any program point. For instance, the type of a byte-sized variable is associated with the safe range $[0, 255]$, which corresponds to all possible bit patterns in the concrete program. Interestingly, a safe range for the variable x_m is $[0, 0]$ since assuming that m does not correspond to any concrete memory region is a safe assumption because every access to m will raise a warning. The idea is to insert the range whenever two polyhedra are joined and a variable is present in one but not the other. While inserting a safe range recovers the precision in the example above, it may lead to unwanted side effects. Assume that the loop above is part of a bigger loop such that the range of x_m at the end of the loop R is propagated back to the beginning of the loop P. If x_m is found to be unstable in the outer loop, all constraints on x_m may be widened away such that P would be updated with a state where x_m is unbounded. A new evaluation of the inner loop in which the memory region m is used after the `malloc` statement is evaluated will result in a warning (since x_m is unbounded) but will also restrict x_m to a finite range (since the user has been warned about the erroneous range of x_m). Upon calculating the join $P \sqcup_N T$, the variable x_m is again unbounded in P but bounded in T such that the safe range $x_m = 0$ is inserted into P before the join is calculated. Hence, the newly calculated P is smaller than the previous value, which indicates that inserting a safe range whenever a variable is unbounded can jeopardise the termination of the fixpoint calculation. Thus, a more principled approach is necessary.

One solution to the non-termination problem is to annotate the polyhedron with a set of variables that it contains. This set might contain a variable even if it is unbounded in the polyhedron. This set of variables can then be propagated around just like a separate domain. Whenever two polyhedra are joined whose sets of variables differ, the missing variables are inserted into the corresponding other polyhedron using the safe range. This approach finesses the problem of reintroducing variables once they become unbounded but requires that an additional set of variables be stored with every polyhedron.

The implementation of the TVPI domain uses a more unconventional approach in that it reinterprets the absence of a variable by assuming that the variable takes on its safe range. In this approach, a single global table of types is necessary that maps every variable to its safe range. A domain with no information therefore maps every variable to the range implied by its type. In contrast, a domain in which all variables are unbounded has to contain an entry for each variable that maps to the unbounded interval. As a consequence, projecting out a variable will insert the variable into the domain with

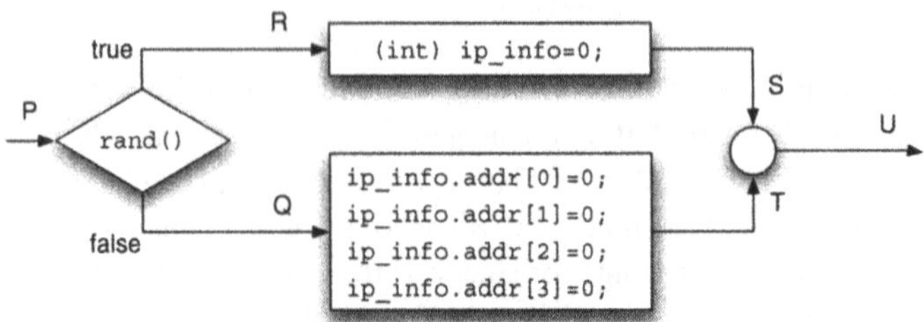

Fig. 10.3. Adding new fields to the F map. The safe ranges $[0, 255]$ are inserted into S for the byte-sized fields, whereas the safe range $[0, 2^{32} - 1]$ is inserted into T for the four-byte integer field if either branch is evaluated with an empty field map.

an unbounded interval. The analysis, in turn, must be careful to distinguish between projecting out a variable and removing a variable. The former operation is used in assignments when a variable must be unbounded since it may be restricted to a value outside its safe range. The latter is used to reduce the size of the domain; for example, when variables go out of scope.

10.2.2 Populating the Field Map

Adding variables on-the-fly not only allows for a context-sensitive treatment of dynamically allocated memory regions but is also key to populating the map of fields $F : \mathcal{M} \cup \mathcal{D} \to \mathcal{P}(\mathbb{Z} \times \mathbb{Z} \times \mathcal{X})$. A possible precision loss was pointed out in Sect. 6.4, where different but overlapping fields were added in two different branches of the following conditional:

```
if (rand()) { (int) ip_info=0; } else {
  ip_info.addr[0]=0; ip_info.addr[1]=0;
  ip_info.addr[2]=0; ip_info.addr[3]=0; }
```

Fig. 10.3 depicts the control-flow graph of the code fragment above with polyhedra $P, Q, R, S, T, U \in \mathit{Num}$ decorating the edges. Suppose that `ip_info` has no fields associated with it before the code is executed. The evaluation of the conditional will add one 4-byte field for the first branch and four 1-byte fields for the second branch. Let these fields be represented by x_4 and by $x_0, \ldots x_3$, respectively. One disadvantage of adding variables on demand is the assumption that every polyhedron includes the new variable as if it were constrained to its safe range; this assumption is too weak when adding fields that overlap with other fields. For instance, if the branch with the polyhedra R, S is analysed first, a 4-byte field is added and $S = R \triangleright x_4 := 0$. When analysing the other branch, the 4-byte field in Q is in the field map but merely constrained to the safe range $[0, 2^{32} - 1]$. The first three byte-sized updates add new variables but fail to refine x_4 since the upper bits are not constant.

When adding the fourth byte-sized field, the upper eight bits of x_4 can be set to zero. Thus, in the join $U = T \sqcup_N S$, $x_0, \ldots x_3$ are zero in T but take on their safe range in S, whereas x_4 is restricted to less than 2^{24} in T and is zero in S. A read access to the 4-byte field in the context of U will propagate some information between the variables using the function *prop* in Fig. 5.2 on p. 95. Without going into detail, this step tightens the ranges of x_4 and x_3 to zero but leaves $x_0, \ldots x_2$ constrained to $[0, 1]$. Interestingly, if the Q, T branch is evaluated first, the byte-sized fields are present when analysing the 4-byte access, and all five fields are set to zero in S. In this case, a 4-byte read access in the context of U will propagate enough information that all fields will be known to be zero. In this case, adding fields on demand is as precise as specifying them up front.

The fact that the iteration strategy determines the outcome of the analysis is not very satisfactory. However, the impact is marginal since overlapping fields are rare and writing overlapping fields in basic blocks that do not dominate each other is even less likely to occur. The problem could be refined by extending the type system to indicate which variables overlap, thereby pushing the task of propagating information between overlapping fields onto the domain. As the propagation of values is based on temporary variables, the implementation as part of the domain seems to be too intrusive. In practice, this effort seems to be exaggerated given the frequency of the phenomenon.

We conclude the topic of adding variables on demand with an overview of the possible variable types.

A Summary of Safe Ranges

In this section, we summarise the various ranges that are used to add variables on demand. In particular, the analysis creates the following types:

- For variables representing the number of dynamically allocated memory regions that are summarised in one address, the safe range is $[0, 0]$.
- For variables representing the size of dynamically allocated memory regions, the safe range is $[0, 2^{32} + 2^{31} - 1]$; i.e., any size between zero and 3 GB is possible.
- For variables representing fields of s bytes, the safe range $[0, 2^{8s} - 1]$ is inserted, where $s = 1, 2, 4, 8$.
- For variables representing NUL positions of memory regions of s bytes, the safe range $[0, s]$ is inserted.
- For flags that specify if an address is in a points-to set, the safe range is $[0, 0]$.

Note that the last two variable types will not be used until Chap. 11 and Chap. 13, respectively. Note further that temporary variables have no type and are projected out when they do not appear in both arguments of a join.

The last topic of this chapter addresses requirements on the analysis with respect to applying widening.

10.3 Applying Widening in Fixpoint Calculations

The widening operator on polyhedra extrapolates changes between loop iterates in order to force the fixpoint computation to terminate. In the context of the TVPI domain, widening amounts to applying the planar widening operator to each planar projection in the domain. The result, however, is not necessarily a widened and closed TVPI domain: A widening operator on sets of inequalities over n dimensions requires its input systems to be non-redundant [62]. This property is satisfied for each planar projection of a closed TVPI system. However, the key idea of the TVPI domain is that the closure operation expresses existing information in terms of inequalities over every pair of variables. These additional inequalities are by definition redundant with respect to the whole system, which jeopardises the correctness of widening. Miné presented an example of how interleaving closure and widening can indeed compromise termination of a fixpoint computation [128]. This non-terminating computation can be generated by the following C fragment:

```
y=z;
if (rand()) z--; else z++;
assert(y<=x)
do {
  if (rand()) x--; else x++;
  i = min(y-x,x-z);
} while (y-x<=i && x-z<=i);
```

Here, the calls to the random number generator **rand()** are used to model non-deterministic behaviour. In order to discuss a fixpoint computation of the program above, consider the control-flow graph in Fig. 10.4, whose edges are decorated with TVPI polyhedra $P, Q, R, S, T \in \mathit{Two}$. For the sake of argument, we assume that program variables have an infinite range such that variables do not wrap. Assuming an unrestricted state space, evaluating the first statement and conditional gives rise to the system $P = \{z-1 \leq y \leq z+1\}$. Evaluating the assertion y<=x and closing the system yields $Q = \{z-1 \leq y \leq z+1, y \leq x, z \leq x+1\}$. The loop is characterised by $R = Q \sqcup_P T$, $S = R \triangleright x := x-1 \sqcup_P R \triangleright x := x+1$, and $T = \{z-1 \leq y \leq y, y \leq x+i, z \leq x+i\}$, where i is the value returned by the statement $min(z-x, y-x)$. In order to examine the computation of a fixpoint, let P_i denote the set of inequalities P in iteration i. Widening is incorporated by replacing $R = Q \sqcup_P U$ with $R_0 = Q$ and $R_{i+1} = R_i \nabla (R_i \sqcup_P T_i)$ for all $i \geq 0$. For best precision, the loop body is always evaluated in terms of the closed system $cl(R)$. The fixpoint calculation proceeds as follows:

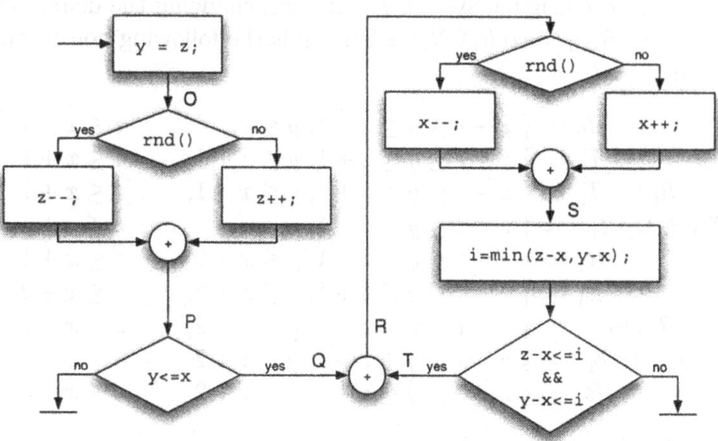

Fig. 10.4. Program to demonstrate the interaction between widening and closure.

$$\begin{aligned}
R_0 = cl(R_0) &= \{\, z-1 \leq y \leq z+1, \quad y \leq x, \quad z \leq x+1 \,\} \\
S_0 &= \{\, z-1 \leq y \leq z+1, y \leq x+1, z \leq x+2 \,\} \\
T_0 &= \{\, z-1 \leq y \leq z+1, y \leq x+1, z \leq x+1 \,\} \\
R_0 \sqcup_P T_0 &= \{\, z-1 \leq y \leq z+1, y \leq x+1, z \leq x+1 \,\} \\
R_1 &= \{\, z-1 \leq y \leq z+1, \qquad z \leq x+1 \,\} \\
cl(R_1) &= \{\, z-1 \leq y \leq z+1, y \leq x+2, z \leq x+1 \,\} \\
S_1 &= \{\, z-1 \leq y \leq z+1, y \leq x+3, z \leq x+2 \,\} \\
T_1 &= \{\, z-1 \leq y \leq z+1, y \leq x+2, z \leq x+2 \,\} \\
R_1 \sqcup_P T_1 &= \{\, z-1 \leq y \leq z+1, \qquad z \leq x+2 \,\} \\
R_2 &= \{\, z-1 \leq y \leq z+1 \,\} \\
cl(R_2) &= \{\, z-1 \leq y \leq z+1 \,\} \\
T_2 &= \{\, z-1 \leq y \leq z+1 \,\}
\end{aligned}$$

At this point, $T_2 \sqsubseteq_P R_2$ and a fixpoint is reached. Note that the variable i is an aid to extracting the smaller of the two constants c_1, c_2 of the two inequalities $z - x \leq c_1$ and $y - x \leq c_2$ from S. An equivalent formulation that only involves a conditional is **x-z<=y-x && y-x<=z-x**. However, these inequalities would be normalised to $2x - z - y \leq 0$ and $-2x + y - z \leq 0$, which cannot be approximated with the methods presented in Sect. 8.2.3 on p. 160.

Widening removes inequalities in certain projections, which creates a non-closed TVPI system if other inequalities exist that, when combined, imply a slightly weaker inequality than the one that had been removed. (No inequalities can exist that imply exactly the same inequality since these would have to have changed and would thereby be removed by widening.) In the example above, the inequality $y \leq x$ in R_0 is relaxed after one iteration to $y \leq x + 1$ and thus removed by widening. However, closing R_1 reintroduces the weaker $y \leq x + 2$. While the loop should be analysed with this closed system, the next widening may not be performed with respect to the closed $cl(R_1)$ but with respect to the original system R_1. In fact, changing the definition of R_i to $R_0 = Q$ and $R_{i+1} = cl(R_i \nabla R_i \sqcup_P T_i)$ yields the following non-terminating calculation:

$$
\begin{array}{rll}
R_0 = & \{\, z-1 \leq y \leq z+1, y \leq x, & z \leq x+1 \,\} \\
T_0 = & \{\, z-1 \leq y \leq z+1, y \leq x+1, & z \leq x+1 \,\} \\
R_0 \sqcup_P T_0 = & \{\, z-1 \leq y \leq z+1, y \leq x+1, & z \leq x+1 \,\} \\
R_0 \nabla (R_0 \sqcup_P T_0) = & \{\, z-1 \leq y \leq z+1, & z \leq x+1 \,\} \\
R_1 = & \{\, z-1 \leq y \leq z+1, y \leq x+2 & z \leq x+1 \,\} \\
T_1 = & \{\, z-1 \leq y \leq z+1, y \leq x+2, & z \leq x+2 \,\} \\
R_1 \sqcup_P T_1 = & \{\, z-1 \leq y \leq z+1, y \leq x+2, & z \leq x+2 \,\} \\
R_1 \nabla (R_1 \sqcup_P T_1) = & \{\, z-1 \leq y \leq z+1, y \leq x+2 & \,\} \\
R_2 = & \{\, z-1 \leq y \leq z+1, y \leq x+2 & z \leq x+3 \,\} \\
& \quad \vdots \qquad\qquad\qquad \vdots & \\
R_{2i} = & \{\, z-1 \leq y \leq z+1, y \leq x+2i & z \leq x+2i+1 \,\} \\
T_{2i} = & \{\, z-1 \leq y \leq z+1, y \leq x+2i, & z \leq x+2i \,\} \\
R_{2i} \sqcup_P T_{2i} = & \{\, z-1 \leq y \leq z+1, y \leq x+2i, & z \leq x+2i \,\} \\
R_{2i} \nabla (R_{2i} \sqcup_P T_{2i}) = & \{\, z-1 \leq y \leq z+1, y \leq x+2i & \,\} \\
R_{2i+1} = & \{\, z-1 \leq y \leq z+1, y \leq x+2i+2 & z \leq x+2i+1 \,\} \\
T_{2i+1} = & \{\, z-1 \leq y \leq z+1, y \leq x+2i+1, & z \leq x+2i+1 \,\} \\
R_{2i+1} \sqcup_P T_{2i+1} = & \{\, z-1 \leq y \leq z+1, y \leq x+2i+1, & z \leq x+2i+1 \,\} \\
R_{2i+1} \nabla (\ldots) = & \{\, z-1 \leq y \leq z+1, & z \leq x+2i+1 \,\}
\end{array}
$$

Closing the result of widening adds $y \leq x+2$ to R_1, which is kept when widening this state with respect to $R_1 \sqcup_P T_1$. Analogously, $z \leq x+3$ is added to R_2 and is not affected by widening in the next iteration. The lower part of the table shows the generalisation for $i \geq 1$. Since the inequalities over x, y and z, x are reintroduced alternately with increasing constants, the computation does not terminate despite widening. Note that termination is guaranteed if widening uses the previously widened state rather than a closure of it. Hence, an analysis needs to keep the previous result of widening but should evaluate the body of the loop with a copy of the state that is closed.

An alternative approach is to perform widening in such a way that all inequalities are removed whose closure could reintroduce inequalities with a different constant [9]. While more complex, such an approach has the potential of reducing the number of iterations necessary to infer a fixpoint. Unfortunately, the algorithm presented is geared towards the Octagon domain and may not carry over to the more general TVPI domain.

Our implementation therefore follows Miné's approach of storing the widened state at the head of a loop without closing it. While the system that is used to analyse the loop body could be closed, our analysis does not do so due to the lack of an implementation of the full TVPI closure. We have not found an example where this leads to a loss of precision.

This concludes the discussion of the TVPI abstract domain. We now focus on mechanisms to improve the overall precision of the analysis.

Part III

Improving Precision

11

Tracking String Lengths

Programs that provide Internet services such as email, web browsing, and remote login communicate over the network by sending streams of bytes. Most of these exchanges are interpreted as strings (sequences of characters) that denote commands or requests. Parsing these commands is a particular challenge in servers written in C since the received byte stream is retrieved in chunks, where each chunk has an explicit size. Using chunks of memory with explicit size stands in contrast to the convention of standard string functions in C, which expect the length of a string to be determined by a NUL character (a zero byte) at the end of a string. Mixing these two conventions can lead to subtle bugs in the program that do not show up until, for example, a malicious attacker sends a request string that contains a NUL character. Another example is the program presented in the introduction (Fig. 1.2), which is incorrect on many platforms yet probably works seamlessly on most inputs since characters larger than 127 are rarely encountered in text files.

To prove the memory management of server software correct, an analysis must be able to express the length of a string in both representations. While the ability to argue about explicitly sized buffers was already given in the value-range analysis presented in previous chapters, tracking the terminating NUL position requires a special abstraction. For the sake of an efficient analysis, it is not possible to model the contents of a memory buffer explicitly. Moreover, dynamically allocated buffers may be of different sizes in separate runs of a program and therefore cannot be represented explicitly. Hence, rather than inferring the possible content of a buffer, we infer information *about* its contents. Specifically, this chapter details how the position of the first NUL character in each memory region can be tracked, which allows a precise analysis of programs that make use of implicitly terminated strings. The decision to track only the first NUL position ensures that only a single variable is needed to argue about the content of an arbitrarily sized memory region. Thus, the number of variables needed to infer information about string buffers is finite.

This chapter presents the modifications to the abstract semantics necessary to track NUL positions of memory regions. It is organised as follows. Section 11.1 presents an example that copies a string using pointer operations. We deduce and solve an inequality system and generalise the approach in Sect. 11.2, where we also refine the abstraction map $\propto$. The chapter concludes with an overview of work related to string buffer analysis.

11.1 Manipulating Implicitly Terminated Strings

Parsing an incoming request for further processing is usually done by a mix of string functions from the C standard library and hand-crafted loops that iterate over incoming string buffers. The way standard string functions such as `strlen` or `strcpy` modify NUL positions can be implemented as primitives. However, the analysis of pointer accesses to buffers within loops requires that the change of the terminating NUL be updated with each pointer access. Consider the following code that uses **char**-pointers to copy a string:

```
char *p;
char t[16] = "Aero";
char *u = "plane␣";
p = t+4;
while (*p=*u) { p++; u++; };
printf("t␣=␣'%s'\n", t);
```

The purpose of the loop is to append the contents of u to t and to print the result. The call to `printf` requires that the passed-in pointer t point to a NUL -terminated string. In order to verify that no out-of-bounds access occurs during the execution of `printf`, the analysis must be able to infer that the 16-element character array that t points to contains a NUL position in one of its 16 elements. The said array is initialised by a 5-byte string constant that includes a terminating NUL character. The buffer is thereafter modified by copying characters pointed to by u, which initially points to a string buffer containing the six characters "plane␣" but is repeatedly incremented until the NUL character at `u[6]` is reached. Observe that t is not modified directly but rather through the pointer p, which is initialised to point to `t[4]`, which corresponds to the initial NUL position of the 16-element character array. Thus, during the very first execution of the loop, the NUL position of t is overwritten. During the next five iterations, t might or might not contain a NUL position. In the final iteration, the NUL position contained in the buffer pointed to by u is copied to t. The task of the analysis is to infer invariants in the form of a single polyhedron that represents these three phases of the loop and furthermore indicates that the loop terminates iff p points to `t[10]`.

All this information can be inferred automatically by manipulating an abstract variable that represents the first NUL position of t. Specifically, let

String Constants.
$[\![\ v = "c_0 c_1 \ldots c_{k-1}"\ ]\!]^\sharp_{\text{Stmt}} \langle N, A\rangle = \langle N' \rhd x_n := k, A'\rangle$
where $x_n = S(L(v))$
$\langle N', A'\rangle = \mathit{clearMem}(v, 0, k, N, A)$

Fig. 11.1. Abstract transfer function for assigning a string constant.

$S : \mathcal{A} \to \mathcal{X}$ assign a variable $x_n^a \in \mathcal{X}$ that indicates the location of the first NUL character relative to the abstract address $a \in \mathcal{A}$, where $x_n^a = S(a)$. Since each memory region $m \in \mathcal{M} \cup \mathcal{D}$ has an associated abstract address $L(m)$, a single NUL position $S(L(m))$ is manipulated for each memory region during the analysis of a program. Thus, the number of variables required to track NUL positions is proportional to the number of declared variables and summarised heap regions. Chapter 13 will detail how to reduce the number of variables representing NUL positions further. In the context of the example, let $t_n = S(L(\mathtt{t}))$ denote the first NUL position of `t`. Note that the NUL positions of the pointers `u` and `p` are irrelevant (they contain an address consisting of 4 bytes, of which some might be zero) and that the NUL position of the buffer pointed to by `u` is always 6.

11.1.1 Analysing the String Loop

We now detail how the previously described static analysis can be extended to successfully prove that the buffer `t` that is passed to `printf` is NUL terminated and hence that the `printf` statement never accesses `t` out of bounds.

In order to infer that the initial NUL position of `t` resides at index 4 (the fifth element), the abstract semantics of assigning a string constant has to be modified as shown in Fig. 11.1. As before, the content of the memory region v is cleared and the individual characters making up the string are ignored. What is retained, however, is the information that the kth byte of the memory region v is NUL by setting the variable $S(L(v))$ to k. In order to make the rest of the abstract semantics aware of NUL positions, it is now merely necessary to redefine the helper functions that access memory regions. Before we redefine the four access functions $\mathit{read}^{F,H}$, $\mathit{write}^{F,H}$, *copyMem*, and *clearMem* from Chap. 5, we illustrate the required changes to these functions by discussing the analysis of the loop in the running example.

To this end, consider the translation of the example into Core C as shown in Fig. 11.2. We associate a polyhedron with each label P, Q, R, S that represents the state space at that program location. Rather than delving into the complexity of the abstract semantics, we merely deduce an equation system from the control-flow graph of the program, which is presented in Fig. 11.3.

For simplicity, let $u, p \in \mathcal{X}$ denote the 4-byte fields of the pointers `u` and `p` that denote the offsets relative to the addresses of `s1` and `t`, respectively.

```
1  var s1 : 7, s2 : 10;
2    s1  = "plane␣";
3    s2  = "t␣=␣'%s'\n";
4  function main
5  local p : 4, q : 4, t : 16, u : 4,
6          tmp1 : 4, tmp2 : 4, tmp3 : 4;
7  P:
8    t   = "Aero";
9    u   = &s1;
10   tmp1  = &t;
11   32 p = tmp1+4;
12   jump R
13 Q:
14   32 p = p+1;
15   32 u = u+1;
16   jump R
17 R:
18   8 c = *u;
19   8 *p = c;
20   if (uint8) c!=0 then jump Q
21   jump S
22 S:
23   tmp2  = &s2;
24   tmp3  = &t;
25   printf((uint32) tmp2, (uint32) tmp3)
26   return
```

Fig. 11.2. Translation of the running example into Core C.

Furthermore, let $c \in \mathcal{X}$ denote the byte-sized field of the corresponding variable `c`. Let $s_n, t_n \in \mathcal{X}$ denote the NUL positions of the buffers `s1` and `t`, respectively. Note that $s_n = 6$ throughout the program. We assume that no fields are tracked for `s1` and `t` such that all information regarding these two buffers is contained in s_n and t_n. The flow graph features additional states, for instance T and U, which reflect the result of executing the basic blocks P and Q before they are joined to yield R. Furthermore, V represents the state at line 20 of the Core C program, before the conditional is evaluated.

The flow equations are given as follows. The initialisation statements in Fig. 11.2 determine P; in particular, $P(s_n) = \{6\}$ due to the statement `s1 = "plane␣";`, which is evaluated by the new rule for string constants in Fig. 11.1. The same rule updates P to $P' = P \rhd t_n := 4$ by evaluating `t = "Aero";`. The address assignments modify both the points-to and the numeric domains. For the remainder of the loop, the points-to domain remains unchanged, and hence we ignore any operations on it. The abstract variables

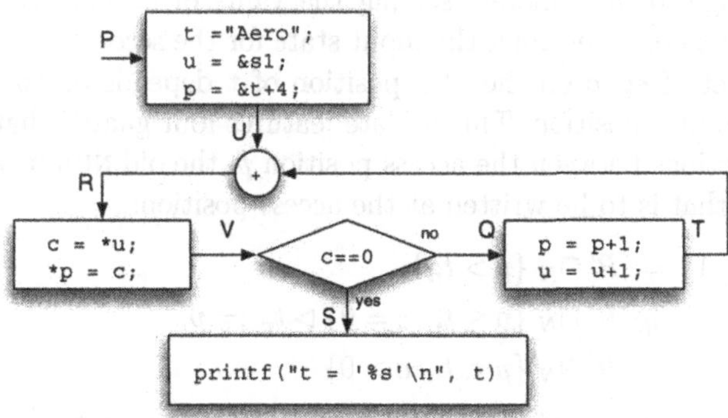

Fig. 11.3. The control-flow graph of the string-copying loop.

u and p corresponding to the pointer offsets of u and p are set to 0 and 4, respectively; thus, $U = P' \rhd u := 0 \rhd p := 4$. The polyhedron $R = U \sqcup_N T$ joins the initial state and the back arc from the loop.

In order to determine the polyhedron V before the evaluation of the conditional, the result c of reading `*u` needs to be calculated. Since the buffer `s1` pointed-to by u has no fields, the value of c is merely defined by the NUL position s_n. Assuming that the access to the buffer `s1` is within bounds, three principal accesses can occur, namely an access in front of the first NUL position if $u < s_n$, an access at the first NUL position if $u = s_n$, and an access beyond the first NUL position if $u > s_n$. In the first case, the read value must be larger than zero since any potential NUL position is located at higher indices. An access at the same index as the NUL position reads a value of 0. An access beyond the first NUL position can be NUL (if the buffer contains more than one NUL character) or any other character. Thus, in this case the read byte can be any value that can be represented as the interval $[0, 255]$.

Since the buffers `s1` and `t` contain no fields, a read access to either buffer via $read^{F,H}$ calls *prop* on an empty access tree that returns a temporary variable restricted to the unsigned range of a byte (that is, $[0, 255]$). For ease of presentation, we merge this behaviour and the refinement of this range with respect to the NUL position to define the semantic equation of the assignment statement `c=*u` as follows:

$$\begin{aligned} R' &= (\exists_c(R) \sqcap_N \{u < s_n, 1 \leq c \leq 255\}) \\ &\sqcup_N (\exists_c(R) \sqcap_N \{u = s_n, c = 0\}) \\ &\sqcup_N (\exists_c(R) \sqcap_N \{u > s_n, 0 \leq c \leq 255\}) \end{aligned}$$

The three different types of accesses are defined as a union of three different behaviours; each behaviour, in turn, is guarded by an (in)equality expressing the relation between access position and NUL position. While these guards partition the state space into three distinct regions, the intersection with the possible range for c is merely setting the value of c. The result R' of the semantics of c=*u determines the input state for the second assignment *p=c.

The effect of *p=c on the NUL position of t depends on the value c and the previous NUL position. This update features four guards that express the possible relations between the access position p, the old NUL position t_n, and the value c that is to be written at the access position:

$$\begin{aligned} V &= R' \sqcap_N \{p > t_n\} \\ &\sqcup_N R' \sqcap_N \{p \leq t_n, c = 0\} \rhd t_n := p \\ &\sqcup_N R' \sqcap_N \{p < t_n, c > 0\} \\ &\sqcup_N \exists_{t_n}(R' \sqcap_N \{p = t_n, c > 0\}) \sqcap_N \{p < t_n \leq 16\} \end{aligned}$$

The first case retains the current NUL position since the access position lies beyond t_n such that the NUL position cannot be affected. The second case sets the NUL position t_n to the access position p since a NUL value c is written in front of or on the current NUL position. In the third case, the NUL position is unaffected since a non-NUL character is written in front of the current NUL position t_n. The fourth case applies when the current NUL position is overwritten by a non-NUL character. The state that corresponds to this case, namely $R' \sqcap_N \{p = t_n, c > 0\}$, is modified by projecting out the old value of the NUL position and setting the new NUL position to be greater than the access position p (since a non-NUL character is written) but no larger than the size of the buffer. Enforcing an upper bound on t_n is done merely to improve the precision of the join operations because a join with a bounded polyhedron is often more precise. Note that this does not affect correctness: Even if the NUL position of t is modelled to be at index 16, this index is never accessed since the bound checks in $read^{F,H}$ will restrict the pointer offset p to the valid range $[0, 15]$ before considering the effect of t_n. Thus, it is irrelevant for correctness whether the NUL position t_n is restricted to 16 or unbounded.

With respect to the semantics of the conditional, the polyhedron V is now intersected with $c = 0$ and $c \neq 0$ to yield the two results of the conditional:

$$\begin{aligned} Q &= R' \sqcap_N \{c > 0\} \\ S &= R' \sqcap_N \{c = 0\} \end{aligned}$$

Here, the negation of $c = 0$ is implemented as an intersection with $c > 0$, which is acceptable since c is always in the range $[0, 255]$ and as such never wraps. Incrementing the two pointers defines the remaining state T, which is given by $T = Q \rhd p := p + 1 \rhd u := u + 1$.

Note that the evaluation of c during the write access assumes that c is within the bounds of an unsigned character. The flow equations presented later ensure that c is within bounds before the NUL position is evaluated or updated. If c is not within $[0, 255]$, the NUL position is updated conservatively.

This completes the equation system for the string-copying loop. The next section calculates a solution using a fixpoint computation.

11.1.2 Calculating a Fixpoint of the Loop

In this section, we infer a solution to the equations above. Given the initial state P, calculating U is only a matter of evaluating the equation above. For convenience, we write elements of *Num* as sets of inequalities such that $U = \{s_n = 6, t_n = 4, u = 0, p = 4\}$. In order to calculate the remaining states, a fixpoint calculation of the cycle R, V, Q, T is necessary. Using Jacobi iteration [58], this fixpoint is inferred by setting $R_0 = V_0 = S_0 = Q_0 = T_0 = \bot_N$ and calculating the iterates $R_{j+1}, S_{j+1}, V_{j+1}, Q_{j+1}$, and T_{j+1} by evaluating the equations using R_j, V_j, S_j, Q_j, and T_j on the right-hand sides. The resulting iterates are shown in Fig. 11.4. For the sake of readability, we omit $\{p = u + 4, s_n = 6\}$ from all inequality sets, as these are valid in every state. Furthermore, $P = \{s_n = 6\}$ and $U = \{s_n = 6, t_n = 4, u = 0, p = 4\}$ are constant and therefore omitted. Since U is constant, $U_i = U$ for all i and the state for R is given by $R_{i+1} = U \sqcup_N T_i$, and, since $T_0 = \bot_N$, $R_1 = U$. Note that the inequalities of states that have changed are shown in bold. The state R_1 is propagated along the control flow, leading to a new value of V. Given that $p = 4$, the omitted equality $p = u + 4$ implies that $u = 0$. Since $s_n = 6$ throughout, the definition of R' implies that c is updated to the range of a non-NUL character, hence $1 \leq c \leq 255$ appears in V_2. Furthermore, writing a non-NUL character at position $p = 4$ triggers the fourth case in the definition of V, which sets the NUL position t_n to $p < t_n \leq 16$; hence $5 \leq t_n \leq 16$ appears in V_2. The conditional does not affect this state since $c \geq 0$, and thus $Q_3 = V_2$. Incrementing both pointers retains the omitted relationship $p = u + 4$ but results in $p = 5$ appearing in the state T_4.

In the iterate $R_5 = U \sqcup_N T_4$, the inequalities $-t_n + 12p \geq 44$ (or equivalently $t_n - 12p \leq -44$) and $p \leq t_n$ are introduced through the join operation. For the sake of a more accessible illustration of inequalities that arise due to the convexity of the state space, we shall illustrate the essential updates of the table using graphs. For instance, the t_n, p projection of the state R_5 is depicted in the first graph of Fig. 11.5. The cross indicates the point $\langle 4, 4 \rangle$, contributed by U, whereas the line $[\![\{p = 5, 5 \leq t \leq 16\}]\!]$, indicated by the line segment, depicts the contribution of T_4. The grey area depicts the join of these states, which constitutes the state R_5.

Note that the most intricate join operations occur in the calculation of R' and V since these states are defined by combining the different behaviours that are chosen based on the input polyhedron. In order to see how an input polyhedron is partitioned and transformed, it is helpful to apply the rules to polyhedra that are as large as possible. To this end, we skip the discussion of iterations 6–20 and only consider how R_{21} is transformed, as this triggers the same behaviours as for R_5, albeit with a larger range for p. Indeed, illustrating the calculation with the range of $4 \leq p \leq 9$ turns out to be simpler.

j	R_j	V_j	S_j	Q_j	T_j
1	$\left\{\begin{array}{l}\mathbf{p=4,}\\ \mathbf{t_n=4}\end{array}\right\}$	$\bot_N$	$\bot_N$	$\bot_N$	$\bot_N$
2	$\left\{\begin{array}{l}p=4,\\ t_n=4\end{array}\right\}$	$\left\{\begin{array}{l}\mathbf{p=4,}\\ \mathbf{5\le t_n\le 16,}\\ \mathbf{1\le c\le 255}\end{array}\right\}$	$\bot_N$	$\bot_N$	$\bot_N$
3	$\left\{\begin{array}{l}p=4,\\ t_n=4\end{array}\right\}$	$\left\{\begin{array}{l}p=4,\\ 5\le t_n\le 16,\\ 1\le c\le 255\end{array}\right\}$	$\bot_N$	$\left\{\begin{array}{l}\mathbf{p=4,}\\ \mathbf{5\le t_n\le 16,}\\ \mathbf{1\le c\le 255}\end{array}\right\}$	$\bot_N$
4	$\left\{\begin{array}{l}p=4,\\ t_n=4\end{array}\right\}$	$\left\{\begin{array}{l}p=4,\\ 5\le t_n\le 16,\\ 1\le c\le 255\end{array}\right\}$	$\bot_N$	$\left\{\begin{array}{l}p=4,\\ 5\le t_n\le 16,\\ 1\le c\le 255\end{array}\right\}$	$\left\{\begin{array}{l}\mathbf{p=5,}\\ \mathbf{5\le t_n\le 16,}\\ \mathbf{1\le c\le 255}\end{array}\right\}$
5	$\left\{\begin{array}{l}\mathbf{4\le p\le 5,}\\ \mathbf{4\le t_n\le 16,}\\ \mathbf{p\le t_n,}\\ \mathbf{-t_n+12p\ge 44}\end{array}\right\}$	$\left\{\begin{array}{l}p=4,\\ 5\le t_n\le 16,\\ 1\le c\le 255\end{array}\right\}$	$\bot_N$	$\left\{\begin{array}{l}p=4,\\ 5\le t_n\le 16,\\ 1\le c\le 255\end{array}\right\}$	$\left\{\begin{array}{l}p=5,\\ 5\le t_n\le 16,\\ 1\le c\le 255\end{array}\right\}$
$\vdots$	$\vdots$	$\vdots$	$\vdots$	$\vdots$	$\vdots$
21	$\left\{\begin{array}{l}\mathbf{4\le p\le 9,}\\ \mathbf{4\le t_n\le 16,}\\ \mathbf{p\le t_n,}\\ \mathbf{-t_n+12p\ge 44}\end{array}\right\}$	$\left\{\begin{array}{l}4\le p\le 8,\\ 5\le t_n\le 16,\\ p<t_n,\\ 1\le c\le 255\end{array}\right\}$	$\bot_N$	$\left\{\begin{array}{l}4\le p\le 8,\\ 5\le t_n\le 16,\\ p<t_n,\\ 1\le c\le 255\end{array}\right\}$	$\left\{\begin{array}{l}5\le p\le 9,\\ 5\le t_n\le 16,\\ p\le t_n,\\ 1\le c\le 255\end{array}\right\}$
22	$\left\{\begin{array}{l}4\le p\le 9,\\ 4\le t_n\le 16,\\ p\le t_n,\\ -t_n+12p\ge 44\end{array}\right\}$	$\left\{\begin{array}{l}\mathbf{4\le p\le 9,}\\ \mathbf{5\le t_n\le 16,}\\ \mathbf{p<t_n,}\\ \mathbf{1\le c\le 255}\end{array}\right\}$	$\bot_N$	$\left\{\begin{array}{l}4\le p\le 8,\\ 5\le t_n\le 16,\\ p<t_n,\\ 1\le c\le 255\end{array}\right\}$	$\left\{\begin{array}{l}5\le p\le 9,\\ 5\le t_n\le 16,\\ p\le t_n,\\ 1\le c\le 255\end{array}\right\}$

j	R_j	V_j	S_j	Q_j	T_j
23	$\left\{\begin{array}{l} 4 \le p \le 9, \\ 4 \le t_n \le 16, \\ p \le t_n, \\ -t_n + 12p \ge 44 \end{array}\right\}$	$\left\{\begin{array}{l} 4 \le p \le 9, \\ 5 \le t_n \le 16, \\ p < t_n, \\ 1 \le c \le 255 \end{array}\right\}$	$\perp_N$	$\left\{\begin{array}{l} \mathbf{4 \le p \le 9,} \\ \mathbf{5 \le t_n \le 16,} \\ \mathbf{p < t_n,} \\ \mathbf{1 \le c \le 255} \end{array}\right\}$	$\left\{\begin{array}{l} 5 \le p \le 9, \\ 5 \le t_n \le 16, \\ p \le t_n, \\ 1 \le c \le 255 \end{array}\right\}$
24	$\left\{\begin{array}{l} 4 \le p \le 9, \\ 4 \le t_n \le 16, \\ p \le t_n, \\ -t_n + 12p \ge 44 \end{array}\right\}$	$\left\{\begin{array}{l} 4 \le p \le 9, \\ 5 \le t_n \le 16, \\ p < t_n, \\ 1 \le c \le 255 \end{array}\right\}$	$\perp_N$	$\left\{\begin{array}{l} 4 \le p \le 9, \\ 5 \le t_n \le 16, \\ p < t_n, \\ 1 \le c \le 255 \end{array}\right\}$	$\left\{\begin{array}{l} \mathbf{5 \le p \le 10,} \\ \mathbf{5 \le t_n \le 16,} \\ \mathbf{p \le t_n,} \\ \mathbf{1 \le c \le 255} \end{array}\right\}$
25	$\left\{\begin{array}{l} \mathbf{4 \le p \le 10,} \\ \mathbf{4 \le t_n \le 16,} \\ \mathbf{p \le t_n,} \\ \mathbf{-t_n + 12p \ge 44} \end{array}\right\}$	$\left\{\begin{array}{l} 4 \le p \le 9, \\ 5 \le t_n \le 16, \\ p < t_n, \\ 1 \le c \le 255 \end{array}\right\}$	$\perp_N$	$\left\{\begin{array}{l} 4 \le p \le 9, \\ 5 \le t_n \le 16, \\ p < t_n, \\ 1 \le c \le 255 \end{array}\right\}$	$\left\{\begin{array}{l} 5 \le p \le 10, \\ 5 \le t_n \le 16, \\ p \le t_n, \\ 1 \le c \le 255 \end{array}\right\}$
26	$\left\{\begin{array}{l} 4 \le p \le 10, \\ 4 \le t_n \le 16, \\ p \le t_n, \\ -t_n + 12p \ge 44 \end{array}\right\}$	$\left\{\begin{array}{l} \mathbf{4 \le p \le 10,} \\ \mathbf{5 \le t_n \le 16,} \\ \mathbf{6p + t_n \le 70,} \\ \mathbf{5p - 6t_n \le -10,} \\ \mathbf{255u + c \le 1530,} \\ \mathbf{u + 6c \ge 6,} \\ \mathbf{0 \le c \le 255} \end{array}\right\}$	$\perp_N$	$\left\{\begin{array}{l} 4 \le p \le 9, \\ 5 \le t_n \le 16, \\ p < t_n, \\ 1 \le c \le 255 \end{array}\right\}$	$\left\{\begin{array}{l} 5 \le p \le 10, \\ 5 \le t_n \le 16, \\ p \le t_n, \\ 1 \le c \le 255 \end{array}\right\}$
27	$\left\{\begin{array}{l} 4 \le p \le 10, \\ 4 \le t_n \le 16, \\ p \le t_n, \\ -t_n + 12p \ge 44 \end{array}\right\}$	$\left\{\begin{array}{l} 4 \le p \le 10, \\ 5 \le t_n \le 16, \\ 6p + t_n \le 70, \\ 5p - 6t_n \le -10, \\ 255u + c \le 1530, \\ u + 6c \ge 6, \\ 0 \le c \le 255 \end{array}\right\}$	$\left\{\begin{array}{l} \mathbf{p = 10,} \\ \mathbf{t_n = 10,} \\ \mathbf{c = 0} \end{array}\right\}$	$\left\{\begin{array}{l} \mathbf{4 \le p \le 9,} \\ \mathbf{5 \le t_n \le 16,} \\ \mathbf{p < t_n,} \\ \mathbf{1 \le c \le 255} \end{array}\right\}$	$\left\{\begin{array}{l} 5 \le p \le 10, \\ 5 \le t_n \le 16, \\ p \le t_n, \\ 1 \le c \le 255 \end{array}\right\}$

Fig. 11.4. Calculating a fixpoint of the string loop example.

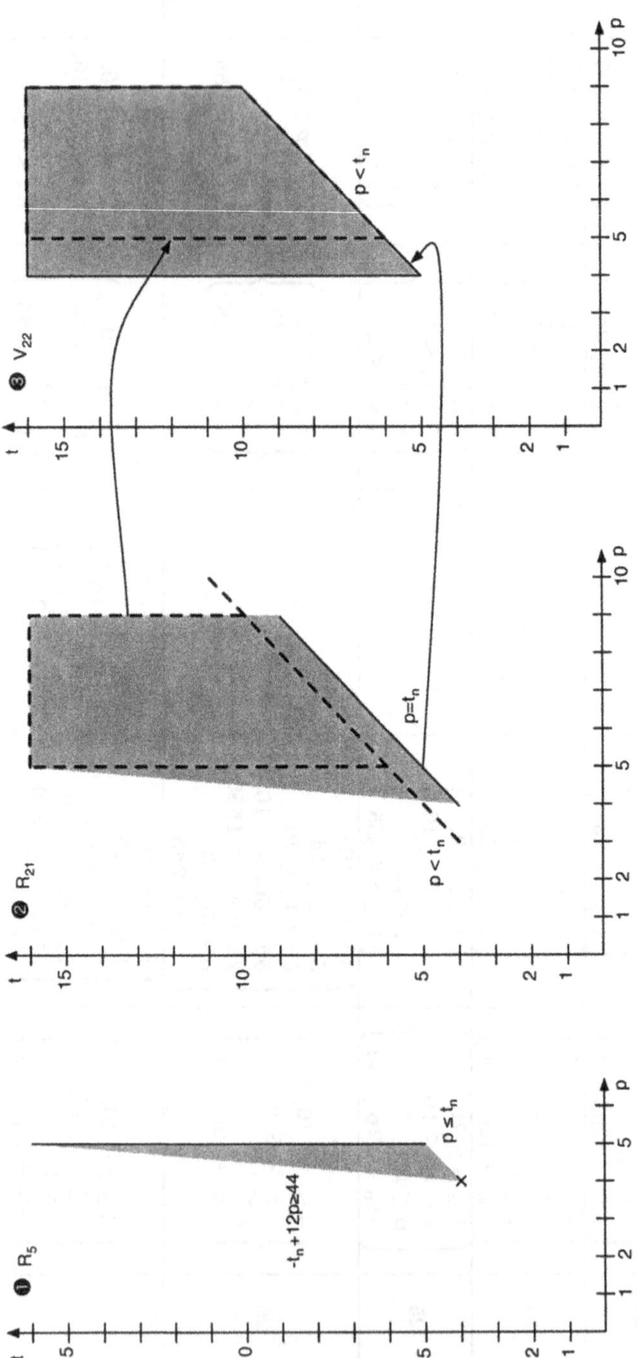

1 R5
-tn+12p≥44
p ≤ tn
2 R21
p=tn
p < tn
3 V22
p < tn

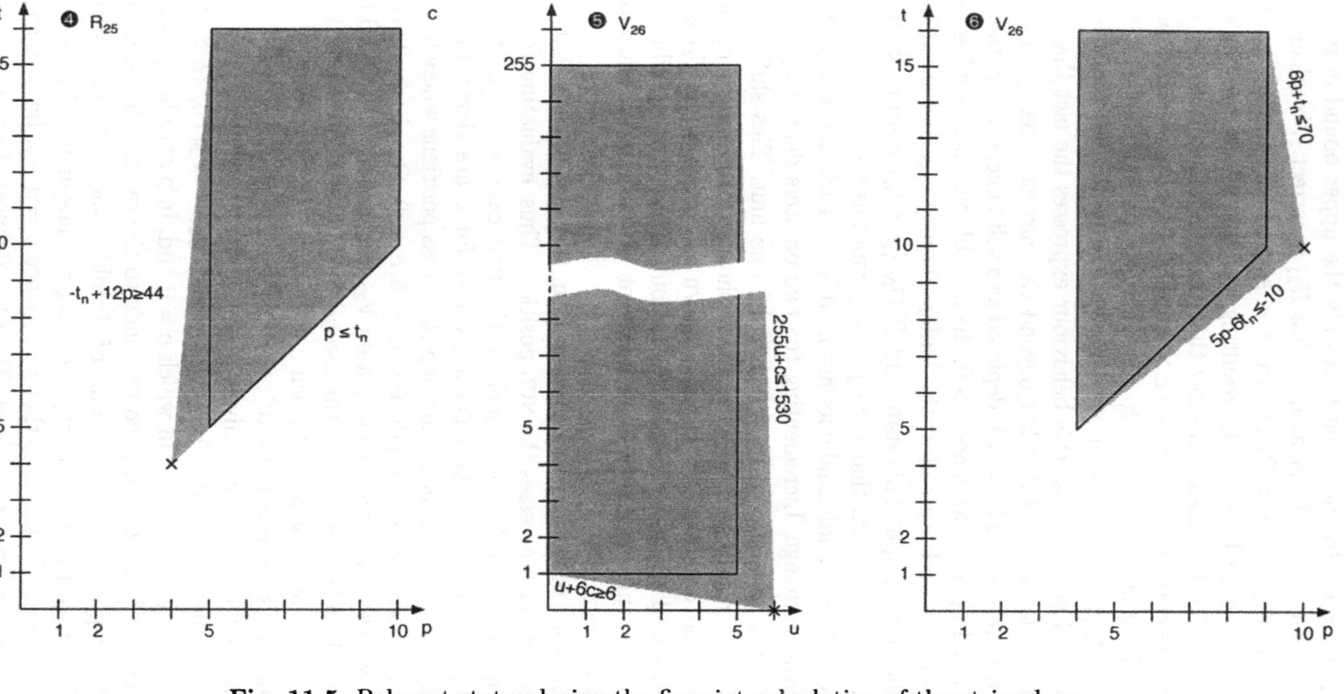

Fig. 11.5. Relevant states during the fixpoint calculation of the string loop.

The second graph in Fig. 11.5 depicts R_{21}, which corresponds to R_5 except that p is relaxed to $4 \leq p \leq 9$. Rather than arguing correctness of all iterates up to R_{21}, we show by induction that only the upper bound of p increases with each loop iteration. Thus, assume that R_{21} is correct. We calculate R' by observing that $0 \leq u \leq 5$ in R_{21}; hence, $u < s_n$ and only the first case in the definition of R' contributes to the resulting state, in which case c is set such that $1 \leq c \leq 255$. It remains to show that the two inequalities $5 \leq t_n \leq 16$ and $p < t_n$ are correct in V_{22}. Since $c > 0$ and $p \leq t_n$, only the last two cases in the definition of V contribute to the result. In particular, the state space $R' \sqcap_N \{p < t_n\}$, which is delineated by the dashed line in graphs two and three, is copied verbatim, as this behaviour expresses the fact that the NUL position t_n is not affected if it lies beyond the current access position p. In contrast, the state $R' \sqcap_N \{p = t_n\}$, depicted as a solid line, changes the current NUL position. In particular, since $c > 0$, the possible NUL position lies past the access position p and the end of the buffer, thereby creating the delineated state in the third graph. This state and $R' \sqcap_N \{p < t_n\}$ both contain $p < t_n$ (that is, $p \leq t_n - 1$), such that their join V_{22} contains $p < t_n$.

Since $c > 0$, the conditional does not modify V_{22} such that it is propagated to Q_{23} without change. Incrementing the two pointers defines T_{24}, in which the state for t_n, p is shifted to the right by one unit. This shifted state is joined with U, resulting in R_{25}, which is shown in the fourth graph. The point $\langle 4, 4\rangle$ that constitutes U is again shown as a cross, while T_{24} is shown as the delineated space. Note that T_{24} is bounded by $p \leq t_n$, which proves our claim that this inequality is an invariant of the loop, at least for loop iterations where $p \leq 10$.

At this point of the fixpoint computation, $0 \leq s \leq 6$, such that dereferencing the pointer u accesses the NUL position. Thus, evaluating R' in terms of R_{25} triggers two behaviours, namely the first case, where $u < s_n$ (that is, $u < 6$) and $u = s_n$. The resulting values for c are shown in the fifth graph of Fig. 11.5. Here, the square depicts access positions where $0 \leq u \leq 5$, $1 \leq c \leq 255$, and the cross indicates $u = 6, c = 0$. Graph six depicts the state V_{26}, which corresponds to the state V_{22} augmented with an additional behaviour, which is expressed by the second case in the definition of V. Since $p = u + 4$, it follows that $p = 10$ when $c = 0$. Thus, the NUL position t_n is set to 10, which adds the point $\langle 10, 10\rangle$, as indicated by the cross in the graph.

For the first time in the analysis of the loop, the conditional partitions the state V_{26} into $S_{27} = V_{26} \sqcap_N \{c = 0\}$ and $Q_{27} = V_{26} \sqcap_N \{c > 0\}$. The former contains a single point in which $c = 0$ and, in particular, $p = 10$. The c, u-projection of the state Q_{27} corresponds to the rectangular area denoted in the fifth graph, which is the result of shrinking the space $V_{26} \sqcap_N \{c > 0\}$ around the integral points, thereby reducing the inequality $255u + c \leq 1530$ to $u \leq 5$. Since $u \leq 5$, it follows that $p \leq 9$ since $p = u + 4$. Integral shrinking of the t_n, p projection therefore results in the delineated space of graph six, which constitutes Q_{27}. Note that the shrunk space of Q_{27} is equal to Q_{26}, so a fixpoint is reached at this point. We conclude the discussion of the example with some interesting observations on the precision of equations.

11.1.3 Prerequisites for String Buffer Analysis

Observe that a precision loss occurs when t_n is allowed to become unbounded rather than restricting it to be no larger than the size of the buffer. In order to illustrate this, reconsider graph six of Fig. 11.5. If t_n were not bounded from above in the definition of V, V_{26} would have no upper bound $t_n \leq 16$ and the join of the second behaviour of V' (which contributes the point $\langle 10, 10\rangle$) and the unbounded state would not contain the inequality $6p + t_n \leq 70$ but merely $p \leq 10$ as an upper bound. An intersection with the conditional would then result in a state Q_{27} in which $t_n \geq 10$; that is, the fact that the first NUL position is at index 10, and hence within `t`, is lost.

A more benign loss of precision can occur when naively adapting the equation to signed **char** types. In this case, the range of a non-NUL character in Core C is $-128, \ldots -1, 1, \ldots 127$. Given this range, R' can be changed to the following:

$$\begin{aligned} R' &= (\exists_c(R) \sqcap_N \{u < s_n, -128 \leq c \leq 127\}]\!]) \\ &\sqcup_N (\exists_c(R) \sqcap_N \{u = s_n, c = 0\}) \\ &\sqcup_N (\exists_c(R) \sqcap_N \{u > s_n, -128 \leq c \leq 127\}) \end{aligned}$$

Thus, an access in front of the NUL position s_n sets c to a range containing 0, and hence the read character can always be zero. This implies that the loop in the example may exit in any iteration, which is imprecise. However, even for platforms where **char** is signed, the original definition of R' is correct since the abstraction relation ensures that $[1, 255]$ maps to all non-zero sequences of 8 bits. The next section justifies the choice of using the unsigned range on all platforms.

11.2 Incorporating String Buffer Analysis

Extending the analysis presented in Chap. 5 to honour NUL positions is a straightforward task, as all memory accesses are ultimately expressed by the four functions $read^{F,H}$, $write^{F,H}$, $copyMem$, and $clearMem$. The notable exception to this is the semantics of assigning a string constant, which was already presented in Fig. 11.1. Thus, in this section we present the functions $readStr^{S,F,H}$, $writeStr^{S,F,H}$, $copyMemStr^{S}$, and $clearMemStr^{S}$, which must be substituted into the abstract semantics in Figs. 6.2, 6.4, and 6.5 in order to make the analysis aware of NUL positions. Here, the additional parameter S denotes the mapping $S : \mathcal{A} \to \mathcal{X}$ that takes each abstract address to the variable representing the NUL position, as discussed in the previous section. The

$$
\begin{aligned}
&\mathit{readStr}^{S,F,H}(m, e_o, s, N, A) = \text{if } s \neq 1 \text{ then } \mathit{read}^{F,H}(m, e_o, s, N, A) \text{ else } \langle N'', x, a\rangle\\
&\quad \text{where } \langle N', x, a\rangle = \mathit{read}^{F,H}(m, e_o, 1, N, A)\\
&\qquad x_n = S(L(m))\\
&\qquad N'' = \text{if } \mathit{inURange}(N', x, 1) \text{ then } N' \sqcap_N \{e_o < x_n, x > 0\} \sqcup_N\\
&\qquad\qquad\qquad\qquad\qquad\qquad\qquad N' \sqcap_N \{e_o = x_n, x = 0\} \sqcup_N\\
&\qquad\qquad\qquad\qquad\qquad\qquad\qquad N' \sqcap_N \{e_o > x_n\}\\
&\qquad\qquad\qquad\qquad\qquad\quad \text{else } N'
\end{aligned}
$$

$$
\begin{aligned}
&\mathit{writeStr}^{S,F,H}(m, e_o, s, e_v, a, N, A) = \langle N'', A'\rangle\\
&\quad \text{where } \langle N', A'\rangle = \mathit{write}^{F,H}(m, e_o, s, e_v, a, N, A)\\
&\qquad x_n = S(L(m))\\
&\qquad e_s = \begin{cases} x_s & \text{if } m \in \mathcal{D} \wedge H(m) = \langle x_m, x_s\rangle\\ \mathit{size}(m) & \text{if } m \in \mathcal{M}\end{cases}\\
&\qquad N'' = N' \sqcap_N \{e_o > x_n\} \sqcup_N \mathit{updateNul}(N' \sqcap_N \{e_o \leq x_n\})\\
&\qquad \mathit{updateNul}(N) =\\
&\qquad\quad \begin{cases} \mathit{writeNonNul}(N \sqcap_N \{e_v > 0\}) \sqcup_N & \\ \quad (N \sqcap_N \{e_v = 0\} \rhd x_n := e_o) & \text{if } s = 1 \wedge \mathit{inURange}(N, e_v, 1)\\ \exists_{x_n}(N) \sqcap_N \{e_o \leq x_n \leq e_s\} & \text{otherwise}\end{cases}\\
&\qquad \mathit{writeNonNul}(N) = N \sqcap_N \{e_o < x_n\} \sqcup_N\\
&\qquad\qquad\qquad\qquad\qquad (\exists_{x_n}(N \sqcap_N \{e_o = x_n\}) \sqcap_N \{e_o < x_n \leq e_s\})
\end{aligned}
$$

Fig. 11.6. Obeying NUL positions when reading from and writing to memory.

new memory access functions make use of the original functions and refine their actions if a character is accessed.

Figure 11.6 presents the $\mathit{readStr}^{S,F,H}$ function, which refines $\mathit{read}^{F,H}$. If more than one byte is accessed, the function reduces to $\mathit{read}^{F,H}$ since the NUL position cannot easily refine the value of a variable that is wider than 1 byte. In contrast, a character-sized value x can be refined if it is in the range of an unsigned byte. While this property could be enforced by wrapping the result to a **uint8**, we refrain from doing so in order not to reduce precision when the program accesses an **int8** variable. Refinements of this tactic are possible; for instance, a variable in the range $[-128, -1]$ can be interpreted as a non-zero character in addition to the range $[1, 255]$. The numeric domain N' is refined by the three cases discussed in the example. The first is guarded by $e_o < x_n$; that is, it deals with accesses in front of the NUL position x_n, in which case the value of x, which is known to be in $[0, 255]$, is restricted to $[1, 255]$. The second case deals with accesses at the NUL position and restricts the return value to zero. The third case handles accesses beyond the first NUL position, in which case no refinement of the read value is possible. Note that an access to a memory region that does not contain any field will return a temporary variable for x that is restricted to $[0, 255]$, even if the statement that triggered the access is reading an **int8** value. This can be observed in the definition of $\mathit{read}^{F,H}$, which in this case calls $\mathit{readAcc}(1, \varepsilon_{AT}, N)$, which is defined in Fig. 5.2 on p. 95. While this behaviour is a prerequisite for $\mathit{inURange}(N', x, 1)$ to hold,

$$
\begin{aligned}
&\mathit{clearMemStr}^S(m, e_o, s, N, A) = \langle \exists_{x_n}(N') \sqcap_N \{0 \leq x_n \leq e_s\}, A' \rangle \\
&\quad \text{where } \langle N', A' \rangle = \mathit{clearMem}(m, e_o, s, N, A) \\
&\qquad\quad x_n = S(L(m)) \\
&\qquad\quad e_s = \begin{cases} x_s & \text{if } m \in \mathcal{D} \wedge H(m) = \langle x_m, x_s \rangle \\ \mathit{size}(m) & \text{if } m \in \mathcal{M} \end{cases}
\end{aligned}
$$

$$
\begin{aligned}
&\mathit{copyMemStr}^S(m_1, o_1, m_2, o_2, s, N, A) = \langle N'', A' \rangle \\
&\quad \text{where } \langle N', A' \rangle = \mathit{copyMem}(m_1, o_1, m_2, o_2, s, N, A) \\
&\qquad\quad x_n^1 = S(L(m_1)) \\
&\qquad\quad x_n^2 = S(L(m_2)) \\
&\qquad\quad e_s = \begin{cases} x_s & \text{if } m_1 \in \mathcal{D} \wedge H(m_1) = \langle x_m, x_s \rangle \\ \mathit{size}(m) & \text{if } m_1 \in \mathcal{M} \end{cases} \\
&\qquad\quad N'' = \begin{cases} N' \rhd x_n^1 := x_n^2 & \text{if } o_1 = o_2 = 0 \\ \exists_{x_n^1}(N') \sqcap_N \{0 \leq x_n^1 \leq e_s\} & \text{otherwise} \end{cases}
\end{aligned}
$$

Fig. 11.7. Obeying NUL positions when clearing and copying memory regions.

it is not as specific to string buffer analysis as it may seem: Returning an unsigned range also allows for more linear relationships between overlapping fields, as illustrated in *fromLower* in the same figure.

Refining a write access to update the NUL position correctly is slightly more involved. In particular, the NUL position has to be updated even if the access is larger than a character or if the value to be written is not in the range $[0, 255]$. For this reason, the function $\mathit{writeStr}^{S,F,H}$ updates the NUL position in several stages. The first stage defines N'' and distinguishes between accesses past the current NUL position, in which case the NUL position is not affected and accesses at or in front of the current NUL position, in which case the function *updateNul* is called. The latter function distinguishes between two cases. If the written value is within the range $[0, 255]$ and a character-sized value is written, the NUL position is updated. If the written value e_v is positive, *writeNonNul* is called; if it is zero, the NUL position x_n is simply set to the access position. If the written value e_v is out-of-range of an unsigned byte or is wider than a byte, the NUL position is set to lie between the access position e_o and the size of the accessed memory region e_s, which is either a constant or the symbolic upper bound of a dynamically allocated memory region. Finally, the function *writeNonNul* is called to update the NUL position when a non-zero character is written. If the access lies in front of the current NUL position, the NUL position does not change. If the access coincides with the NUL position, it is pushed to the right; that is, the new NUL position lies beyond the current NUL position. In both cases, the NUL position is restricted to be no larger than the buffer size e_s, thereby ensuring that join operations do not lose precision due to unbounded inputs.

The two functions that handle the assignment of whole structures, namely *clearMem* and *copyMem*, have to be replaced by the NUL position-aware functions $\mathit{clearMemStr}^S$ and $\mathit{copyMemStr}^S$, shown in Fig. 11.7. The function *clearMemStr*^{S} sets the NUL position of the memory region to any possible position in the buffer and one element beyond the end of the buffer if the buffer itself contains no NUL . The *copyMemStr*^{S} function performs the same step if the offsets of the target and source region are not both zero. Otherwise, the NUL position of the source is simply assigned to the target. A more precise implementation is possible when the offsets are non-zero, but the approach presented is sufficient in practice.

The next section refines the abstraction relation to express information on NUL positions, thereby showing that our analysis can be extended in a modular way.

11.2.1 Extending the Abstraction Relation

The analysis presented in Chap. 5 uses abstract variables to express the values of fields – that is, the possible bit patterns of small, fixed consecutive regions in memory. In contrast, variables denoting NUL positions express a property of a memory region as a whole. A given memory region may contain fields as well as information in the polyhedral variable representing its NUL position. Thus, the abstraction relation $\propto$ must combine the information contained in fields with that representing the NUL positions of the variables. We alter the relation $\propto$ by redefining the concretisation map $\gamma_\rho : \mathit{Num} \times \mathit{Pts} \to \mathcal{P}(\Sigma)$. The original definition of γ_ρ was given in Section 5.4 as follows:

$$\gamma_\rho(N, A) = \{\mathit{mem}_\rho(\mathbf{v}, A) \mid \mathbf{v} \in [\![N]\!]\}$$

The set $\gamma_\rho(N, A) \subseteq \Sigma$ denotes all memory configurations for a given numeric domain $N \in \mathit{Num}$ and a points-to domain $A \in \mathit{Pts}$. Specifically, for each vector of values $\mathbf{v} \in [\![N]\!]$, the function mem_ρ is called to create the concrete memory states that correspond to the vector of values $\mathbf{v}$. In this section, we redefine γ_ρ to express the information contained in each NUL position variable $S(a)$, $a \in \mathcal{A}$. In particular, the concrete state spaces that correspond to a given vector $\mathbf{v}$ are refined by the states $\mathit{string}_\rho(\mathbf{v})$ such that the concretisation can be redefined to incorporate the information on NUL positions as follows:

$$\gamma_\rho(N, A) = \{\mathit{mem}_\rho(\mathbf{v}, A) \cap \mathit{string}_\rho(\mathbf{v}) \mid \mathbf{v} \in [\![N]\!]\}$$

Here, $\mathit{string}_\rho : \mathit{Num} \to \mathcal{P}(\Sigma)$ is defined as an intersection of all memory states, in which the NUL positions of the memory regions are set to a specific location within the region. Thus, the intersection above refines the memory states returned by mem_ρ with the additional information available from the abstract variables denoting the NUL positions. The function string_ρ itself is defined as follows:

$$\mathit{string}_\rho(\mathbf{v}) = \bigcap_{m \in \mathcal{M}} \left(\bigcap_{a \in \rho(L(m))} \{\mathit{nul}_a^{\mathit{size}(m)}(\pi_n(\mathbf{v})) \mid x_n = S(L(m))\} \right)$$

$$\cap \bigcap_{m \in \mathcal{D}} \left(\bigcap_{a \in \rho(L(m))} \left\{ nul_a^{\pi_s(\mathbf{v})}(\pi_n(\mathbf{v})) \,\middle|\, \begin{array}{l} x_n = S(L(m)) \wedge \\ H(m) = \langle x_i, x_s \rangle \end{array} \right\} \right)$$

The definition distinguishes the case of imposing a NUL position onto a declared memory region $m \in \mathcal{M}$ and onto a dynamically allocated region $m \in \mathcal{D}$. In the first case, the size *size*(m) is fixed by the program, whereas in the second case the size is determined by the current value of x_s, where x_s is the abstract variable denoting the size of the dynamic memory region m. The actual memory configurations are created by $nul_a^s(n)$, which defines a set of stores $\overline{\sigma} \subseteq \Sigma$ in which the bytes $a, \ldots (a + s - 1)$ are restricted to reflect the first NUL position at the index n. This function is defined as follows:

$$\begin{aligned} nul_a^s(n) = & \bigcap_{i \in [0, n-1], i < s} \{ bits_{a+i}^1(v) \mid v \in [1, 255] \} \\ & \cap \bigcap_{n < s} bits_{a+n}^1(0) \end{aligned}$$

The definition above uses the function $bits_{a+i}^s$ from Sect. 4.4 to create stores where the byte at address $a + i$ is restricted to $[1, 255]$, where i ranges over all indices of the buffer that lie in front of the NUL position. The second line represents the set of stores in which the byte at address $a + n$ is set to NUL. As before, we require that any intersection ranging over an empty set, or in this case an unsatisfiable condition, is equated to Σ such that the set of states is not restricted. In particular, the condition in both cases ensures that no stores are created in which bytes outside of the buffer $a, \ldots (a+s-1)$ are restricted. For instance, if the abstract state contains a state in which the NUL position lies beyond the size of the buffer, the second line reduces to an intersection with Σ.

This concludes the presentation of the refined abstraction map. We finish the chapter with a discussion of related work. The next chapter presents an extrapolation strategy that is a prerequisite for an efficient analysis of string buffers. Limitations of our analysis are presented later in Sect. 14.4.

11.3 Related Work

Wagner was the first to attempt an analysis of string buffers by tracking the NUL position within memory regions [184,185]. The implemented tool creates a constraint system from the input C program, which is then solved using intervals. Separating the C-to-constraints translation from the solving process means that pointer operations cannot be analysed precisely. The constraints generated by a pointer access depend on what fields are accessed, which in turn depends on the pointer offset, which is not known until the constraint system is solved. Hence, a precise analysis that aims to be sound with respect to pointer operations needs to alternate between generating and solving constraints. As a consequence, Wagner cannot precisely model modifications of string buffers through pointer accesses but merely models how library functions interact with the NUL position in buffers. The inability to treat pointer accesses renders his analysis unsound. Furthermore, the use of intervals in the constraint solver employed leads to imprecise results and thereby to a large number of incorrect warnings, thereby prohibiting practical applications.

Dor et al. were the first to propose explicit tracking of NUL positions in the presence of pointer operations [71]. The NUL position and the size of string buffers are stored on a per-pointer basis such that a write through a pointer p might change another pointer q if both point to the same underlying buffer. In order to argue about relationships of the NUL position between different pointers, Dor et al. introduce special *p_overlaps_q* variables to quantify the pointer offset and thereby model string length interaction. This tactic potentially increases quadratically the number of abstract variables in the number of program variables. Since their tool is based on the domain of polyhedra, the large number of variables restricts the applicability of their analysis to small examples. Worse, their analysis does not cater to weak updates; that is, it writes through pointers that can possibly point to several buffers. This renders their tool unsound.

In [167], we show that a single abstract variable is enough to infer information on the NUL position. Furthermore, we present a sound framework for analysing a subset of C. In particular, we combine the value-range analysis of variables with points-to analysis, which enables us to correctly distinguish between weak and strong updates. However, the functions presented for reading and writing buffers are given by case distinctions on how the interval that denotes the access position and the interval of the NUL position overlap, thereby losing relational information, which would lead to a precision loss when different iterates of a loop are joined. The analysis in this chapter is a refinement of this earlier work in that the language is generalised to full C, NUL positions are associated with the string buffers rather than with the pointers that point to them, and, most importantly, the abstract transfer functions are simpler and more precise by modelling NUL positions in a relational way.

Dor et al. revised their approach to an analysis called CSSV that is sound on some C programs but uses program annotations [72]. However, as in the work of Wagner, string manipulations are expected to be performed by the standard C string functions and not through direct pointer accesses.

Many tools exist that use heuristics to find buffer overflows and are thus unsound. On the positive side, tools such as Archer [189] can be applied to large code bases, as the underlying analysis method is more lightweight. The results of a heuristic analysis can be tuned between missing potential errors and reporting too many false positives. Since it is unknown how many errors these analyses miss, the precision of unsound tools is essentially incomparable. However, the authors point out that the inability to argue about NUL positions

and the use of string functions from the C library that rely on NUL positions incur a substantial precision loss.
Other unsound approaches for analysing string operations, such as LClint [115, 116], can increase their precision by incorporating user-supplied annotations.

12

Widening with Landmarks

An inherent challenge in program analysis is the requirement to reason about an arbitrary number of loop iterations in finite time. Polyhedra provide a powerful way to summarise an arbitrary number of loop iterations using a finite representation. However, in order to quickly analyse a large or potentially infinite number of iterations, special acceleration techniques are required. A formal basis for accelerating a fixpoint computation is the widening/narrowing approach that was developed in the context of abstract interpretation [59,62].

12.1 An Introduction to Widening/Narrowing

In order to illustrate the widening/narrowing approach on the domain of polyhedra and to discuss the implications of applying narrowing in an analyser, consider the control-flow graph of `for (i=0; i<100; i++) {/*empty*/}`:

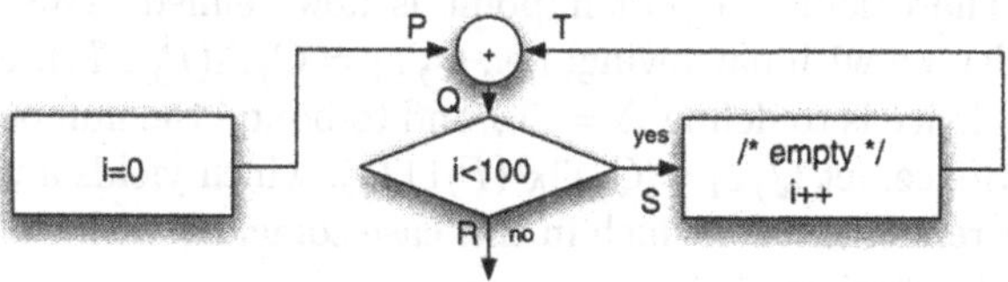

The analysis amounts to characterising the values that can arise on the edges of the control-flow graph. For simplicity, assume that each edge is decorated with an element of the numeric domain *Num* rather than a tuple $\langle N, A\rangle$ of numeric and points-to domains. Given that the program contains only a single variable i, the polyhedra $P, Q, R, S, T \in Num$ are represented as possibly unbounded intervals over integers. In the example, the polyhedron $P = \{0 \leq i \leq 0\}$ describes the value of i at the beginning of the program. The $+$ node joins the polyhedra P and T to obtain $Q = P \sqcup_N T$. Due to the integrality of i, the polyhedra that characterise the two outcomes of the test

j	P_j	Q_j	R_j	S_j	T_j
1	$\mathbf{[0,0]}$	$\bot_N$	$\bot_N$	$\bot_N$	$\bot_N$
2	$[0,0]$	$\mathbf{[0,0]}$	$\bot_N$	$\bot_N$	$\bot_N$
3	$[0,0]$	$[0,0]$	$\bot_N$	$\mathbf{[0,0]}$	$\bot_N$
4	$[0,0]$	$[0,0]$	$\bot_N$	$[0,0]$	$\mathbf{[1,1]}$
5	$[0,0]$	$\mathbf{[0,\infty]}$	$\bot_N$	$[0,0]$	$[1,1]$
6	$[0,0]$	$[0,\infty]$	$\mathbf{[100,\infty]}$	$\mathbf{[0,99]}$	$[1,1]$
7	$[0,0]$	$[0,\infty]$	$[100,\infty]$	$[0,99]$	$\mathbf{[1,100]}$
8	$[0,0]$	$[0,\infty]$	$[100,\infty]$	$[0,99]$	$[1,100]$
$1'$	$[0,0]$	$\mathbf{[0,100]}$	$[100,\infty]$	$[0,99]$	$[1,100]$
$2'$	$[0,0]$	$[0,100]$	$\mathbf{[100,100]}$	$[0,99]$	$[1,100]$

Fig. 12.1. Calculating the fixpoint of a simple **for**-loop using Jacobi iteration.

$i < 100$ are $R = Q \sqcap_N \{i \geq 100\}$ and $S = Q \sqcap_N \{i \leq 99\}$. The last polyhedron T is characterised by the update $T = S \triangleright i := i + 1$.

A solution to the equations can be found by applying Jacobi iteration [58], which calculates new polyhedra $P_{j+1}, Q_{j+1}, R_{j+1}, S_{j+1}, T_{j+1}$ from the polyhedra of the previous iteration P_j, Q_j, R_j, S_j, T_j. To ensure rapid convergence, a widening point must be inserted into the Q, S, T cycle. Widening at Q amounts to replacing the equation for Q with $Q_{j+1} = Q_j \nabla (P_j \sqcup T_j)$, where ∇ is a widening operator that removes unstable bounds [59]. The possible values of i are given in Fig. 12.1 as intervals; the updated entries are shown in bold.

In iteration 5, the output of the + node is $P_4 \sqcup_N T_4 = \{0 \leq i \leq 1\}$. The widening operator compares $P_4 \sqcup_N T_4$ against $Q_4 = \{0 \leq i \leq 0\}$ and removes the unstable upper bound, yielding $Q_5 = \{0 \leq i\}$. Stability is reached in iteration 8. The calculated post-fixpoint is now refined. This is realised by replacing widening with narrowing; i.e., $Q_{j+1} = Q_j \Delta (P_j \sqcup T_j)$. For polyhedra, the common choice is to define $\Delta = \sqcap_N$ and to bound the number of iterations [59, p. 290]. Hence, let $Q_{j+1} = Q_j \sqcap_N (P_j \sqcup T_j)$, which yields a refined state $1'$ and a further refinement $2'$, which in this case coincides with the least fixpoint of the original equations.

12.1.1 The Limitations of Narrowing

In order to illustrate one drawback of narrowing, consider a reanalysis of the example above where the widening is applied on S rather than on Q. Thus, let $S_{j+1} = S_j \nabla (Q_i \sqcap_N \{i \leq 99\})$. Figure 12.2 shows a difference after iteration 4.

In the first analysis, only the polyhedra Q and R are larger before narrowing commences. In the second analysis, S and T are also larger before narrowing. In order to illustrate the impact of this in the context of verification of memory accesses, suppose `/*empty*/` is replaced by `b = array[i]`, where

j	P_j	Q_j	R_j	S_j	T_j
1	$\mathbf{[0,0]}$	$\bot_N$	$\bot_N$	$\bot_N$	$\bot_N$
2	$[0,0]$	$\mathbf{[0,0]}$	$\bot_N$	$\bot_N$	$\bot_N$
3	$[0,0]$	$[0,0]$	$\bot_N$	$\mathbf{[0,0]}$	$\bot_N$
4	$[0,0]$	$[0,0]$	$\bot_N$	$[0,0]$	$\mathbf{[1,1]}$
5	$[0,0]$	$\mathbf{[0,1]}$	$\bot_N$	$[0,0]$	$[1,1]$
6	$[0,0]$	$[0,1]$	$\bot_N$	$\mathbf{[0,\infty]}$	$[1,1]$
7	$[0,0]$	$[0,1]$	$\bot_N$	$[0,\infty]$	$\mathbf{[1,\infty]}$
8	$[0,0]$	$\mathbf{[0,\infty]}$	$\bot_N$	$[0,\infty]$	$[1,\infty]$
9	$[0,0]$	$[0,\infty]$	$\mathbf{[100,\infty]}$	$[0,\infty]$	$[1,\infty]$
10	$[0,0]$	$[0,\infty]$	$[100,\infty]$	$[0,\infty]$	$[1,\infty]$
1′	$[0,0]$	$[0,\infty]$	$[100,\infty]$	$\mathbf{[0,99]}$	$[1,\infty]$
2′	$[0,0]$	$[0,\infty]$	$[100,\infty]$	$[0,99]$	$\mathbf{[1,100]}$
3′	$[0,0]$	$\mathbf{[0,100]}$	$[100,\infty]$	$[0,99]$	$[1,100]$
4′	$[0,0]$	$[0,100]$	$\mathbf{[100,100]}$	$[0,99]$	$[1,100]$

Fig. 12.2. The placement of the widening point affects the fixpoint calculation.

`array` has 100 elements. To avoid an avalanche of false warning messages, it is common practice to intersect S with the legal range of the index i [30], in this case $0 \leq i \leq 99$, yielding the polyhedron S', and thereafter use S' instead of S. Moreover, since the out-of-bounds check amounts to the subsumption test $S \not\sqsubseteq_N S'$, it is straightforward to perform the check during fixpoint calculation; the test could be postponed until a fixpoint is reached, but this would require S' to be recalculated unnecessarily. However, this technique does not combine well with narrowing since a warning is issued if S is nominated for widening rather than Q, and thus the placement of the widening point can determine whether a warning is issued or not.

Another implication of reducing a post-fixpoint with narrowing relates to the way separate domains interact. Assume that the array above is embedded into a C structure declared as `struct { int[100] array; int* p } s;` and that the loop body is changed to `b = s.array[i]`. Consider again the second analysis, in which S is widened to $\{0 \leq i\}$ so that the upper bound of the array index i is temporarily lost. In this case, a points-to analysis would generate a spurious l-value flow from `s.p` to `b`. Once narrowing infers $0 \leq i < 100$, it is desirable to remove this spurious flow. However, points-to analyses are typically formulated in terms of either closure operations [99] or union-find algorithms [176], none of which support the removal of flow information. Thus, even if narrowing can recover precision in one domain, the knock-on precision loss induced in other domains may be irrecoverable.

Furthermore, narrowing on polyhedra [59] cannot recover precision if the loop invariant is expressed as a disequality [30]. For instance, narrowing has no effect if the loop invariant in the example is changed from `i<100` to the semantically equivalent test `i!=100`. Since it is unrealistic to modify the program under test, narrowing is unable to recover precision in programs with disequalities as loop conditions.

12.1.2 Improving Widening and Removing Narrowing

Rather than recovering inequalities through narrowing that were widened away, our contribution is to use unsatisfiable inequalities as oracles to guide the fixpoint acceleration. Specifically, we propose widening with landmarks, which records inequalities that were found to be unsatisfiable in two consecutive iterates. We then extrapolate to the first iterate that makes any of these inequalities satisfiable. If this extrapolation is not a fixpoint, we continue until no unsatisfiable inequalities remain, at which point standard widening is applied [10, 91]. The rationale for observing unsatisfiable inequalities is that the transition from unsatisfiable to satisfiable indicates a change in the behaviour of a program. Widening with landmarks is similar in spirit to widening with thresholds [30]. In this related approach, the value of an unstable variable is extrapolated to the next threshold from a set of user-supplied values. Rather than guiding widening with thresholds on individual variables, our approach automatically observes linear inequalities that arise during the analysis of a program, which are then used to bound the degree of extrapolation. As a consequence, no narrowing is needed once a fixpoint is reached, which simplifies the implementation of a fixpoint engine considerably.

The remainder of this chapter is organised as follows. Section 12.2 details an example motivated by the last chapter on iterating over a NUL-terminated string, thereby presenting the ideas behind widening with landmarks. Sections 12.3 and 12.4 formalise the notion of landmarks, which are used in Sect. 12.5 to define an extrapolation strategy. We discuss our approach and related work in Sect. 12.6.

12.2 Revisiting the Analysis of String Buffers

In this section, we explain the ideas behind widening with landmarks. Consider the task of advancing a pointer `s` to the end of a string buffer by executing the loop `while (*s) s++;`. For ease of presentation, we expand the loop as follows:

```
char s[32] = "the␣string";
int i = 0;
while (true) {
  c = s[i];
  if (c==0) break;
  i = i+1;
};
```

The task is to check that the string buffer `s` is only accessed within bounds. This program is challenging for automatic verification because the loop invariant is always satisfied and the extra exit condition within the loop does not mention the loop counter `i`. In order to simplify the presentation, we merely show the polyhedral operations resulting from applying the abstract semantics. In particular, let n represent the index of the first NUL position in `s`. The control-flow graph of the string buffer example is decorated with polyhedra P, Q, R, S, T, U as follows:

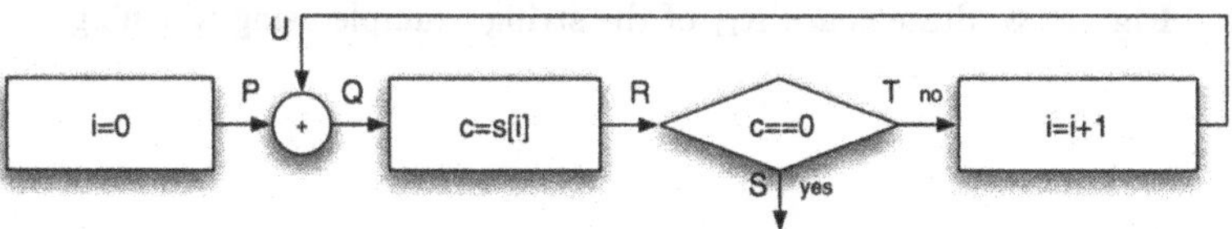

The initial values of the program variables are described by $P = \{i = 0, n = 10\}$. The merge of this polyhedron and the polyhedron on the back edge, U, defines $Q = P \sqcup_N U$. To verify that the array access `s[i]` is within bounds, we compute $Q' = Q \sqcap_N \{0 \leq i \leq 31\}$ and issue a warning if $Q \not\sqsubseteq_N Q'$. The analysis continues under the premise that the access was within bounds and hence R is defined in terms of Q' rather than Q, following the definition of $readStr^{S,F,H}$ in Fig. 11.6:

$$\begin{aligned} R = & (\exists_c(Q') \sqcap_N \{i < n, 1 \leq c \leq 255\}) \\ & \sqcup (\exists_c(Q') \sqcap_N \{i = n, c = 0\}) \\ & \sqcup (\exists_c(Q') \sqcap_N \{i > n, 0 \leq c \leq 255\}) \end{aligned}$$

By assuming that the string `s` has no fields, the corresponding empty access tree will return c such that $0 \leq c \leq 255$. This range is thereafter refined by $readStr^{S,F,H}$, depending on where the read access occurs. The definition of R above summarises these steps by removing all information pertaining to c in Q' and restricting c to the desired range. Specifically, the value of c is restricted to $[1, 255]$ if $i < n$ and is set to 0 if $i = n$ and to $[0, 255]$ if $i > n$. The last three equations that comprise the system are given by the following definitions:

$$\begin{aligned} S &= R \sqcap_N \{c = 0\} \\ T &= (R \sqcap_N \{c < 0\}) \sqcup_N (R \sqcap_N \{c > 0\}) \\ U &= T \triangleright i := i + 1 \end{aligned}$$

Note that the disequality check `c!=0` is implemented by the join of the two state spaces $R \sqcap_N \{c < 0\}$ and $R \sqcap_N \{c > 0\}$ and that $R \sqcap_N \{c < 0\} = \bot_N$ in this example, as the definition of R confines c to the range $[0, 255]$. Now consider using widening to calculate a fixpoint of this system.

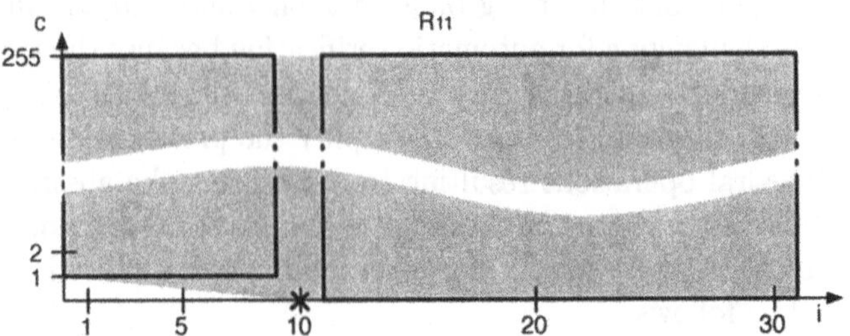

Fig. 12.3. State space R_{11} of the string example using widening.

12.2.1 Applying the Widening/Narrowing Approach

As before, we solve these equations iteratively, nominating Q as the widening point to ensure convergence in the cycle Q, R, T, U. Thus, when the equations are reinterpreted iteratively, the equation Q is replaced with $Q_{j+1} = Q_j \nabla (P_j \sqcup_N U_j)$. Applying the standard widening/narrowing approach results in the iterates shown in Fig. 12.4. Again, we apply widening when Q is evaluated the third time, so that widening is applied on Q_9 and $P_9 \sqcup_N U_9$ to obtain Q_{10}. The resulting polyhedron $Q_{10} = [\![\{0 \leq i\}]\!]$ is intersected with the verification condition to yield $Q'_{10} = [\![\{0 \leq i \leq 31\}]\!]$, thereby raising a warning since $Q_{10} \neq Q'_{10}$. Before proceeding to the evaluation of R_{11}, observe that $\exists_c(Q'_j) = Q'_j$ in all iterations j since P_j does not constrain c and consequently neither does $Q_j = U_j \sqcup_N P_j$. Given that Q'_{10} allows i to take on any value in $[0, 31]$, the three cases in the definition of R, which are guarded by $i < n$, $i = n$, and $i > n$, all contribute to the result R_{11}. This result is depicted as the grey region in Fig. 12.3, which shows the relationship between i and c. The three regions whose join forms the polyhedron R_{11} are marked with two rectangles and a small cross for the $c = 0$ case. Observe that applying narrowing (that is, replacing $Q_{j+1} = Q_j \nabla (U_j \sqcup_N P_j)$ with $Q_{j+1} = Q_j \Delta (U_j \sqcup_N P_j)$) yields another iterate $1'$ in which the value of i ranges over $[0, 32]$, which still violates the array bound check since $Q'_{1'} \neq Q_{1'}$, where $Q'_{1'} = Q_{1'} \sqcap_N \{0 \leq i \leq 31\}$ corresponds to $Q_{1'}$ restricted to valid array indices.

12.2.2 The Rationale behind Landmarks

Consider the same fixpoint calculation using widening with landmarks as shown in Fig. 12.6. We omit the first nine iterates before widening is applied since they coincide with those given in Fig. 12.4. While landmarks are gathered throughout the fixpoint calculation, we focus on the calculation of the polyhedron R, as it gives rise to the only landmarks that are of relevance in this example. The three graphs in Fig. 12.5 depict the relation between i and c in the polyhedra R_3, R_7, R_{11}, which are the three iterates in which R_j

j	Q_j	R_j	S_j	T_j	U_j
1	$\bot_N$	$\bot_N$	$\bot_N$	$\bot_N$	$\bot_N$
2	$\{\mathbf{0 \le i \le 0}\}$	$\bot_N$	$\bot_N$	$\bot_N$	$\bot_N$
3	$\{0 \le i \le 0\}$	$\{\mathbf{0 \le i \le 0, 1 \le c \le 255}\}$	$\bot_N$	$\bot_N$	$\bot_N$
4	$\{0 \le i \le 0\}$	$\{0 \le i \le 0, 1 \le c \le 255\}$	$\bot_N$	$\{\mathbf{0 \le i \le 0, 1 \le c \le 255}\}$	$\bot_N$
5	$\{0 \le i \le 0\}$	$\{0 \le i \le 0, 1 \le c \le 255\}$	$\bot_N$	$\{0 \le i \le 0, 1 \le c \le 255\}$	$\{\mathbf{1 \le i \le 1, 1 \le c \le 255}\}$
6	$\{\mathbf{0 \le i \le 1}\}$	$\{0 \le i \le 0, 1 \le c \le 255\}$	$\bot_N$	$\{0 \le i \le 0, 1 \le c \le 255\}$	$\{1 \le i \le 1, 1 \le c \le 255\}$
7	$\{0 \le i \le 1\}$	$\{\mathbf{0 \le i \le 1, 1 \le c \le 255}\}$	$\bot_N$	$\{0 \le i \le 0, 1 \le c \le 255\}$	$\{1 \le i \le 1, 1 \le c \le 255\}$
8	$\{0 \le i \le 1\}$	$\{0 \le i \le 1, 1 \le c \le 255\}$	$\bot_N$	$\{\mathbf{0 \le i \le 1, 1 \le c \le 255}\}$	$\{1 \le i \le 1, 1 \le c \le 255\}$
9	$\{0 \le i \le 1\}$	$\{0 \le i \le 1, 1 \le c \le 255\}$	$\bot_N$	$\{0 \le i \le 1, 1 \le c \le 255\}$	$\{\mathbf{1 \le i \le 2, 1 \le c \le 255}\}$
10	$\{\mathbf{0 \le i}\}$	$\{0 \le i \le 1, 1 \le c \le 255\}$	$\bot_N$	$\{0 \le i \le 1, 1 \le c \le 255\}$	$\{1 \le i \le 2, 1 \le c \le 255\}$
11	$\{0 \le i\}$	$\{\mathbf{0 \le i \le 31, 0 \le c \le 255, i + 10c \ge 10}\}$	$\bot_N$	$\{0 \le i \le 1, 1 \le c \le 255\}$	$\{1 \le i \le 2, 1 \le c \le 255\}$
12	$\{0 \le i\}$	$\{0 \le i \le 31, 0 \le c \le 255, i + 10c \ge 10\}$	$\{\mathbf{10 \le i, i \le 31, 0 \le c \le 0}\}$	$\{\mathbf{0 \le i \le 31, 1 \le c \le 255}\}$	$\{1 \le i \le 2, 1 \le c \le 255\}$
13	$\{0 \le i\}$	$\{0 \le i \le 31, 0 \le c \le 255, i + 10c \ge 10\}$	$\{10 \le i, i \le 31, 0 \le c \le 0\}$	$\{0 \le i \le 31, 1 \le c \le 255\}$	$\{\mathbf{1 \le i \le 32, 1 \le c \le 255}\}$
14	$\{0 \le i\}$	$\{0 \le i \le 31, 0 \le c \le 255, i + 10c \ge 10\}$	$\{10 \le i, i \le 31, 0 \le c \le 0\}$	$\{0 \le i \le 31, 1 \le c \le 255\}$	$\{1 \le i \le 32, 1 \le c \le 255\}$
1′	$\{\mathbf{0 \le i \le 32}\}$	$\{0 \le i \le 31, 0 \le c \le 255, i + 10c \ge 10\}$	$\{10 \le i, i \le 31, 0 \le c \le 0\}$	$\{0 \le i \le 31, 1 \le c \le 255\}$	$\{1 \le i \le 31, 1 \le c \le 255\}$

Fig. 12.4. Fixpoint calculation of the string loop. The column P_j is omitted since $P_j = \{i = 0\}$ for all iterations j. Furthermore, we omit $n = 10$ from all polyhedra.

changes. The polyhedron R_3 is derived from $\exists_c(Q'_3) = Q'_3 = \{i = 0\}$. During this computation, Q'_3 is intersected with $\{i < n, 1 \le c \le 255\}$, $\{i = n, c = 0\}$, and $\{i > n, 0 \le c \le 255\}$, which represent three different behaviours of the program. As the fixpoint calculation progresses, polyhedra grow and new behaviours are incrementally enabled. A behaviour can only change from being disabled to being enabled when one of its constituent inequalities is initially unsatisfiable and becomes satisfiable. A fixpoint may exist in which not all behaviours of a program are enabled; that is, these behaviours still contain

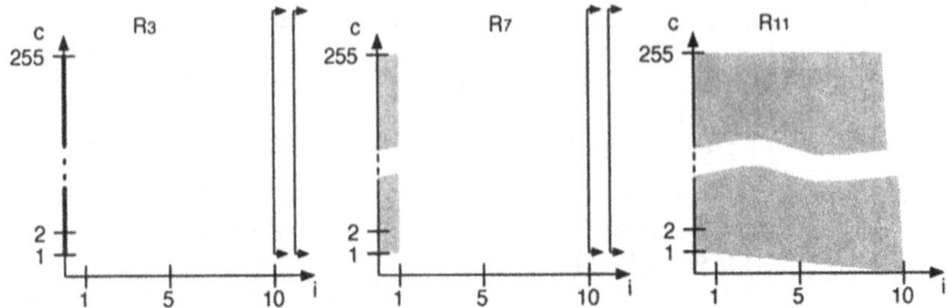

Fig. 12.5. State space R_{11} of the string example using widening with landmarks.

j	Q_j	R_j	S_j	T_j	U_j
10	$\left\{\begin{array}{l}\mathbf{0 \leq i,} \\ \mathbf{i \leq 10}\end{array}\right\}$	$\left\{\begin{array}{l}0 \leq i \leq 1, \\ 1 \leq c \leq 255\end{array}\right\}$	$\bot_N$	$\left\{\begin{array}{l}0 \leq i \leq 1, \\ 1 \leq c \leq 255\end{array}\right\}$	$\left\{\begin{array}{l}1 \leq i \leq 2, \\ 1 \leq c \leq 255\end{array}\right\}$
11	$\left\{\begin{array}{l}0 \leq i, \\ i \leq 10\end{array}\right\}$	$\left\{\begin{array}{l}\mathbf{0 \leq i,} \\ \mathbf{c \leq 255,} \\ \mathbf{255i + c \leq 2550,} \\ \mathbf{i + 10c \geq 10}\end{array}\right\}$	$\bot_N$	$\left\{\begin{array}{l}0 \leq i \leq 1, \\ 1 \leq c \leq 255\end{array}\right\}$	$\left\{\begin{array}{l}1 \leq i \leq 2, \\ 1 \leq c \leq 255\end{array}\right\}$
12	$\left\{\begin{array}{l}0 \leq i, \\ i \leq 10\end{array}\right\}$	$\left\{\begin{array}{l}0 \leq i, \\ c \leq 255, \\ 255i + c \leq 2550, \\ i + 10c \geq 10\end{array}\right\}$	$\left\{\begin{array}{l}\mathbf{10 \leq i,} \\ \mathbf{i \leq 10,} \\ \mathbf{0 \leq c,} \\ \mathbf{c \leq 0}\end{array}\right\}$	$\left\{\begin{array}{l}\mathbf{0 \leq i \leq 9,} \\ \mathbf{1 \leq c \leq 255}\end{array}\right\}$	$\left\{\begin{array}{l}1 \leq i \leq 2, \\ 1 \leq c \leq 255\end{array}\right\}$
13	$\left\{\begin{array}{l}0 \leq i, \\ i \leq 10\end{array}\right\}$	$\left\{\begin{array}{l}0 \leq i, \\ c \leq 255, \\ 255i + c \leq 2550, \\ i + 10c \geq 10\end{array}\right\}$	$\left\{\begin{array}{l}10 \leq i, \\ i \leq 10, \\ 0 \leq c, \\ c \leq 0\end{array}\right\}$	$\left\{\begin{array}{l}0 \leq i \leq 9, \\ 1 \leq c \leq 255\end{array}\right\}$	$\left\{\begin{array}{l}\mathbf{1 \leq i \leq 10,} \\ \mathbf{1 \leq c \leq 255}\end{array}\right\}$
14	$\left\{\begin{array}{l}0 \leq i, \\ i \leq 10\end{array}\right\}$	$\left\{\begin{array}{l}0 \leq i, \\ c \leq 255, \\ 255i + c \leq 2550, \\ -i - 10c \leq -10\end{array}\right\}$	$\left\{\begin{array}{l}10 \leq i, \\ i \leq 10, \\ 0 \leq c, \\ c \leq 0\end{array}\right\}$	$\left\{\begin{array}{l}0 \leq i \leq 9, \\ 1 \leq c \leq 255\end{array}\right\}$	$\left\{\begin{array}{l}1 \leq i \leq 10, \\ 1 \leq c \leq 255\end{array}\right\}$

Fig. 12.6. Fixpoint calculation using widening with landmarks.

unsatisfiable inequalities. The rationale for widening with landmarks is to find these fixpoints by systematically considering the inequalities that prevent a behaviour from being enabled. These inequalities are exactly those inequalities in the semantic equations that are unsatisfiable in the context of the current iterate. In the example, the last two behaviours contain the inequalities $n \leq i$ (arising from $i = n$) and $n < i$ that are responsible for enabling the second and third behaviours. These inequalities are unsatisfiable in Q'_3 and are therefore stored as landmarks. The leftmost graph in Fig. 12.5 indicates the position of the two inequalities $n \leq i$ and $n < i$, which define the landmarks we record for R_3.

12.2.3 Creating Landmarks for Widening

A landmark is a triple comprised of an inequality and two distances. On creation, the first distance is set to the shortest straight-line distance to which the inequality must be translated so as to touch the current iterate. In this example, translations by 10 and 11 units are required for $n \leq i$ and $n < i$, respectively, to touch R_3.

In the seventh iteration, when R_j is updated again, a second measurement is taken between the inequality and the new iterate. This distance is recorded as the second distance in the existing landmark. In the example, the second distance for $n \leq i$ and $n < i$ is set to 9 and 10 units, respectively.

By iteration 8, both landmarks have acquired a second measurement; however, it is not until widening is applied in iteration 10 that the landmarks are actually used. The difference between the two measurements of a particular landmark indicates how fast the iterates R_j are approaching the as yet unsatisfiable inequality of that landmark. From this difference, we estimate how many times R_j must be updated until the inequality becomes satisfiable. In the example, the difference in distance between the two updates R_3 and R_7 is one unit for each landmark. Thus, at this rate, R_j would be updated nine more times until the closer inequality, namely $n \leq i$, becomes satisfied. Rather than calculating all these intermediate iterates, we use this information to perform an extrapolation step when the widening point Q is revisited.

12.2.4 Using Landmarks in Widening

From the perspective of the widening operator, the task is, firstly, to gather all landmarks that have been generated in the traversal of the cycle in which the widening operator resides. Secondly, the widening operator ranks the landmarks by the number of iterations needed for the corresponding inequality to become satisfied. Thirdly, the landmark with the smallest rank determines the amount of extrapolation the widening operator applies. In the example, recall that the unsatisfiable inequality $n \leq i$ in R_7 would become satisfiable after nine more updates of R, whereas the other unsatisfiable inequality $n < i$ becomes satisfiable after ten updates. Hence, $n \leq i$ constitutes the nearest inequality and, rather than applying widening when calculating $Q_{10} = Q_9 \nabla (P_9 \sqcup_N U_9)$, extrapolation is performed. Specifically, the changes between $Q_9 = \{0 \leq i \leq 1\}$ and $P_9 \sqcup_N U_9 = \{0 \leq i \leq 2\}$ are extrapolated nine times to yield $Q_{10} = \{0 \leq i \leq 10\}$. The new value of Q_{10} forces a reevaluation of R, yielding R_{11}, as shown in Fig. 12.6. In the next iteration, the semantic equation for T yields $\{0 \leq i, 1 \leq c \leq 255, 255i + c \leq 2550\}$. By applying integral tightening techniques discussed in Chap. 8, this polyhedron is refined to the entry T_{12}, as shown in the table. A final iteration leads to a fixpoint.

Note that it is possible to apply extrapolation as soon as a single landmark acquires its second measurement. However, to ensure that the state is extrapolated only to the point where the first additional behaviour is enabled,

Algorithm 10 Adding or tightening a landmark.

procedure *updateLandmark*(N, ι, L), $N \in \mathit{Num}, \iota \in \mathit{Ineq}, L \subseteq \mathit{Lin} \times \mathbb{Z} \times (\mathbb{Z} \cup \{\infty\})$

1: $e \leq c \leftarrow \iota$
2: $c' \leftarrow \mathit{minExp}(N, e)$
3: **if** $c < c'$ **then** /* $N \sqcap_N \{\iota\}$ *is empty* */
4:     $\mathit{dist} \leftarrow c' - c$ /* *calculate the distance between* N *and* $e = c$ */
5:     **if** $\exists \mathit{dist}_c, \mathit{dist}_p \,.\, \langle e, \mathit{dist}_c, \mathit{dist}_p \rangle \in L$ **then**
6:         **return** $(L \setminus \{\langle e, \mathit{dist}_c, \mathit{dist}_p \rangle\}) \cup \{\langle e, \min(\mathit{dist}, \mathit{dist}_c), \mathit{dist}_p \rangle\}$
7:     **else**
8:         **return** $L \cup \{\langle e, \mathit{dist}, \infty \rangle\}$
9:     **end if**
10: **end if**
11: **return** L

the extrapolation step should be deferred until all landmarks have acquired their second value; that is, when no new landmarks were created in the last iteration. Observe that new landmarks cannot be added indefinitely, as at most one landmark is created for each inequality that occurs in the semantic equations, which are in turn finite.

The following sections formalise these ideas by presenting algorithms for gathering landmarks and performing extrapolation using landmarks.

12.3 Acquiring Landmarks

This section formalises the intuition behind widening with landmarks by giving a more algorithmic description of how landmarks are acquired. Algorithm 10 presents *updateLandmark*, which is invoked whenever a state N is intersected with an inequality ι that arises from a semantic equation. In line 2, the distance between the boundary of ι and N is measured by calculating $c' = \mathit{minExp}(N, e)$. Intuitively, $e \leq c'$ is a parallel translation of ι that has a minimal intersection with N. Line 3 compares the relative location of ι and its translation, thereby ensuring that lines 4 to 9 are only executed if ι is unsatisfiable and thus can yield a landmark. If ι is indeed unsatisfiable, line 4 calculates its distance to N. Given this distance, line 5 determines if a landmark is to be updated or created. An update occurs whenever different semantic equations contain the same unsatisfiable inequality. In this case, line 6 ensures that the smaller distance is stored in the landmark. The rationale for storing the smaller distance is to choose the closer inequality as the landmark. In particular, if extrapolation to the closer inequality does not lead to a fixpoint, the closer inequality is satisfiable in the extrapolated space and cannot induce a new landmark. At this point, the inequality that is farther away can become a landmark. Hence, tracking distances to closer inequalities ensures that all landmarks are considered in turn. When creating a new

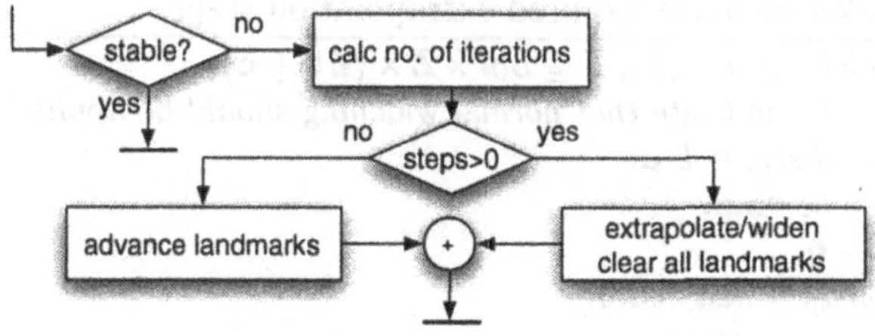

Fig. 12.7. Operations performed at a widening point.

Algorithm 11 Advance a landmark.

procedure *advanceLandmarks*(L), $L \subseteq Lin \times \mathbb{Z} \times (\mathbb{Z} \cup \{\infty\})$

$L' \leftarrow \{\langle e, dist_c, dist_c\rangle \mid \langle e, dist_c, dist_p\rangle \in L\}$

return L'

landmark, line 8 sets the second distance to infinity, which indicates that this new landmark is not yet ready to be used in extrapolation.

The next section details how the acquired landmarks are manipulated.

12.4 Using Landmarks at a Widening Point

The semantic equations of the program induce cyclic dependencies between the states at each program point. A widening point must be inserted into each cycle to ensure that the fixpoint computation eventually stabilises. In the case of nested cycles, a fixpoint is calculated on each inner cycle before moving on to the containing cycle [34]. Figure 12.7 schematically shows the actions taken when a semantic equation at a widening point is evaluated. If stability has not yet been achieved, all landmarks gathered in the current cycle (excluding those in inner cycles) are passed to the algorithm *calcIterations*, which estimates the number of times the cycle needs to be traversed until a state is reached at which the first as yet unsatisfiable inequality becomes satisfiable. This count is denoted as *steps* in Fig. 12.7. Two special values are distinguished: 0 and ∞. A value of zero indicates that new landmarks were created during the last traversal of the cycle. In this case, the left branch of Fig. 12.7 is taken and the algorithm *advanceLandmarks*, which is presented in Alg. 11, is called. Normal fixpoint computation is then resumed, allowing landmarks to acquire a second measurement. The call to *advanceLandmarks* propagates the most recently calculated distance to the third element of each landmark so that *updateLandmark* can update the second element of the landmark tuple during the next iteration.

Algorithm 12 Calculate required extrapolation steps.

procedure *calcIterations*(L), $L \subseteq Lin \times \mathbb{Z} \times (\mathbb{Z} \cup \{\infty\})$
1: $steps \leftarrow \infty$ /* *indicate that normal widening should be applied* */
2: **for** $\langle e, dist_c, dist_p \rangle \in L$ **do**
3:   **if** $dist_p = \infty$ **then**
4:     $steps \leftarrow 0$
5:   **else if** $dist_p > dist_c$ **then**
6:     **if** $steps = \infty$ **then**
7:       $steps \leftarrow \lceil dist_c/(dist_p - dist_c) \rceil$
8:     **else**
9:       $steps \leftarrow \min(steps, \lceil dist_c/(dist_p - dist_c) \rceil)$
10:     **end if**
11:   **end if**
12: **end for**
13: **return** *steps*

The right branch of Fig. 12.7 is selected whenever *calcIterations* returns a non-zero value for *steps*, which indicates that all landmarks have acquired two measurements. This is the propitious moment for extrapolation, as only now can all landmarks participate in predicting the number of cycles until the first as yet unsatisfiable inequality is reached. Algorithm 12 shows how this number is derived. The function *calcIterations* calculates an estimate of the number of iterations necessary to satisfy the nearest landmark stored in *steps*. This variable is initially set to ∞, which is the value returned if no landmarks have been gathered. An infinite value in *steps* indicates that widening, rather than extrapolation, has to be applied. Otherwise, the loop in lines 2–12 examines each landmark in turn. For any landmark with two measurements (i.e., those for which $dist_p \neq \infty$), lines 7–8 calculate the number of steps after which the unsatisfiable inequality that gave rise to the landmark $\langle e, dist_c, dist_p \rangle$ becomes satisfiable. Specifically, $dist_p - dist_c$ represents the distance traversed during one iteration. Given that $dist_c$ is the distance between the boundary of the unsatisfiable inequality and the polyhedron in that iteration, the algorithm computes $\lceil dist_c/(dist_p - dist_c) \rceil$ as an estimate of the number of iterations required to make the inequality satisfiable. This number is stored in *steps* unless another landmark has already been encountered that can be reached in fewer iterations. If $steps = \infty$, no landmark has been found so far towards which the state space grows and *steps* is always set.

The next section presents an algorithm that extrapolates the change between two iterates by a given number of steps. It thereby completes the suite of algorithms necessary to realise widening with landmarks.

Algorithm 13 Extrapolate the change between P_1 and $P_1 \sqcup_P P_2$.

procedure *extrapolate*$(P_1, P_2, steps)$, $P_1, P_2 \in Poly$, $steps \in \mathbb{N} \cup \{\infty\}$

1: $[\![\{\iota_1, \ldots \iota_n\}]\!] \leftarrow P_1$ /* $\iota_1, \ldots \iota_n$ *is a non-redundant description of* P_1 */
2: $P \leftarrow P_1 \sqcup_P P_2$
3: **if** $steps = 0$ **then**
4:     **return** P
5: **else**
6:     $Ires \leftarrow \emptyset$
7:     **for** $i = 1, \ldots n$ **do**
8:         $e \leq c \leftarrow \iota_i$
9:         $c' \leftarrow minExp(P, e)$
10:         **if** $c' \leq c$ **then**
11:             $Ires \leftarrow Ires \cup \{e \leq c\}$ /* *since* $P \sqsubseteq_P [\![\iota_i]\!]$ */
12:         **else if** $steps \neq \infty$ **then**
13:             $Ires \leftarrow Ires \cup \{e \leq (c + (c' - c)steps)\}$
14:         **end if**
15:     **end for**
16:     **return** $Ires$
17: **end if**

12.5 Extrapolation Operator for Polyhedra

In contrast to standard widening, which removes inequalities that are unstable, extrapolation by a finite number of steps merely relaxes inequalities until the next landmark is reached. Algorithm 13 presents a simple extrapolation algorithm that performs this relaxation based on two iterates, namely P_1 and $P_2 \in Poly$. This extrapolation is applied by replacing any semantic equation of the form $Q_{i+1} = Q_i \nabla (Q_i \sqcup_P R_i)$ with $Q_{i+1} = extrapolate(Q_i, R_i, steps)$, where $steps = calcIterations(L)$ and L is the set of landmarks relevant to this widening point. Thus the first argument to *extrapolate*, namely P_1, corresponds to the previous iterate Q_i, while P_2 corresponds to R_i. Line 2 calculates the join P of both P_1 and P_2, which forms the basis for extrapolating the polyhedron P_1. Specifically, bounds of P_1 that are not preserved in the join are extrapolated. The loop in lines 7–15 implements this strategy, which resembles the original widening on polyhedra [62], which can be defined as $I_{res} = \{\iota_i \mid P \sqsubseteq_N [\![\{\iota_i\}]\!]\}$, where $\iota_1, \ldots \iota_n$ is a non-redundant set of inequalities such that $[\![\{\iota_1, \ldots \iota_n\}]\!] = P_1$; see [10]. Note that this widening might not be well defined if the dimensionality of P_1 is smaller than that of $P = P_1 \sqcup_N P_2$; other inequalities from P can be added to I_{res} to remedy this [10, 91]. The latter is not necessary in the context of the TVPI domain when implemented as a reduced product of intervals and TVPI inequalities due to the unique representation of planar polyhedra; see Sect. 8.2.5.

The extraction of stable inequalities in the definition of I_{res} is implemented as follows. The entailment check $P \sqsubseteq_N [\![\{\iota_i\}]\!]$ for $\iota_i \equiv e \leq c$ is implemented in line 9 by calculating the smallest c' such that $P \sqsubseteq_N [\![\{e \leq c'\}]\!]$. If $c' \leq c$,

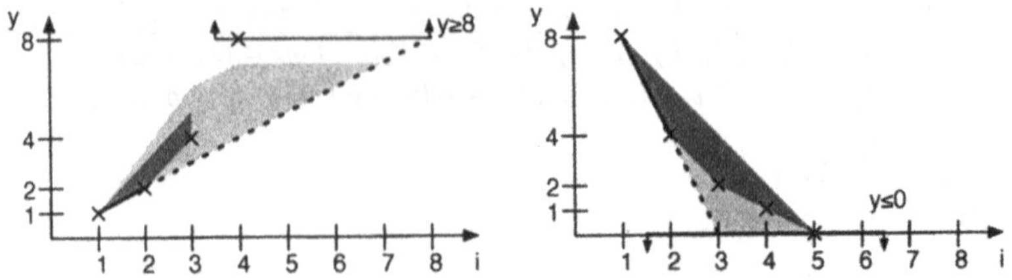

Fig. 12.8. Illustrating non-linear growth.

the entailment holds and line 11 adds the inequality to the result set. If the entailment does not hold, the inequality is discarded whenever *steps* $= \infty$. In this case, *extrapolate* reduces to a simple widening. If *steps* is finite, line 13 translates the inequality, thereby anticipating the change that is likely to occur during the next *steps* loop iteration.

The algorithm presented performs a linear translation of inequalities. Since array accesses are typically linear, this approach is well suited for verifying that indices fall within bounds. However, a non-linear relationship such as that arising in the C loop `int i=1; for(int y=1; y<8; y=y*2) i++;` is not amenable to linear extrapolation and thus leads to a loss of precision. The loop creates successive values for `i`, `y` that correspond to the points $\langle 1,1\rangle$, $\langle 2,2\rangle$, $\langle 3,4\rangle$, and finally, at the exit of the loop, the point $\langle 4,8\rangle$. These are indicated as crosses in the left graph of Fig. 12.8. The best polyhedral approximation of these points restricted by the loop invariant $y < 8$ is shown in dark grey. However, extrapolating the first two iterates, namely the polyhedron $\{\langle 1,1\rangle\}$ and the polyhedron that additionally contains $\langle 2,2\rangle$, predicts that the shown landmark $y \geq 8$ becomes satisfiable after seven additional loop iterations. The extrapolation results in the polyhedron depicted as a dashed line; continuing with the fixpoint calculation leads to the light grey area for the state within the loop, which is a coarser approximation than the optimal polyhedron, which is depicted as the dark grey area.

A loss of precision also occurs in the case of sub-linear growth as in the loop `int i=1; for(int y=8; y>0; y=y>>1) i++;`, shown in the right graph. After intersecting the extrapolated state with $y > 0$ and performing integral tightening, the estimated state for the loop body will merely include the incorrect point $\langle 2,1\rangle$ in addition to the best abstraction, shown in dark grey. However, in this case the linear extrapolation is too conservative and the exit condition of the loop becomes satisfied, although the state space within the loop has not yet stabilised. In this case, normal widening might be applied, thereby incurring an even greater precision loss. Note that termination of widening with landmarks is guaranteed in both cases.

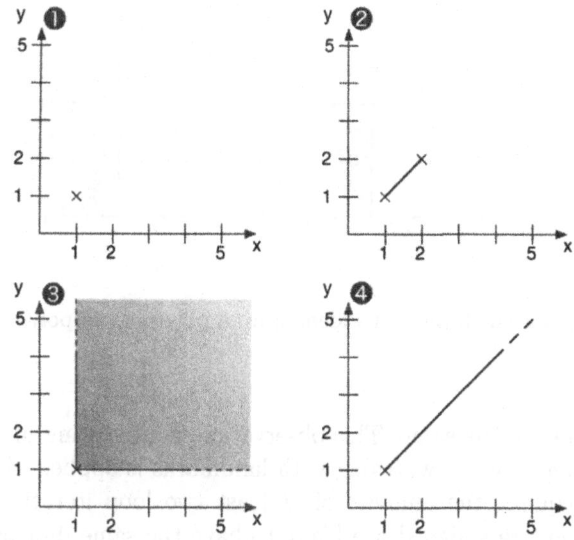

Fig. 12.9. Standard widening versus revised widening.

12.6 Related Work

Although the foundations of widening and narrowing were laid three decades ago [55], the value of widening was largely unappreciated until comparatively recently [59]. In the last decade, there has been a resurgence of interest in applying polyhedral analysis and, specifically, polyhedral widenings [10,25,28]. The original widening operator in [62] discards linear relationships that result from joining the state of the previous loop iteration with the current loop iteration. This causes a loss of precision, especially when widening is applied in each loop iteration. In order to illustrate this shortcoming, consider the polyhedron $P_1 = [\![\{x = 1, y = 1\}]\!]$ in the first graph of Fig. 12.9. Assuming that this polyhedron changes to $P_2 = [\![\{x = 2, y = 2\}]\!]$ after executing the loop body, an over-approximation of the loop invariant can be calculated by widening P_1 with respect to $P_1 \sqcup_P P_2$, the latter being depicted in the second graph. The original widening will calculate $P' = P_1 \nabla (P_1 \sqcup_P P_2) = [\![\{x \geq 1, y \geq 1\}]\!]$, shown in the third graph. This result is imprecise in that the linear relationship $x = y$, which is present in $P_1 \sqcup_P P_2$, is not included in P'. The so-called revised widening [91] remedies this by adding additional inequalities from the join $P_1 \sqcup_P P_2$ that fulfill certain properties. In the example, the equality $x = y$ can be added to P', yielding the polyhedron depicted in the fourth graph of Fig. 12.9. Given that checking whether an inequality from the join can be added to P' is expensive, it is interesting to note that Benoy [25] showed that no inequalities need to be added if the dimensionalities of the

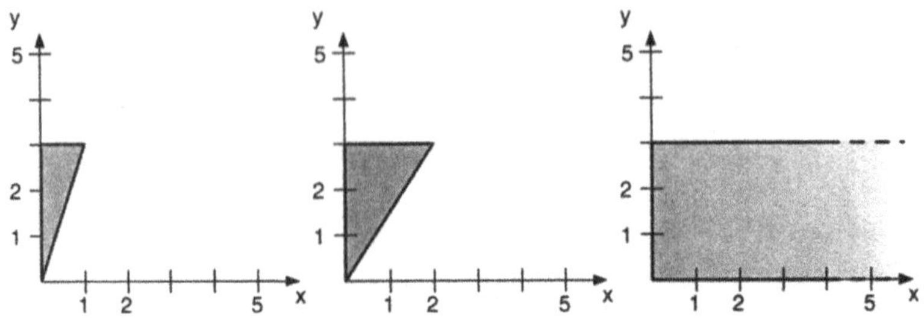

Fig. 12.10. Improved widening from polytopes to polyhedra.

two polyhedra are the same. This observation is interesting in the context of the TVPI domain when widening with landmarks is applied. Since gathering landmarks requires the analysis of at least two loop iterations, the planar polyhedra that are widened are likely to have the same dimension since the topological dimension of a planar polyhedron can be at most two.

Besson et al. [28] present widenings that are especially precise when widening polytopes into polyhedra. For instance, the iterates shown in Fig. 12.10 feature an inequality with changing coefficients that standard widening would remove. Instead, this inequality is widened to $y \geq 0$, thereby retaining a lower bound on y. Extending our extrapolation function to include inequalities with changing coefficients is an interesting research question. Bagnara et al. [10] combine the techniques of Besson et al. and other widenings with extrapolation strategies that delay widening. More closely related is work on extrapolation using information from the analysed equation system. For instance, widening with thresholds [30] uses a sequence of user-specified values (thresholds) on individual variables up to which the state space is extrapolated in sequence. Halbwachs et al. [93] deduce thresholds automatically from guards in the semantic equations. However, they observe redundant inequalities rather than unsatisfiable inequalities, thereby possibly extrapolating to thresholds where no fixpoint can exist, such as redundant inequalities that express verification conditions. The restriction of inferring thresholds on single variables is lifted by lookahead widening, which uses standard widening and narrowing operators and thereby is able to find bounds that are expressed with more than one variable [81]. It uses a pilot polyhedron on which widening and narrowing are performed alongside a main polyhedron. Once the pilot value has stabilised after narrowing, it is promoted to become the main value. By using the main value to evaluate effects in other domains, the problems of domain interaction as discussed in the introduction of this chapter do not occur. Furthermore, by discarding behaviours that are enabled after widening but are disabled with respect to the main value, lookahead widening is able to find fixpoints in which not all behaviours are enabled, as is the case in the example on string buffers. While lookahead widening solves essentially the same problem as widening with landmarks, the analysis operates on two polyhedra instead of one. In many cases, the pilot and main value coincide such that only a single polyhedron needs to be propagated, which reduces the overhead of using two polyhedra.

The need for a pilot polyhedron can be refined by temporarily removing all unsatisfiable behaviours from equation systems [82]. The idea is to find a fixpoint using widening and narrowing only on the enabled behaviours. Once a fixpoint is found, behaviours are enabled and another iteration is performed. Again, behaviours that are still disabled are removed and another widening/narrowing cycle is performed. One of the downsides of this approach is the complexity of the implementation when behaviours are not branches in the control-flow graph but part of the abstract transfer functions of statements as in the string buffer example. Note that both approaches require a narrowing step for full precision. Implementing narrowing in an analyser can require a substantial implementation effort. For instance, the authors of lookahead widening point out that narrowing was not implemented in their WPDS++ tool, as it would have required a major redesign [81].

Further afield is the technique of counterexample-driven refinement, which has recently been adapted to polyhedral analysis [89]. This approach is in some sense orthogonal to narrowing that refines a single fixpoint. In counterexample-driven refinement, the fixpoint computation is repeatedly restarted, guided by a backwards analysis from the point of a false warning to some widening point. Finally, it has been shown that widening and narrowing can be avoided altogether in a relational analysis if the semantic equations are affine [177]. Incredibly, for this restricted class of equations, least fixpoints can be found in polynomial time.

13

Combining Points-to and Numeric Analyses

This chapter addresses a loss of precision that is due to the lack of interaction between the numeric domain and the points-to domain. Chapter 11 on string buffer analysis demonstrated that, within the domain of polyhedra, an invariant such as $p < n$ (the access position lies in front of the NUL position) can be recovered through the relational information in the domain merely by intersection with the loop invariant $c \neq 0$ (the read character is not NUL). Since no relational information exists between the polyhedral domain and the points-to domain, the intersection with loop invariants cannot recover any information in the points-to domain. This deficiency can compromise precision, as demonstrated by the following example, which is a modification of the running example in the chapter on string buffer analysis:

```
char *p;
char t[16] = "Aero";
char v[8] = "boat";
char *u = "plane␣";
p = t+4;
if (rand()) u = &v;
while (*p=*u) { p++; u++; };
printf("t␣=␣'%s'\n", t);
```

The difference from the original example lies in the conditional statement `if (rand())`, which, depending on a random number, updates the pointer u to point to the string buffer v. Thus, during the analysis of the **while**-loop, the pointer u has a points-to set of $\{a_p, a_v\}$, where a_p denotes the abstract address of the buffer containing `"plane␣"` and a_v denotes the address of v. Hence, whenever the read access `*u` is evaluated, the analysis evaluates each buffer access separately and joins the results. Suppose that c denotes the result of the read access `*u`. The first graph in Fig. 13.1 shows the value of c with respect to the pointer offset u when dereferencing the abstract address a_p;

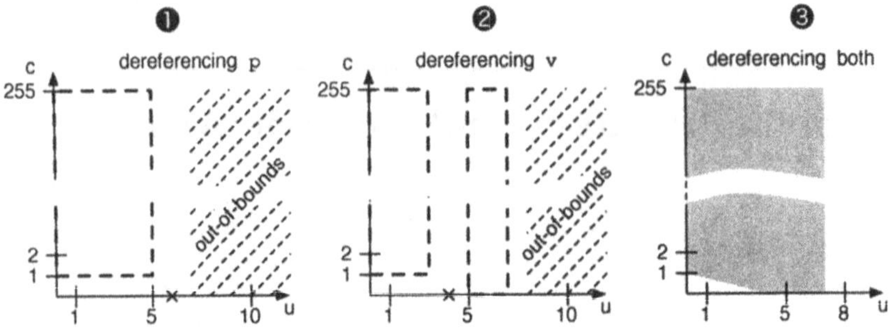

Fig. 13.1. Finding the NUL position on more than one buffer at a time is impossible without stating that a points-to set changes from a certain iteration onwards.

the second graph shows values of c when accessing v. The loss of precision becomes apparent in the third graph, which depicts the join of the first two graphs. While the join still expresses that c is zero and thus that the loop exits at $u = 4$, the information that the loop only exits if *u accesses v is lost. Worse, since a_v is not removed from the points-to set of u, subsequent iterations still read from the buffer v even though u cannot point to v for loop iterations with $u > 4$. As a result, the analysis assumes that *u accesses v past the first known NUL position so that c remains non-zero even at $u = 6$, where the NUL position of the string constant "plane␣" is reached. Hence, it is not possible to infer that the loop always terminates at $u = 6$, and a warning is raised that the string buffers "plane␣" and "boat" accessed out-of-bounds.

In order to prove the correctness of the example above, the analysis must be able to express that *u does not access the buffer v once the pointer offset of u exceeds 4.[1] If this fact can be inferred in loop iterations $u > 4$, only the string constant "plane␣" would be read when evaluating *u and a fixpoint could be detected at $u = 6$. Relating the content of the points-to set of u to the pointer offset u can only be accomplished by introducing a relationship in the polyhedral domain. Specifically, the idea is to keep Boolean variables in the polyhedron that indicate if a specific abstract address is present in the points-to set of a variable. In this chapter, we will describe how a vector of Boolean flags is kept for each pointer-sized variable, where each flag in the vector is an abstract variable. The value of that variable in the polyhedron then indicates if an abstract address is part of the points-to set.

The chapter is structured as follows. The next section illustrates the theory of Boolean flags and how they can be used to improve points-to analysis. Section 13.2 presents a modification of the analysis in which the presence of an l-value in a points-to set is governed by a flag in the numeric domain. The practical implementation and other applications are discussed in Sect. 13.3.

[1] Note that this requires inequalities with more than two variables, for which the TVPI domain is not expressive enough on its own.

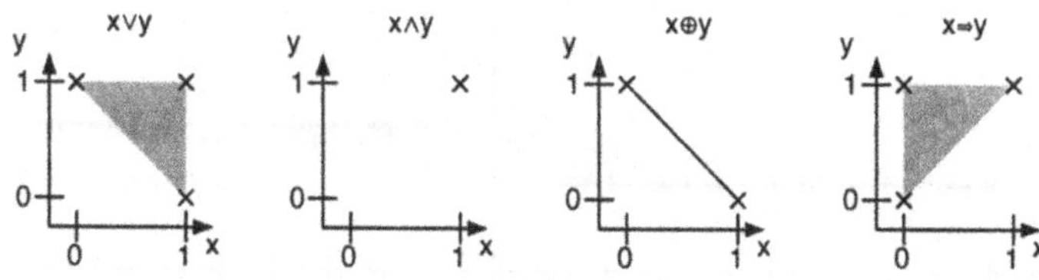

Fig. 13.2. Boolean functions can be expressed exactly in the polyhedral domain when satisfiable assignments of variables are modelled as vertices in the polyhedron.

13.1 Boolean Flags in the Numeric Domain

An interesting property of polyhedra is that it is possible to express binary formulas when Boolean truth values are stored as 0 when the variable is false and as 1 otherwise. For instance, Fig. 13.2 depicts four polyhedra that model the truth values of common Boolean functions. Moreover, it is possible to freely mix Boolean variables and variables that represent the ranges of program variables. In order to illustrate this, consider the evaluation of the following code fragment when given the state $P \in Poly$:

```
int r=MAX_INT;
if (d!=0) r=v/d;
```

Suppose this block of code is executed with a value of $[-9, 9]$ for d. Rather than analysing the division twice, once with positive values of d and once with negative values of d, Fig. 13.3 shows how the states P^+ and P^- can be stored in a single state without introducing an integral point where $d = 0$. Specifically, the figure shows the state $(P^- \sqcap_P [\![f = 0]\!]) \sqcup_P (P^+ \sqcap_P [\![f = 1]\!])$, which collapses to the empty $\mathbb{Z}$-polyhedron when intersected with $d = 0$, where $[\![d = 0]\!]$ is the state for which the division is erroneous. The prerequisites for merging two states without loss are that both states be represented by polytopes (polyhedra in which all variables are bounded) and that techniques for integral tightening be present. This is formalised in the following proposition.

Proposition 11. *Let* $P_0, P_1 \in Poly$, *let* $P = (P_0 \sqcap_P [\![f = 0]\!]) \sqcup_P (P_1 \sqcap_P [\![f = 1]\!])$ *and let* $P'_i = P \sqcap_P [\![f = i]\!]$ *for* $i = 0, 1$. *Then* $P_i \sqcap_P [\![f = i]\!] \cap \mathbb{Z}^n = P'_i \cap \mathbb{Z}^n$ *for all* $i = 0, 1$.

Proof. Without loss of generality, assume $i = 0$. Suppose that $\mathcal{X}$ is arranged as $\langle x_1, \ldots x_{n-1}, f \rangle = \mathbf{x}$, which we abbreviate as $\langle \bar{\mathbf{x}} | f \rangle = \mathbf{x}$. Consider two cases:

"soundness": Let $\langle \bar{\mathbf{a}} | f \rangle \in P_0 \sqcap_P [\![f = 0]\!] \cap \mathbb{Z}^n$. Then $\langle \bar{\mathbf{a}} | f \rangle \in P$ by the definition of $\sqcup_P$. Since $f = 0$ in $\langle \bar{\mathbf{a}} | f \rangle$, it follows that $\langle \bar{\mathbf{a}} | f \rangle \in P \sqcap_P [\![f = 0]\!]$ and thus $\langle \bar{\mathbf{a}} | f \rangle \in P'_0$. We chose the vector such that $\langle \bar{\mathbf{a}} | f \rangle \in \mathbb{Z}^n$ and hence $\langle \bar{\mathbf{a}} | f \rangle \in P'_0 \cap \mathbb{Z}^n$.

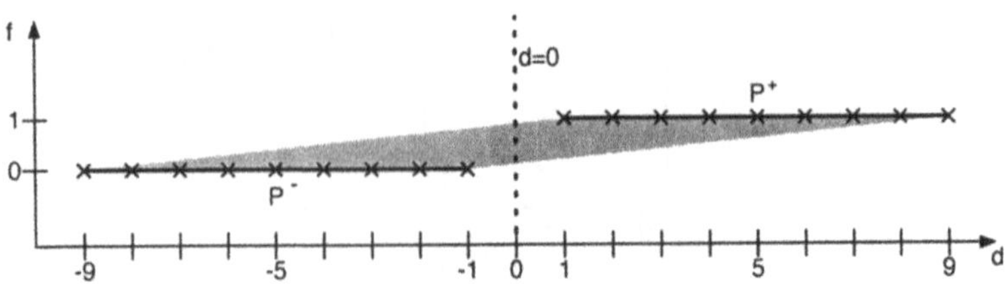

Fig. 13.3. Using a Boolean flag to perform control-flow splitting. Feasible integral points are indicated by crosses, the dashed line indicating the polyhedron $[\![d = 0]\!]$.

"completeness": Let $\langle \bar{\mathbf{a}} | f \rangle \in P'_0 \cap \mathbb{Z}^n$. Then $\langle \bar{\mathbf{a}} | f \rangle \in P \sqcap_P [\![f = 0]\!]$ and hence $f = 0$. For the sake of a contradiction, suppose that $\langle \bar{\mathbf{a}} | 0 \rangle \notin P_0$. Since $P \in \mathit{Poly}$ is convex, $\langle \bar{\mathbf{a}} | 0 \rangle = \lambda \langle \bar{\mathbf{a}}_1 | f_1 \rangle + (1 - \lambda) \langle \bar{\mathbf{a}}_2 | f_2 \rangle$ for some $0 \leq \lambda \leq 1$. Observe that the join $P = (P_0 \sqcap_P [\![f = 0]\!]) \sqcup_P (P_1 \sqcap_P [\![f = 1]\!])$ is defined as an intersection of all states $\hat{P}$ such that $(P_i \sqcap_P [\![f = i]\!]) \sqsubseteq_P \hat{P}$ for $i = 0, 1$; this holds in particular for $\hat{P} = [\![f \geq 0]\!]$ and $\hat{P} = [\![f \leq 1]\!]$. Thus, $0 \leq f \leq 1$ for all $\langle \bar{\mathbf{a}} | f \rangle \in P$. Hence, $0 \leq f_1 \leq 1$ and $0 \leq f_2 \leq 1$ must hold. Given the constraints $0 = \lambda f_1 + (1 - \lambda) f_2$ and $\bar{\mathbf{a}} = \lambda \bar{\mathbf{a}}_1 + (1 - \lambda) \bar{\mathbf{a}}_2$, either $f_0 = 0$ or $f_1 = 0$ so that one vector, say $\langle \bar{\mathbf{a}}_1 | f_1 \rangle$, must lie in P_0, which implies $\lambda = 1$. In particular, with $\bar{\mathbf{a}} = \lambda \bar{\mathbf{a}}_1 + (1 - \lambda) \bar{\mathbf{a}}_2$ and $\bar{\mathbf{a}}_2 \in \mathbb{Z}^{n-1}$ having only finite coefficients, $\bar{\mathbf{a}} = \bar{\mathbf{a}}_1$ follows, which contradicts our assumption of $\langle \bar{\mathbf{a}} | 0 \rangle \notin P_0$.

We briefly comment on the requirements of P_0, P_1 being polytopes and how integral tightening methods affect the precision of Boolean flags.

13.1.1 Boolean Flags and Unbounded Polyhedra

With respect to the first requirement, namely that the polyhedra that describe the state space must be bounded, Fig. 13.4 shows that a precision loss occurs for unbounded polyhedra. Specifically, taking convex combinations of points in the state $P^+ = [\![\{d \geq 1, f = 1\}]\!]$ and those in $P^- = [\![\{-9 \leq d \leq -1, f = 0\}]\!]$ leads to the grey state. Even though the line $[\![\{f = 0, d > -1\}]\!]$ is not part of the grey state, the polyhedral join $P^+ \sqcup_P P^-$ approximates this state with sets of non-strict inequalities, thereby including the line. Hence, the definition of $\sqcup_P$ automatically closes the resulting space and thereby introduces points $\langle \bar{\mathbf{a}}, f \rangle \in P^+ \sqcup_P P^-$, where $f = 0$, even though $\langle \bar{\mathbf{a}}, 0 \rangle \notin P^+$. Note that allowing strict inequalities $\mathbf{a} \cdot \mathbf{x} < c$ [12] cannot counteract this precision loss, as the state $P^+ \sqcup_P P^-$ cannot be described exactly, even when allowing both strict and non-strict inequalities.

The imprecise handling of unbounded polyhedra is generally not a problem in verifiers that perform a forward reachability analysis, as program variables

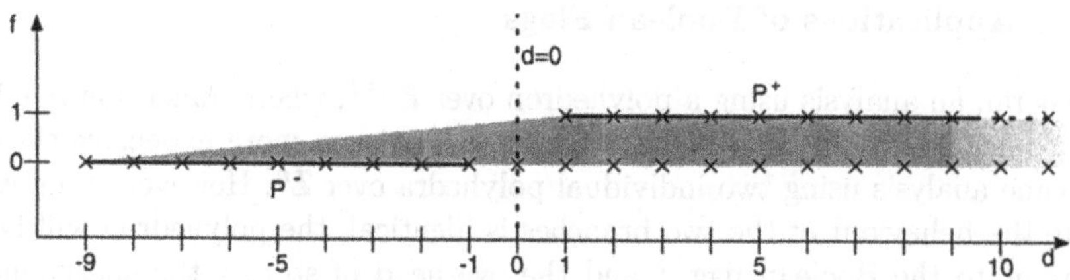

Fig. 13.4. Unbounded polyhedra cannot be fully distinguished using a Boolean flag.

are usually finite and wrap when they exceed their limit. Making wrapping explicit as proposed in Chap. 4 effectively restricts the range of a variable. Thus, states in a forward analysis are usually bounded. Polyhedra have also been used to infer an input-output relationship of a function, thereby achieving a context-sensitive analysis by instantiating this input-output behaviour at various call sites [83]. In this application, the inputs are generally unbounded and a Boolean flag does not distinguish any differences between states. However, even in this application it might be possible to restrict the range of input variables to the maximum range that the concrete program variable may take on, thereby ensuring that input-output relationships are inferred using polytopes.

13.1.2 Integrality of the Solution Space

A second prerequisite for distinguishing two states within a single polyhedron is that the polyhedron be reduced to the contained $\mathbb{Z}$-polyhedron upon each intersection. Tightening a polyhedron to a $\mathbb{Z}$-polyhedron is an exponential operation, which can be observed by translating a Boolean function f over n variables to a $\mathbb{Z}$-polyhedron over $\mathbb{Z}^n$ by calculating the convex hull of all Boolean vectors (using 0 and 1 for false and true, respectively) for which f is true, as shown in Fig. 13.2 for $n = 2$. An argument similar to Prop. 11 is possible to show that joining all n-ary vectors for which f is true leads to a polyhedron that expresses f exactly. The integral meet operation $\sqcap_P^{\mathbb{Z}}$ therefore becomes a decision procedure for satisfiability of n-ary Boolean formulas. As an interesting consequence, observe that octagons [130] together with the complete algorithm for $\sqcap_P^{\mathbb{Z}}$ presented in [13] provide an efficient decision procedure for 2-SAT.

While calculating a full $\mathbb{Z}$-polyhedron from a given polyhedron with rational intersection points is expensive, a cheap approximation often suffices in practice. For instance, the abstract transfer function of the division operation will add the constraint $d = 0$ to the state space shown in Fig. 13.3, with the result that possible values of f lie in $[0.1, 0.9]$. Rounding the bounds to the nearest feasible integral value yields the empty interval $[1, 0]$, which indicates an unreachable state, thereby proving that a division by zero cannot happen.

We now point out an application where using Boolean flags seems to be a good compromise between cost and precision.

13.1.3 Applications of Boolean Flags

In general, an analysis using a polyhedron over $\mathbb{Z}^{n+1}$, where dimension $n+1$ is a Boolean variable, may be as expensive as or even more expensive than the same analysis using two individual polyhedra over $\mathbb{Z}^n$. However, in cases where the behaviour of the two branches is identical, the polyhedron will be invariant to the Boolean flag f and the overhead of storing the constraint $0 \leq f \leq 1$ will be negligible. In practice, a behaviour somewhere in between these two extremes is likely. In the context of the polynomial TVPI domain, adding a Boolean flag in order to split a path is always faster than analysing the path twice. On the downside, TVPI inequalities can only express that the range of one variable changes with respect to a Boolean flag; that is, it cannot be stated that a relationship between two variables changes. However, we found that even TVPI inequalities are valuable to express certain idioms in C. Suppose the flag x_p indicates if the pointer **p** points to **s** in the following block of code:

```
extern s;
int f(struct s **p) {
  if (rand()) return 1; /* error */
  *p = &s; return 0; /* success */
}
```

The purpose of the function **f** is to calculate a result and return a pointer to one of many global structures on success, in which case the constant one is returned. If the actual parameter passed to **f** is not initialised, then **p** can take on any value in the range $[0, 2^{32}-1]$ if one is returned. This and the fact that (the offset of) **p** is zero if zero is returned can readily be inferred by the analysis. However, **f** is likely to be called as follows:

```
struct s *p;
int r;
r = f(&p);
if (!r) printf("value:␣%x", *p);
```

While it is known that the offset of **p** is zero if the return value is zero, it is also necessary to know that **p** definitely points to **s** and is not `NULL`. Setting the Boolean flag x_p to one iff **p** contains the address of **s** is sufficient to convey this information. In particular, the analysis of **f** infers that $x_p = -x_r$, where x_r is the return value. Thus, testing that **r** is zero restricts the points-to set of **p** as expected. In other words, introducing Boolean flags to state whether a certain address is in the points-to set of a variable is a particularly expressive way to make a (possibly flow-insensitive) points-to analysis flow-sensitive.
The next section shows how the presented idea is implemented in our analysis.

13.2 Incorporating Boolean Flags into Points-to Sets

In this section, we detail how to extend the analysis to use Boolean flags in the numeric domain that model the content of points-to sets. With each pointer-sized field whose value is represented by some abstract variable $x \in \mathcal{X}$, we associate a vector of Boolean flags $\langle f_1, \ldots f_n \rangle$, where each $f_i \in \mathcal{X}$. Here, $n = |\mathcal{A}|$; that is, with each pointer-sized field, the polyhedron also tracks $|\mathcal{A}|$ abstract variables that represent the flags. The idea is to set the flag f_i of the field x to one whenever the points-to set of x includes the ith abstract address and to set $f_i = 0$ otherwise. In this new model, the points-to domain is redefined to $Pts = \mathcal{X} \to \mathcal{X}^{|\mathcal{X}|}$. In particular, $A \in Pts$ assigns a vector of Boolean flags $\mathbf{f} = A(x)$ to each pointer-sized field x. For all other abstract variables, the map A is undefined. For the sake of a simpler presentation, we assume that the set of pointer-sized fields and that of abstract addresses $\mathcal{A}$ is fixed such that A is fixed. Section 13.3 will detail how to infer these sets on-the-fly, thereby also reducing the number of flag variables.

In the remainder of this section, we shall present the necessary modifications to the analysis. After revising the functions on access trees and presenting the read and write functions in Sect. 13.2.1, we hint at the required changes to the abstract semantics by presenting the transfer functions for expressions and assignments in Sect. 13.2.2. In order to unveil the full expressiveness of the new abstraction, Sect. 13.2.3 discusses the semantics of conditionals.

13.2.1 Revising Access Trees and Access Functions

While the switch to a new way of tracking points-to sets affects most parts of the analysis presented so far, the propagation functions for overlapping fields remain unchanged, as they only deal with value fields. Figure 13.5 depicts the changed functions for managing l-values in access trees. A new function $hasLVals^A : (\{1, 2, 4, 8\} \times \mathcal{AT} \times Num) \to \{true, false\}$ determines if a given field holds a pointer variable. Most cases simply ascend or descend towards the 4-byte, pointer-sized field. Once there, it is tested if x is a value, which is the case when all points-to flags $\langle f_1, \ldots f_n \rangle = A(x)$ are zero in N. The test proceeds by intersecting N with the assumption and checking that the state has not gotten smaller. The functions $getLVals^A$ and $setLVals^A$ are only defined on access trees that pivot in a pointer-sized field. In particular, $getLVals^A$ merely returns the vector of flags $A(x)$, while $setLVals^A$ uses the notation $N \triangleright A(x) := \mathbf{f}$ as shorthand for assigning each individual flag in $\mathbf{f}$ to those returned by $A(x)$. The two functions *getOfs* and *setOfs* remain unchanged from Fig. 5.4 on p. 102 and are thus omitted here. The last function on access trees is $clear^{F,A}$, which is applied before accessing a field that overlaps with a pointer. The function recurses until the pointer-sized field is the pivot node. As in the case of $hasLVals^A$, it is tested if the points-to set of x is empty – that is, if all points-to flags are zero in N. If not, the value of the pointer-sized

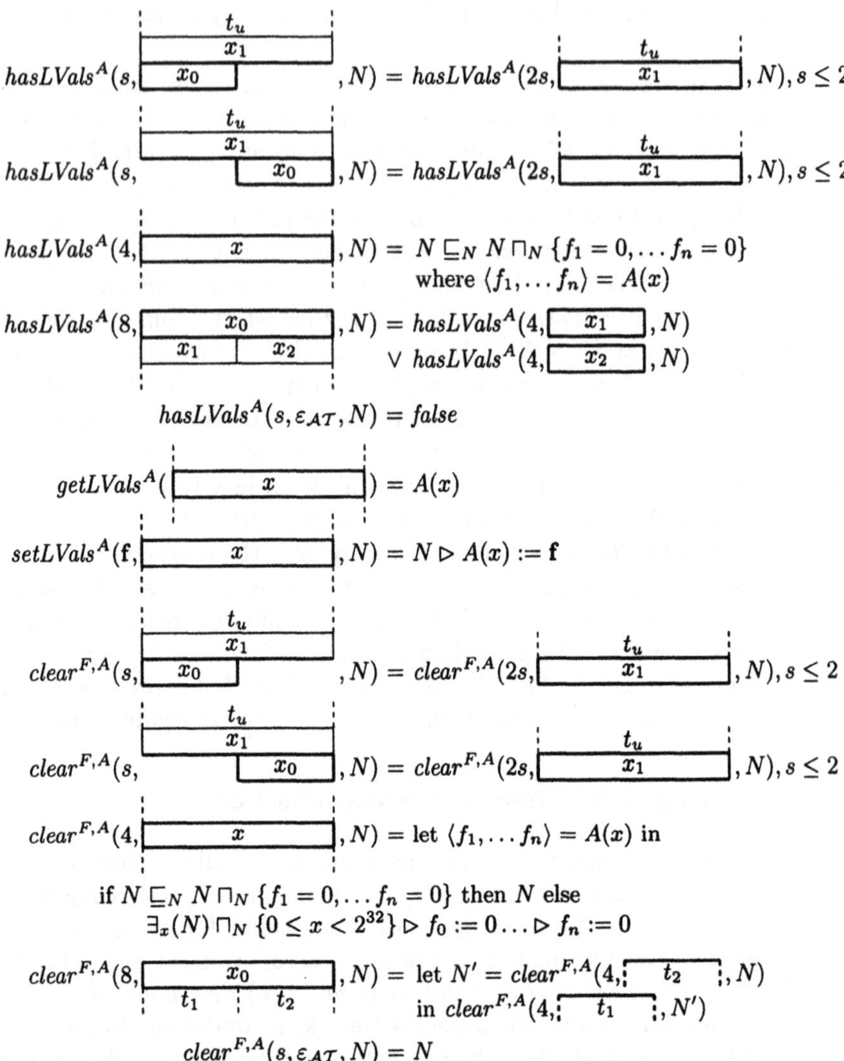

Fig. 13.5. Revised functions for setting and getting l-values. The variable t_u represents a wider tree than the dotted borders suggest.

field is set to its maximum bounds, thereby interpreting the set of l-values as a pure value. Note that the overlapping fields do not have to be set to their maximum value, as this is done by *setOfs* when setting the l-values of x.

The functions on access trees are invoked by the read and write functions presented in Figure 13.6. For brevity, we do not show the wrapper function $read^{F,H,A} : \mathcal{M} \cup \mathcal{D} \times Lin \times \{1,2,4,8\} \times Num \to Num \times \mathcal{X} \times \mathcal{X}^{|\mathcal{X}|}$, which

$$readChk^{F,A}(m, [l,u]_{\equiv d}, s, N) = \langle \bigsqcup_{i=0}^{n} N_i \triangleright x := x_i \triangleright \mathbf{f} := \mathbf{f}_i, x, \mathbf{f}\rangle$$
$$\text{where } x \in \mathcal{X}^T,\ \mathbf{f} \in \mathcal{X}^{T|A|} \text{ fresh}$$
$$\{at_1, \ldots at_n\} = \bigcup_{o \in \{l, l+d, \ldots u\}} access^F(o, s, m)$$
$$\langle N_i, x_i, \mathbf{f}_i\rangle = readTree^A(s, at_i, N),\ i \in [1, n]$$

$$readTree^A(s, at, N) = \langle N', x, \mathbf{f}\rangle$$
$$\text{where if } hasLVals^A(s, at, N) \text{ then}$$
$$\text{if } s = 4 \text{ then } \langle N, getOfs(at), getLVals^A(at)\rangle$$
$$\text{else } \langle N \sqcap_N \{0 \le t < 2^{8s}\}, x, \mathbf{0}\rangle,\ t \in \mathcal{X}^T \text{ fresh}$$
$$\text{else } \langle N', x, \mathbf{0}\rangle \text{ where } \langle N', x\rangle = prop(s, at, N)$$

$$writeChk^{F,A}(m, [l,u]_{\equiv d}, s, e_v, \mathbf{f}, N) = \bigsqcup_{i=1}^{n} writeTree^A(s, e_v, \mathbf{f}, at_i, N)$$
$$\text{where } \{at_1, \ldots at_n\} = \bigcup_{o \in \{l, l+d, \ldots u\}} access^F(o, s, m)$$

$$writeTree^A(s, e_v, \mathbf{f}, at, N) =$$
$$\text{if } s = 4 \text{ then if } N \sqsubseteq_N N\{f_0 = 0, \ldots f_n = 0\}$$
$$\text{then } update(s, e_v, at, N)$$
$$\text{else } setOfs(s, e_v, at, setLVals^A(\mathbf{f}, at, N))$$
$$\text{else } update(s, e_v, at, clear^{F,A}(s, at, N))$$

Fig. 13.6. Accessing abstract memory regions.

can be derived in a straightforward way by omitting the points-to domain A in the definition of $read^{F,H}$ on p. 103. The same holds for the wrapper $write^{F,H,A} : \mathcal{M} \cup \mathcal{D} \times Lin \times \{1,2,4,8\} \times \mathcal{X} \times \mathcal{X}^{|\mathcal{X}|} \times Num \to Num$. In both cases, a vector $\mathbf{f} \in \mathcal{X}^{|\mathcal{X}|}$ replaces the points-to set in the original definition.

The function $\langle N', x, \mathbf{f}\rangle = readChk^{F,A}(m, [l,u]_{\equiv d}, s, N)$ performs the actual read access by calculating a value x and a vector of points-to flags $\mathbf{f}$ that indicates to which l-values x is the offset. Specifically, the function calculates the possible access trees from the given interval $[l,u]_{\equiv d}$, calls $readTree^A$ on each tree, and joins the results. In $readTree^A$, the predicate $hasLVals^A$ is used to check if the access tree overlaps with a field that contains a pointer. If so, the offset and points-to flags are read if the field is as wide as a pointer; otherwise, a safe range $0 \le x \le 2^{8s}$ and the points-to set $\mathbf{0}$ (representing `NULL`) is returned. If the access tree does not cover a field with points-to information, *prop* is called on each tree in order to propagate information from overlapping fields. On return, $readChk^{F,A}$ sets the result variable x to the pivot node of each access tree and joins the resulting states.

Analogously, evaluating $N' = writeChk^{F,A}(m, [l,u]_{\equiv d}, s, e_v, \mathbf{f}, N)$ calculates all access trees in the range $[l,u]_{\equiv d}$ and updates them with the value of e_v and the points-to set $\mathbf{f}$ using *writeTree*. Within *writeTree*, it is checked if the points-to set $\mathbf{f}$ is `NULL`, in which case the tree is updated as a value using *update*. In the case of a pointer, the l-values are set using $setLVals^A$ before the offset is set by *setOfs*. If a field is written that is not as wide as a pointer, the *update* function is called as before, but any l-values of a potentially overlapping pointer field are turned into a value using $clear^{F,A}$.

The read and write functions are applied in the next section when defining the semantics of expressions and assignments between unstructured variables.

13.2.2 The Semantics of Expressions and Assignments

In this section, we present the revised abstract transfer functions for evaluating linear expressions and assignments between variables and pointers. We omit all other functions, as their definition can be adapted analogously from those in Chap. 6. Interestingly, tracking flags for the points-to sets of a variable streamlines much of the definition, in particular the evaluation of linear expressions. For instance, a constant is simply treated as a pure value and a zero vector for the points-to flags. Figure 13.7 shows that the semantics of a term is hardly more elaborate. Specifically, once the value x and the vector of points-to flags $\mathbf{f}$ are calculated by $read^{F,H,A}$, the semantics of the whole expression is simply given by multiplying the value x by n and multiplying each of the points-to flags in $\mathbf{f}$. Each result is then added separately to the remaining expression. Unlike the abstract semantics on p. 117, no warnings are emitted if a multiple or a negative pointer is calculated. In particular, the first statement merely assigns the value and the points-to flags to the target. Warnings are only emitted whenever a pointer is dereferenced, as demonstrated in the last two semantic functions, which write through a pointer and read from a pointer.

Both accesses through pointers use an auxiliary function *derefForMem* that calculates, for each possible memory region m_i, a state N_i, in which the corresponding flag f_i is set. If $f_i = 1$ is not feasible in the passed-in state, the returned domain N_i is itself bottom. Before we detail *derefForMem*, observe that writing through a pointer reduces to a join of the different states that represent the result of each write. Similarly, a write access through a pointer merely calculates the join of possible right-hand sides by assigning the result to the temporary value x_t and the temporary points-to vector $\mathbf{f}_t$.

In order to unravel the definition of *derefForMem*, observe that the three states N', N'', and N''' are consecutive refinements of the input state N calculated by removing state space that indicates an erroneous pointer operation. For instance, it is guaranteed in N' that the pointer variable contains no negation of an address by intersecting N with $f_i \geq 0$. Furthermore, the restriction $f_1 + \ldots + f_n \leq 1$ ensures that the pointer variable contains at most one l-value at a time and not, say, the sum of two pointers or a multiple of one pointer. Symmetrically, N'' is calculated by enforcing that at least one flag is one, thereby ensuring that the dereferenced pointer cannot be `NULL` in the actual program. The next line defines the indices $k_1, \ldots k_m$ that correspond to abstract addresses of functions. These flags are restricted to zero in N''', thereby ensuring that the pointer access does not write to or read from the program code. Finally, the memory region m_i for each abstract address a_i is returned together with the domain N_i, in which the ith flag is restricted to

Expressions.

$[\![\, n \,]\!]^{\sharp,s}_{\text{Expr}} N = \langle N, n, \mathbf{0} \rangle$

$[\![\, n * v.o + exp \,]\!]^{\sharp,s}_{\text{Expr}} N = \langle N'', nx, n \cdot \mathbf{f} + \mathbf{f}' \rangle$
where $\langle N', x, \mathbf{f} \rangle = read^{F,H,A}(v, o, s, N)$
$\langle N'', e, \mathbf{f}' \rangle = [\![\, exp \,]\!]^{\sharp,s}_{\text{Expr}} N'$

Assignments.

$[\![\, s\ v.o = exp \,]\!]^{\sharp}_{\text{Stmt}} N = \exists_{\mathcal{X}^T}(N'')$
where $\langle N', e, \mathbf{f} \rangle = [\![\, exp \,]\!]^{\sharp,s}_{\text{Expr}} N$
$N'' = write^{F,H,A}(v, o, s, e, \mathbf{f}, N')$

$[\![\, s\ v \to o = exp \,]\!]^{\sharp}_{\text{Stmt}} N = \exists_{\mathcal{X}^T}(N''')$
where $\langle N', e, \mathbf{f}_e \rangle = [\![\, exp \,]\!]^{\sharp,s}_{\text{Expr}} N$
$\langle N'', x_o, \mathbf{f} \rangle = read^{F,H,A}(v, 0, 4, N')$
$\langle\, \langle N_1, m_1 \rangle, \ldots \langle N_n, m_n \rangle \,\rangle = derefForMem(N'', \mathbf{f})$
$N''' = \bigsqcup_{i=1}^{n} write^{F,H,A}(m_i, x_o + o, s, e, \mathbf{f}_e, N_i)$

$[\![\, s\ v_1.o_1 = v_2 \to o_2 \,]\!]^{\sharp}_{\text{Stmt}} N = \exists_{\mathcal{X}^T}(N'')$
where $x_t \in \mathcal{X}^T, \mathbf{f}_t \subseteq \mathcal{X}^T$ fresh
$\langle N', x_o, \mathbf{f} \rangle = read^{F,H,A}(v_2, 0, 4, N)$
$\langle\, \langle N_1, m_1 \rangle, \ldots \langle N_n, m_n \rangle \,\rangle = derefForMem(N, \mathbf{f})$
$\langle N_i', x_i, \mathbf{f}_i \rangle = read^{F,H,A}(m_i, x_o + o_2, s, N_i),\ i \in [1, n]$
$N'' = write^{F,H,A}(v_1, o_1, s, x_t, \mathbf{f}_t, \bigsqcup_{i=1}^{n} N_i' \rhd x_t := x_i \rhd \mathbf{f}_t := \mathbf{f}_i)$

$derefForMem(N, \langle f_1, \ldots f_n \rangle) = \langle\, \langle N_1, m_1 \rangle, \ldots \langle N_n, m_n \rangle \,\rangle$
where $N' = N \sqcap_N \{f_1 \geq 0, \ldots, f_n \geq 0, f_1 + \ldots + f_n \leq 1\}$
$N'' = N' \sqcap_N \{f_1 + \ldots + f_n \geq 1\}$
$\{k_1, \ldots k_m\} = \{i \mid \langle a_1, \ldots a_n \rangle = \mathcal{A}, a_i \in \mathcal{A}^{\mathcal{F}}\}$
$N''' = N'' \sqcap_N \{f_{k_1} = 0, \ldots, f_{k_m} = 0\}$
$\langle m_1, \ldots m_n \rangle = \langle L^{-1}(a_1), \ldots L^{-1}(a_n) \rangle$ where $\langle a_1, \ldots a_n \rangle = \mathcal{A}$
$N_i = N''' \sqcap_N \{f_1 = 0, \ldots f_{i-1} = 0, f_i = 1, f_{i+1} = 0, \ldots f_n = 0\},\ i \in [1, n]$
warn "L-value is not a single pointer." **if** $N \not\sqsubseteq_N N'$
warn "Dereferencing a NULL pointer." **if** $N' \not\sqsubseteq_N N''$
warn "Dereferencing a function pointer." **if** $N'' \not\sqsubseteq_N N'''$

Fig. 13.7. Revised abstract semantics for expressions and assignments.

one. For better precision in the context of the weaker TVPI domain, the other flags are restricted to zero, which is implicit when using a general polyhedral domain.

The ability to work with a single domain simplifies most of the abstract transfer functions. As a by-product, incorporating the points-to domain into the numeric domain also improves the expressiveness. The next section discusses the implementation of the conditional statement, which is key in leveraging the expressiveness gained into a more precise analysis.

13.2.3 Conditionals and Points-to Flags

Conditionals in a program are crucial to a static analysis that aims to verify programs in that they constitute the main way of restricting the state space and thereby proving the correctness of a program. Indeed, the reason for expressing the points-to analysis in terms of the numeric domain is to improve the precision of the transfer function of conditionals, which is shown in Fig. 13.8. While the transfer function looks overwhelming at first, recall that the previous conditional was defined in terms of a function *cond* for which only a few useful cases were given and that ignored certain combinations of points-to sets and values. In contrast, the revised conditional treats all combinations precisely using a three-tier approach consisting of *cond*, which, as before, implements the semantics of an **if**-statement; *addAddress*, which incorporates the information in the points-to domain into the comparison as offset; and *intersect*, which performs wrapping and calculates the actual restriction of the state space. We shall detail each function in turn.

The transfer function itself evaluates both sides of the condition to a value and a vector of points-to flags. The latter are then examined to ensure that the variable does not contain the sum of several pointers or a negated pointer. The result N''' is passed to *cond*, which calculates the effect of the comparison when using the operator *op* and when using the opposing operator *neg*(*op*).

In *cond*, different combinations of points-to sets are generated. Specifically, if $n = |\mathcal{A}|$ abstract addresses exist, $(n+1)^2$ states are calculated such that N_{ij} represents the state in which the left-hand side points to a_i, and the right-hand side points to a_j where $\langle a_1, \ldots a_n \rangle = \mathcal{A}$. As a special case, if i or j is zero, the corresponding points-to set of that side is empty; that is, all points-to flags are zero. The result of applying the condition is the join over all $(n+1)^2$ states after calling the *addAddress* function on each. The two possible addresses a_i and a_j are passed into this function as arguments, and the special tag NULL is passed if the corresponding side is a pure value.

Before detailing the *addAddress* function, observe that the *intersect* function merely calls *wrap* on the two linear expressions e_x and e_y using the type $t\ s$ before intersecting the passed-in state N with the constraint $e_x\ op\ e_y$. This function forms the building block for the *addAddress* function.

The *addAddress* function is defined using four different patterns; the first matching pattern determines the result. For instance, given two NULL flags as l-values of the two sides of the condition, *addAddress* passes control straight to *intersect*. In other cases, an offset is added that reflects the possible value of the pointer in the concrete program, namely a value between 4096 (the first location past the first virtual memory page) and $2^{30} + 2^{31}$ (the 3-GB barrier above which lie the reserved 1-GB of the operating system). In particular, the second pattern applies if a pointer on the left side of the condition is compared with a value. The temporary variable o is restricted to the above-mentioned range and added to the pointer offset, creating a value of the left-hand side that

$$
\begin{array}{l}
[\![\textbf{if } t\ s\ v.o\ op\ exp \textbf{ then jump } l\ ;\ nxt]\!]^\sharp_{\text{Next}} \langle N, l_0 \cdots l_s\rangle = \\
\quad \{\langle \exists_{\mathcal{X}^T}(N^{then}), l_0 \cdots l_s \cdot l\rangle\} \cup [\![nxt]\!]^\sharp_{\text{Next}} \langle \exists_{\mathcal{X}^T}(N^{else}), l_0 \cdots l_s\rangle \\
\quad \text{where } \langle N', x, \langle f_1^x, \ldots f_n^x\rangle\ \rangle = read^{F,H,A}(v, o, s, N) \\
\qquad \langle N'', e, \langle f_1^e, \ldots f_n^e\rangle\ \rangle = [\![exp]\!]^{\sharp, s}_{\text{Expr}} N' \\
\qquad N''' = N'' \sqcap_N \{f_1^x \geq 0, \ldots f_n^x \geq 0, f_1^e \geq 0, \ldots f_n^e \geq 0\} \\
\qquad\qquad\qquad \sqcap_N \{f_1^x + \ldots + f_n^x \leq 1, f_1^e + \ldots + f_n^e \leq 1\} \\
\qquad N^{then} = cond(N''', t\ s, x, \langle f_1^x, \ldots f_n^x\rangle, e, \langle f_1^e, \ldots f_n^e\rangle, op) \\
\qquad N^{else} = cond(N''', t\ s, x, \langle f_1^x, \ldots f_n^x\rangle, e, \langle f_1^e, \ldots f_n^e\rangle, neg(op)) \\
\qquad \textbf{warn } \text{“L-values are not single pointers.” } \textbf{if } N'' \not\sqsubseteq_N N'''
\end{array}
$$

$$
\begin{array}{l}
cond(N, t\ s, x, \langle f_1^x, \ldots f_n^x\rangle, e, \langle f_1^e, \ldots f_n^e\rangle, op) = \exists_{\mathcal{X}^T}(N') \\
\quad \text{where } \delta_{ij} = \begin{cases} 1 & \text{if } i = j \\ 0 & \text{otherwise} \end{cases} \\
\qquad N_{ij} = N \sqcap_N \{f_1^x = \delta_{i1}, \ldots f_n^x = \delta_{in}, f_1^e = \delta_{1j}, \ldots f_n^e = \delta_{nj}\},\ i, j \in [0, n] \\
\qquad a_0 = \text{NULL},\ \langle a_1, \ldots a_n\rangle = \mathcal{A} \\
\qquad N' = \bigsqcup_{i,j \in [0,n]} addAddress(N_{ij}, t\ s, x, a_i, e, a_j, op)
\end{array}
$$

$$
\begin{array}{l}
addAddress(N, t\ s, x, \text{NULL}, e, \text{NULL}, op) = intersect(N, t\ s, x, e, op) \\
addAddress(N, t\ s, x, a_x, e, \text{NULL}, op) = \exists_{\mathcal{X}^T}(N'') \\
\quad \text{where } N' = N \sqcap_N \{4096 \leq o < 2^{30} + 2^{31}\},\ o \in \mathcal{X}^T \text{ fresh} \\
\qquad N'' = intersect(N', t\ s, x + o, e, op) \\
addAddress(N, t\ s, x, \text{NULL}, e, a_e, op) = \exists_{\mathcal{X}^T}(N'') \\
\quad \text{where } N' = N \sqcap_N \{4096 \leq o < 2^{30} + 2^{31}\},\ o \in \mathcal{X}^T \text{ fresh} \\
\qquad N'' = intersect(N', t\ s, x, e + o, op) \\
addAddress(N, t\ s, x, a_x, e, a_e, op) = \exists_{\mathcal{X}^T}(N''') \\
\quad \text{where } N' = N \sqcap_N \{4096 \leq o_x < 2^{30} + 2^{31}\},\ o_x \in \mathcal{X}^T \text{ fresh} \\
\qquad N'' = N' \sqcap_N \{4096 \leq o_e < 2^{30} + 2^{31}\},\ o_e \in \mathcal{X}^T \text{ fresh} \\
\qquad N''' = \begin{cases}
intersect(N'', t\ s, x + o_x, e + o_e, op) \\
\qquad\qquad\qquad\qquad \text{if } L(a_x) \in \mathcal{D} \vee L(a_e) \in \mathcal{D} \\
intersect(N'' \sqcap_N \{o_x < o_e\}, t\ s, x + o_x, e + o_e, op) \sqcup_N \\
intersect(N'' \sqcap_N \{o_x > o_e\}, t\ s, x + o_x, e + o_e, op) \\
\qquad\qquad\qquad\qquad \text{if } a_x \neq a_e \\
intersect(N'' \sqcap_N \{o_x = o_e\}, t\ s, x + o_x, e + o_e, op) \\
\qquad\qquad\qquad\qquad \text{if } a_x = a_e
\end{cases}
\end{array}
$$

$$
\begin{array}{l}
intersect(N, t\ s, e_x, e_y, op) = \exists_{\mathcal{X}^T}(N''') \\
\quad \text{where } N' = wrap(N \triangleright x := e_x, t\ s, x),\ x \in \mathcal{X}^T \text{ fresh} \\
\qquad N'' = wrap(N' \triangleright y := e_y, t\ s, y),\ y \in \mathcal{X}^T \text{ fresh} \\
\qquad N''' = \begin{cases} N'' \sqcap_N \{x < y\} \sqcup_N N'' \sqcap_N \{x > y\} & \text{if } op = \text{'}\neq\text{'} \\ N'' \sqcap_N \{x\ op\ y\} & \text{otherwise} \end{cases}
\end{array}
$$

Fig. 13.8. Abstract transfer function for the revised conditional.

lies in $[4096, 2^{30} + 2^{31} + s]$, where s is the maximum value of the pointer offset x. Observe that the two cases presented so far are key to implementing the common test p==NULL for some pointer p. Consider the calculation of N_{00}; that is, the state in which it is assumed that both sides have no l-values. The first case applies since $a_x = a_e = \text{NULL}$. The state space remains unchanged since *intersect* intersects the state N_{00} with the tautologous $\{0 = 0\}$. Now consider the calculation of N_{i0} – that is, the fact that the left-hand side contains a pointer to a_i. The second case of *addAddress* applies, which replaces the l-value with the range $o \in [4096, 2^{30} + 2^{31}]$. In *intersect*, the state is restricted to $N_{i0} \sqcap_N \{o = 0\} = \bot_N$. Thus, all states N_{i0} of the join operation in *cond* are empty so that no state in which a flag of p is set contributes to the final state. As a result, the upcoming statements are evaluated with flags that are all zero for p implying an empty points-to set. The dual test p!=0 is analogous.

The third pattern of *addAddress* implements the case that is symmetric to the second pattern and requires no further explanation. The fourth case implements the comparison of two non-NULL pointers. In this case, two variables o_x and o_e are created and restricted to the range of possible concrete addresses. However, three cases are distinguished when comparing the two expressions. In the first case, two abstract addresses are compared of which at least one is a dynamically allocated heap region, which may exist several times in the concrete program. Adding o_x and o_e to the expressions effectively inhibits the inference of any useful information. This solution is probably as precise as possible, considering that even an equality test between pointers cannot refine any points-to relationship, as both pointers already point to a single abstract address. In the second case, the abstract addresses are different, implying that their concrete addresses must be different, too. Since a disequality $o_x \neq o_e$ cannot be expressed in the polyhedral domain, the two states $N''\{o_x < o_e\}$ and $N''\{o_x > o_e\}$ are intersected with the condition separately and the results are joined. Given that pointers can only be meaningfully compared using the operators $\{=, \neq\}$, these two states are sufficiently precise to return either an empty result (if $op \equiv {}'{=}'$) or the unchanged state (if $op \equiv {}'{\neq}'$). The third case applies when pointers contain the same l-value. These pointers have the same concrete address, and thus the refined state $N'' \sqcap_N \{o_e = o_x\}$ is passed to *intersect*. Note that if the pointer offsets x and e are in $[0, 2^{30}]$, *wrap* does not alter the state, so that, for example, $x + o_x < e + o_e$ reduces to $x < e$ since $o_x = o_e$. Thus, the last case effectively compares the pointer offsets.

An interesting observation is that the variables o_x and o_e represent nothing but the address of a memory region in the actual program. The transfer function specifies these addresses in the form of temporary variables that are projected out after *intersect* returns. While this strategy is prudent to keep the number of polyhedral variables low, the scheme can be generalised by introducing a polyhedral variable for every abstract address that is always present in the polyhedron. This variable representing the address could then be added instead of the temporaries o_x and o_e. Tracking the possible value of every abstract address enables an analysis of programs that examine the relative locations of memory regions. While calculations on pointers to different memory regions are undefined according to the C standard, such operations can be useful in the context of embedded systems, where certain memory regions lie at fixed addresses. In this case, the polyhedral variables that represent abstract addresses could be fixed to the known address of the memory region. Note that in this case other functions, namely $clear^{F,A}$, which converts an abstract address to a value, and the function *wrap*, need adjusting, too. Differentiating actual addresses in memory is more challenging in the presence of heap-allocated regions which are summarised by a single abstract address.

Note further that expressions involving pointers must always occur over **uint32** and that a pointer expression should never wrap. Hence, a practical implementation will replace the call to *wrap* with a simple test that raises a warning if one of the pointer expressions exceeds the **uint32** range.

The technique of combining a pointer offset with the range $[4096, 2^{30}+2^{31}]$ of possible addresses also needs to be used to adapt the semantics of the cast statement. Rather than elaborating on this and various other transfer functions, we conclude with the presentation of the revised abstraction relation.

13.2.4 Incorporating Boolean Flags into the Abstraction Relation

This section considers the necessary changes to the abstraction, consisting of the relation $\propto$, the concretisation function $\gamma_\rho : (Num \times Pts) \to \Sigma$, the address map $\rho : \mathcal{A} \to [0, 2^{32}-1]$, and the function $mem_\rho : (\mathbb{Z}^{|\mathcal{X}|} \times Pts) \to \mathcal{P}(\Sigma)$ synthesising concrete memory states. Changes in the first three definitions are straightforward when considering that the points-to domain $A \in Pts$ is constant and can therefore be removed from the analysis. Only the function mem_ρ, presented on p. 100, needs to be redefined as $mem_\rho^A : \mathbb{Z}^{|\mathcal{X}|} \to \mathcal{P}(\Sigma)$:

$$mem_\rho^A(\mathbf{v}) = \bigcap_{\substack{m\in\\ \mathcal{MUD}}} \left(\bigcap_{\substack{a\in\\ \rho(L(m))}} \left(\bigcap_{\substack{\langle o,s,x_i\rangle\in\\ F(m)}} \{bits^s_{a+o}(\pi_i(\mathbf{v}) + \mathbf{v}\cdot\mathbf{p}) \mid \mathbf{p} \in f_\rho^A(s, x_i)\} \right)\right)$$

Here, $f_\rho^A(s, x_i)$ creates a vector of concrete addresses that is multiplied with the value vector $\mathbf{v} \in [\![N]\!]$. For all field sizes $s \neq 4$, the function simply returns $f_\rho^A(s, x_i) = \mathbf{0}$, and hence mem_ρ^A simply restricts the field to a pure value determined by $\pi_i(\mathbf{v})$, the ith element of $\mathbf{v}$. For pointer-sized fields, $f_\rho^A(4, x_i)$ is defined in terms of the set of abstract addresses $\langle a_1, \ldots a_n\rangle = \mathcal{A}$ and the flags $\mathbf{f} = A(x_i)$. Specifically, let $\mathbf{f} = \langle x_{i_1}, \ldots x_{i_n}\rangle$; that is, the n flags correspond to the domain variables $x_{i_1}, \ldots x_{i_n}$. We define $f_\rho^A(4, x_i)$ as follows:

$$f_\rho^A(4, x_i) = \left\{ \langle 0, \ldots, 0, \underbrace{p_1}_{i_1}, 0, \ldots, 0, \underbrace{p_n}_{i_n}, 0, \ldots 0\rangle \;\middle|\; \begin{array}{l} p_j \in \rho(a_j), j = 1, \ldots n \\ \langle x_{i_1}, \ldots x_{i_n}\rangle = A(x_i) \end{array} \right\}$$

In the vector above, the annotations $i_1, \ldots i_n$ denote the index position of $p_1, \ldots p_n$ in the vector. Thus, the function returns all possible address vectors in which the concrete addresses of the abstract addresses $a_1, \ldots a_n$ are arranged in such a way that the product $\mathbf{v} \cdot \mathbf{p}$ in mem_ρ^A multiplies these addresses with the values of the flags $\mathbf{f}$ that determine if a certain address is added to the field or not. In order to illustrate this process, consider a program variable v with $\langle 0, 4, x_v \rangle \in F(\mathrm{v})$ and $A(x_v) = \langle f_p^1, f_p^2 \rangle$ that determines if v points to one of the two abstract addresses $\mathcal{A} = \langle a_p^1, a_p^2 \rangle$. Let $\rho(a_p^1) = \{p_1, p_1'\}$ and $\rho(a_p^2) = \{p_2\}$, and let the set of abstract variables be $\mathcal{X} = \langle x_v, f_p^1, f_p^2 \rangle$. In order to determine the set of concrete stores, observe that $f_\rho^A(4, x_v) = \{\langle 0, p_1, p_2 \rangle, \langle 0, p_1', p_2 \rangle\}$. Given an abstract state N such that $[\![N]\!] = \{\langle 0,0,0 \rangle, \langle 0,1,0 \rangle, \langle 0,0,1 \rangle\}$, four concrete values of v are possible, namely 0 (from $\langle 0,0,0 \rangle$), p_1, p_1' (from $\langle 0,1,0 \rangle$), and p_2 (from $\langle 0,0,1 \rangle$). Note that Boolean flags are mere abstract domain variables. As such, they may take on other values besides zero and one. In particular, it is possible to store the negation of a pointer in a variable and add the same pointer (possibly with a different offset) later on. Such an operation is undefined according to the C standard [51] but unlikely to be evaluated differently from our model. In particular, an optimising compiler may reorder a linear expression `d=o+p-q` to `d=o-q+p`, which resembles the example above if `p` and `q` are pointers. The evaluation of expressions in Fig. 13.7 makes use of this property in that the points-to vectors of variables are simply added.

The implementation described so far uses a vector of flags for each pointer-sized field, and each vector has an element for every variable in the program. Thus, incorporating the points-to analysis adds a quadratic number of variables to the numeric domain, which is prohibitive for a practical analysis. Thus, it is prudent to remedy this presentational artifact of using vectors of flags for each pointer, which is the topic of the next section.

13.3 Practical Implementation

In this section, we comment on the implementation of the concepts presented so far. In particular, we focus on improving efficiency and point out restrictions of our approach. With respect to efficiency, the next section details how to finesse the use of Boolean vectors for modelling points-to sets by inferring a sufficient set of flags for each variable. Section 13.3.2 extends this automatic inference to abstract addresses and generalises the technique by allowing several abstract addresses per memory region, thereby improving the precision of the analysis with respect to string buffer analysis. Instead of a Boolean flag, it is also possible to distinguish several states using a numeric variable. Section 13.3.3 demonstrates this by relating the elements of an array with the index of each element. We show that a numeric value cannot fully distinguish individual elements and hence that it is weaker than a Boolean flag. Section 13.3.4 concludes with an alternative to model points-to information.

13.3.1 Inferring Points-to Flags on Demand

Chapter 10 introduced the concept of typed domain variables, in which a default value is assigned to a variable that is not mentioned in a domain. Typed domain variables are key to populating the map of fields on-the-fly. Not surprisingly, this concept can easily be applied to other maps in the analyser. Specifically, rather than associating a fixed vector $\mathbf{f} \in A(x)$ with every pointer-sized field x, the points-to map $A \in Pts$ is continuously augmented such that it only holds a flag for l-values that have at one point been assigned to this field. By typing these points-to flags such that their default value is zero, this map can always be extended without the need to update other polyhedra, as the interpretation of the new flag f is that the corresponding abstract address is not part of the variable's points-to set. To this end, the analysis keeps a single map $A \in Pts = \mathcal{A} \to \mathcal{P}(\mathcal{A} \times \mathcal{X})$, which stores a set of tuples $\{\langle a_{i_1}, f_1, \rangle, \dots \langle a_{i_n}, f_n \rangle\}$ for each pointer-sized field, where $\{a_{i_1}, \dots a_{i_n}\}$ is usually a very small subset of $\mathcal{A}$. In particular, the quadratic number of case distinctions in the abstract transfer function of the conditional is finessed since the right-hand side is usually a constant rather than an expression involving l-values.

Note that augmenting a single global points-to map $A \in Pts$ weakens the flow-sensitive points-to analysis used in the early chapters to a flow-insensitive analysis in which flow sensitivity is recovered by consulting the flags in the numeric domain. Furthermore, even the flow-insensitive information that is present in the addresses of the tuples of A can be more precise than a classical subset-based points-to analysis [3,99]. In particular, the result is more precise if, in an assignment from the field $x \in \mathcal{X}$ to $y \in \mathcal{X}$, the abstract address a with $\langle a, f \rangle \in A(x)$ is not added to the set of tuples $A(y)$ whenever f is zero.

Populating the map of fields and the points-to map on-the-fly has the obvious advantage of reducing the number of variables the analysis needs to track, thereby improving the efficiency. The next section discusses the merit of creating abstract addresses on-the-fly.

13.3.2 Populating the Address Map on Demand

The ability to add fields and points-to flags on-the-fly significantly reduces the number of abstract variables used in the analysis. A further reduction in the number of tracked variables can be achieved by restricting the number of manipulated NUL positions. The string buffer analysis in Chap. 11 was defined to infer a NUL position for each abstract address $a \in \mathcal{A}$. Here, the set of abstract addresses was given by the bijective address map $L : \mathcal{M} \cup \mathcal{D} \to \mathcal{A}$. Hence, since one abstract variable is tracked for each NUL position, the number of variables required is equal to the number of memory regions $m \in \mathcal{M} \cup \mathcal{D}$. The crucial observation is that no abstract address is necessary for memory regions whose address is never taken, as these cannot be accessed through a pointer. Since the address of most variables is never taken, the number of variables representing NUL positions is significantly reduced.

Populating the address map on demand opens up an interesting opportunity for improving the precision of the analysis. A loss of precision in the analysis presented so far occurs when arguing about string buffers that are embedded in a structure such as the following:

```
struct {
  char firstname[80];
  char surname[120];
} client;
```

Recall that translating a program to Core C removes all information on data structures, and hence the structure above is represented as a memory region of 200 bytes. Suppose that s represents a variable of the type above. A program that writes a NUL-terminated string to `s.surname` cannot be analysed precisely since the only NUL position for this buffer is tracked relative to the beginning of the whole structure. Worse, suppose a flag `f` decides whether the first or the last name is supposed to be edited. Then the statement

```
if (f) p = &s.firstname; else p = &s.surname;
```

might precede a statement that writes a new NUL-terminated string to p. Even in the revised model, this example cannot be analysed precisely since for both branches p points to the beginning of the memory region s, while the offset of p lies in $[0, 80]$, the best approximation of the offsets 0 and 80 of the two arrays. As a result, evaluating the equations for string buffer accesses cannot infer a definite NUL position within `firstname` since the access position $[0, 80]$ implies that any zero byte might be written beyond the end of the first array.

In order to preempt the precision loss above, observe that the rule for taking an address in Core C is v . n = & v . n ; that is, the rule allows for an offset within the memory region whose address is taken. We exploit this fact by allowing several abstract addresses per memory region. In particular, each time an address-of operator is evaluated, a new abstract address is created at the given offset unless an address already exists at that offset, in which case it is reused. To this end, redefine the bijection $L : \mathcal{M} \cup \mathcal{D} \to \mathcal{A}$ to the partial map $L : (\mathcal{M} \cup \mathcal{D}) \times \mathbb{N} \to \mathcal{A}$, which, for certain offsets, associates an abstract address with a memory region. Using this approach, the evaluation of the conditional above results in the pointer p having a constant offset of zero, albeit having a points-to set containing the abstract address $L(\mathtt{s}, 0)$ for `s.firstname` and the address $L(\mathtt{s}, 80)$ for `s.surname`. As a consequence, two NUL positions are tracked for the structure, one for each array. Allowing several abstract addresses per memory region requires cross-cutting changes to the analysis. For instance, the updates of the NUL position in Fig. 11.6 on p. 210 need to ignore accesses in front of the NUL position since these are not erroneous if

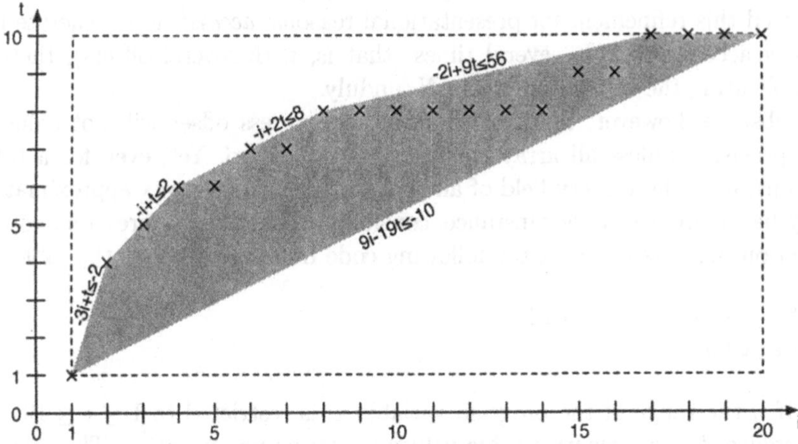

Fig. 13.9. Arrays containing constants can be accessed in a way that creates a linear relationship between the access position and the constant.

the abstract address has an offset. Furthermore, the function *addAddress* in Fig. 13.8 may no longer assume that two different abstract addresses denote two non-overlapping memory regions. In particular, whenever two abstract addresses correspond to the same memory region, the pointers are compared as if the abstract addresses were the same except that the offset difference between the two pointers has to be incorporated. Since these changes are mostly technical, they are omitted for simplicity.

The following section discusses an orthogonal generalisation of using Boolean flags. In particular, we consider how and if a single variable can be used to distinguish more than one state.

13.3.3 Index-Sensitive Memory Access Functions

While it is clear that the precision of an analysis depends not only on the underlying domain but also on the abstract transfer functions, it is often less clear if and how the transfer functions can be refined. This section points out a possible refinement for the functions that model the access to memory. In particular, the precision of the $readChk^{F,A}$ function can be improved by making it sensitive to the offset of the access. For instance, suppose a table containing constants is given whose values correspond to the crosses in Fig. 13.9. The $readChk^{F,A}$ function will create an access tree for each constant and join the result of all the extracted values, yielding the state space indicated by the dashed box. A more precise result can be obtained by intersecting the state N_i in the first line of the function with a constraint that expresses that the current access position is that of the access tree that is about to be assigned to x and f. This approach would yield the grey state space in Fig. 13.9. We omitted this refinement for presentational reasons: $access^F$ may generate the empty access tree $\varepsilon_{\mathcal{AT}}$ several times (that is, with several offsets), thereby complicating the restriction of the N_i unduly.

Observe, however, that incorporating the access offset will not enhance the precision unless all array elements are populated. Yet, even for a table of constants where every field of an array is given, the convex approximation may be too imprecise. For instance, consider a table t that stores a monotone function, and assume that the following code fragment accesses the table:

```
int d = t[i+1]-t[i];
assert(d>=0);
```

If i, the range of the program variable i, is restricted to $1 \leq i \leq 19$, the difference d of consecutive table values is always non-negative. The original $readChk^{F,A}$ function infers the range $[4, 10]$ for t[i+1] and $[1, 10]$ for t[i], resulting in $[4, 10] - [1, 10] = [-14, 9]$ for a bound on the difference d.

Figure 13.10 shows how the precision of this answer improves to $[-3, 4]$ when incorporating the access position. Evaluating t[i+1]-t[i] in the polyhedral domain corresponds to calculating the difference between the state space in Fig. 13.9 and the same state shifted one unit to the right (for t[i+1]). The convex approximation of the table values leads to the overlapping state in dark grey in Fig. 13.10, for which it cannot be shown that the difference is non-negative. This is depicted in the lower graph, which shows d in relation to the index i after integral tightening is applied. The crosses in the lower graph indicate the exact result, which is always non-negative. Thus, even when the index position is incorporated into the array access, it is not possible to prove that d is non-negative, although the maximum range of d is now $[-3, 4]$ and thereby considerably more precise than the solution $[-14, 9]$, which ignored the access position. While incorporating the access position into $readChk^{F,A}$ may increase the precision, this added precision does not seem sufficient to be relevant in practice, as only linear relationships between the index and the content of the array can be expressed. Since a simple linear function would hardly be stored in a table of constants, incorporating the index position into array accesses never yields precise results. The Astrée analyser is able to prove the assertion by analysing the code fragment separately for every value of i, a behaviour triggered by a heuristic or an annotation in the analysed program [69]. This behaviour could be simulated by adding a Boolean flag for each array index. A less costly solution is to partition the indices of the constant table into strictly increasing ranges and constant ranges and to split the control-flow path for every such range. In the example, the ranges $[1, 1], [2, 4], [5, 6], [7, 8], [9, 13], [14, 15], [16, 17], [18, 20]$ constitute such a partitioning. The relations between each index range and the constants of the table are linear, so no precision is lost due to convexity, and monotonicity can

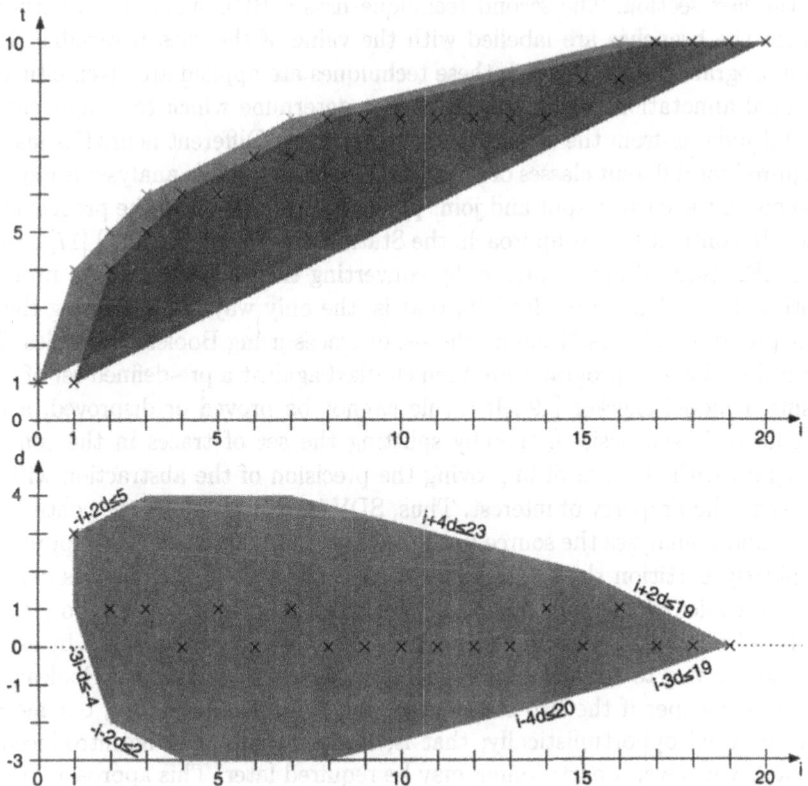

Fig. 13.10. Incorporating the access position cannot prove monotonicity of a table.

be proved. In general, such a refinement can hardly be done automatically, as it requires an understanding of the property to be proved and the reason why precision is lost.

13.3.4 Related Work

When using a convex approximation to the possible state space of a program, a separate analysis of the same piece of code is often an easier solution to recover from a loss of precision in the join operator than designing a new domain. For instance, analysing the division operation in Sect. 13.1 twice is easier than adding a new domain that expresses disequalities of the form $d \neq 0$ and ensuring the propagation of information from this to the numeric domain.

The Astrée analyser [60] distinguishes two ways of partitioning traces. The first is to keep two control-flow paths separate when they join. The second is a partitioning according to the value of a variable. An application of the first technique is the division example and also any kind of loop unrolling. The second technique is used for problems such as the example on accessing arrays in the last section. The second technique uses a BDD-like data structure in which the branches are labelled with the value of the chosen variable [123]. The program points at which these techniques are applied are given either by manual annotations or by heuristics that determine which technique can be useful judging from the structure of the program. Different heuristics may be required for different classes of programs. Thus, the Astrée analyser is required to choose the correct split and joint points in order to prove the program correct. In contrast to this approach, the Static Driver Verifier (SDV) [17] verifies the API usage of device drivers by converting their C source into a program with only Boolean variables [20]; that is, the only way of abstracting the input program is by partitioning the set of traces using Boolean variables. The resulting Boolean programs are then checked against a pre-defined set of rules using a model checker [19]. If a rule cannot be proved or disproved, a new predicate is synthesised, thereby splitting the set of traces in the concrete program with the aim of improving the precision of the abstraction with respect to the property of interest. Thus, SDV iteratively splits the control-flow path and reanalyses the source until the rules can be proved. Both approaches explicitly partition the set of traces so that the cost of the analysis depends on the choice of the split and join points. In this chapter, we propose to use polyhedral analysis but implement the split of a control-flow path by adding a Boolean flag to the domain. The observation is that adding a Boolean flag may be cheaper if the two states it separates are similar. Thus, our method can be used opportunistically; that is, Boolean flags may be introduced by default whenever a partitioning may be required later. This approach may be cheaper than a reanalysis of the same code as long as the states separated by the flag are reasonably similar.

Alternative Approaches to Improving Points-to Information

Using Boolean flags to indicate whether a given l-value is part of a points-to set provides a very expressive points-to analysis but requires a domain that can express these Boolean relationships. Furthermore, the number of Boolean flags required to analyse a program may become large if a program has variables with large points-to sets. In this case, the points-to analysis presented in Chaps. 5 and 6 may be used. Although the string-copying loop in the example cannot be proved using the set-based points-to domain, it might be possible to prove other program properties. Furthermore, the set-based points-to analysis can be refined in order to mend some of its deficiencies. To this end, consider the following variation on the string-copying example:

```
char *p;
char s[16] = "Propeller";
char t[16] = "Aero";
char *u = "plane␣";
```

```
char *o;
p = t+4;
if (rand()) p = &s+5;
o=p; /* Safe p before it is modified. */
while (*p=*u) { p++; u++; };
printf("copied␣%i␣characters\n", p-o);
```

As with the original example in Chap. 11 on string buffer analysis, a single string constant u is appended to the buffer pointed to by p. However, p may point either to the end of the string "Aero" or "Propeller", depending on a random number. Suppose that $a_s \in \mathcal{A}$ and $a_t \in \mathcal{A}$ represent the abstract addresses of the buffers s and t, respectively, and that the abstract variables $p, o \in \mathcal{X}$ represent the values of p and o, respectively. Observe that the points-to set of p after the conditional is $A(p) = \{a_s, a_t\}$. Since p is saved in o before it is modified in the loop, the variable o will contain the same pointer and, hence, $A(o) = \{a_s, a_t\}$. However, when evaluating the difference p-o, the points-to sets do not convey enough information to ensure that o points to exactly the same buffer as p. Hence, it has to be assumed that the two variables may point to different buffers such that the result of p-o is dominated by the difference between the address of s and the address of t (or vice versa). In practice, this implies that p-o cannot be bounded using simple points-to sets.

In the context of program analysis with dependent types [94], this problem is circumvented. Whenever a points-to set contains two or more abstract addresses, it is summarised by a single abstract address that is a representative of this set. For instance, in the example above, the join after the **if** statement merges two control-flow branches in which, on the one hand, $A(p) = \{a_t\}$ and, on the other hand, $A(p) = \{a_s\}$, a new abstract address $\overline{a} \in \mathcal{A}$ is created that represents the set $\{a_s, a_t\}$, and A is updated such that $A(p) = \{\overline{a}\}$. During the evaluation of the loop, when the actual set of memory regions of $\overline{a}$ is required, the set $\{a_s, a_t\}$ that $\overline{a}$ represents is used. For pointer arithmetic, assignments, and comparisons, however, the representative $\overline{a}$ is used. Thus, the assignment o=p; updates A such that $A(o) = \{\overline{a}\}$, and hence the pointer difference p-o is calculated as $p - o$ since p and o both point to $\overline{a}$.

Only minute changes are necessary in order to adapt the points-to domain *Pts* from Chap. 3 to operate on representative addresses. The entailment check can exploit that two domains A and A' are equal if $A(x) = A'(x)$ for all $x \in \mathcal{X}$, even if a points-to set $A(x)$ represents a set of abstract addresses. If $A(x) \neq A'(x)$, the underlying sets must be retrieved and compared such that the overall semantics of the entailment check is unaffected by the introduction of representative addresses. With respect to the join operation, the result for any given variable $x \in \mathcal{X}$ is $A(x)$ if $A(x) = A'(x)$. Otherwise, the result is a new representative abstract address $\overline{a}$ that corresponds to the set $A(x) \cup A'(x)$. Note that this definition of the join operator may create a new representative in every evaluation of a block. However, as the analyser will find a fixpoint in finite time, the number of newly created representatives is finite, too. Termination is guaranteed, as the semantics of the entailment check reduces to subset tests on the underlying sets.

An interesting challenge is the combination of representative addresses and Boolean flags. In particular, it might be possible to reduce the number of Boolean flags necessary to track points-to sets if a single flag can be used for a representative address rather than several flags for the individual addresses that constitute the representative address. An obstacle to this approach is that creating new representative addresses repeatedly for the same pointer is incompatible with adding Boolean flags to the domain, which is a process that can only be done a finite number of times. Even if these problems can be solved, observe that representative addresses merge the numeric information for the pointers that are summarised, so that the termination of the loop discussed at the beginning of the chapter cannot be shown.

14

Implementation

The building blocks of a static analyser for a programming language resemble those of an interpreter or virtual machine, except that the operations performed for each program statement are expressed in terms of the abstract domain rather than the concrete store. While implementing the semantics of a program statement in the context of an interpreter is a clear-cut task, implementing the semantics in the context of a static analysis provides a plethora of possibilities, partly because it involves a trade-off between precision, efficiency, and simplicity of implementation, the latter possibly affecting the correctness. The design of the analysis presented in the last chapter is the result of trying several approaches, including a staged approach in which an off-the-shelf points-to analysis is run on the code [99] before a constraint system is deduced, which is then solved using polyhedral operations. This approach is similar to that of Wagner [184] and shares the inability to generate precise constraints for pointer dereferences since the offset of a pointer is not known until the constraint system is solved. The idea behind the analysis presented is therefore to generate the operations that manipulate polyhedra by determining which fields a pointer may access. Thus, manipulating polyhedra is interleaved with querying the range of certain variables in the polyhedron in order to derive the next polyhedral operation. While this approach leads to more complex transfer functions, it seems like the only viable approach to a precise analysis.

In this chapter, we present an overview of our analyser, some technical aspects that are important in practice, and an extension that we deem to be important to make the analysis precise enough to verify off-the-shelf C programs. We explain where the abstractions presented are too imprecise and how this inhibits the analysis of larger programs. In particular, Sect. 14.1 provides an overview of our prototype analyser, Sect. 14.2 comments on checking for erroneous state space, and Sect. 14.3 discusses the efficient calculation of a fixpoint in a practical analysis. With respect to the precision of the analysis, Sect. 14.4 discusses inherent limitations of the string buffer analysis, whereas Sect. 14.5 suggests a possible extension to the analysis.

14.1 Technical Overview of the Analyser

This section presents the prototype implementation of the static analysis described. The design of the analyser was guided by the task of analysing the C program `qmail-smtp`, which is part of the `qmail` mail transfer agent, a program to handle emails. The chosen program receives incoming mail over a network connection and is thus prone to buffer-overflow attacks. The program is deemed to be bug free, which makes it a prime candidate for designing a static analysis in that any warning can be attributed to the imprecision of the analyser. Furthermore, the program is reinvoked for every new email such that it is single-threaded and does not use long-lived dynamic data structures.

The basic structure of the analyser is shown in Fig. 14.1. In the context of translating the program `qmail-smtp`, several source files have to be translated, as shown schematically in the top left of the diagram. Using the build infrastructure of the `qmail` suite, each source file is compiled by a modified version of the GNU C compiler, dubbed *gcc* in the figure. The compiler is augmented with a *treewriter* module, which emits an image of the intermediate representation called *tree* as part of its assembler output. Each assembler file is then turned into an object file by the standard assembler *as*. The resulting object files contain three new segments that hold the intermediate structure: one for declarations of memory regions, one for string constants, and another for the actual code in *tree* representation. Running the linker on these object files creates a single binary file with the additional segments. These segments can be examined by a tool dubbed *tree browser*, which proved to be important in understanding the intermediate structure of the compiler. For the sake of the analysis, the extra segments in the binary file are translated into Core C. The translation to Core C simplifies the intermediate structure of GCC and creates a single initialisation block that contains assignment statements for initialised variables and, in particular, an array declaration and initialisation for each string constant. Any function that is not defined in the binary file itself is treated as a primitive, which has to be implemented as such when calculating a fixpoint of the semantics of the program. The utility *Core C printer* converts the binary Core C representation into human-readable form. The Core C examples shown throughout this book are the result of running this tool.

In the common case, the binary Core C file is read by the static analyser, dubbed "fixpoint calculation" in the figure, which can be invoked either with an interactive command-line prompt or with a graphical user interface. Both interfaces allow the user to run the fixpoint computation to completion or to evaluate single blocks or single statements. Furthermore, it is possible to query the TVPI domain, the points-to domain, the current set of landmarks, the fields and addresses of memory regions, and the current work list, which contains the list of blocks that are yet to be evaluated.

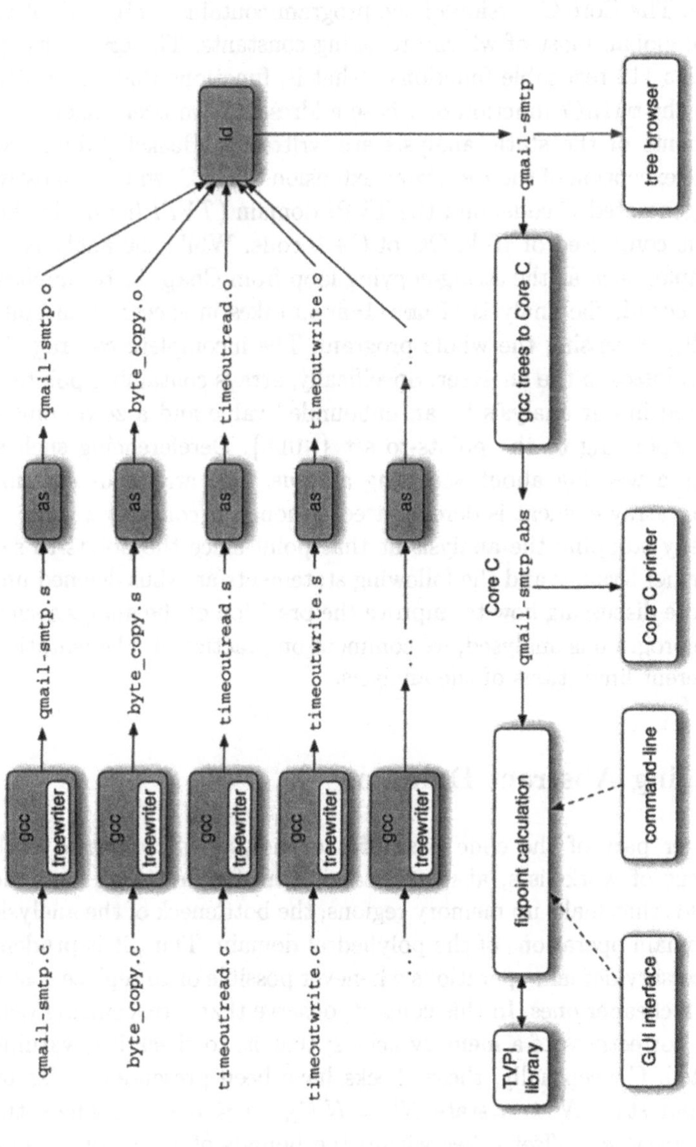

Fig. 14.1. Conceptual overview of the analysis. Grey boxes denote existing components, while white boxes denote components that are specific to the analysis. Note that the information created by the *treewriter* extension is treated as data by the assembler and the linker.

For the example application `qmail-smtp`, which consists of about 3 kLOC (lines of code), the intermediate structure created by GCC consists of 404 kB of code and 17 kB in strings. The intermediate representation translates to 27 kB of variable declarations, 55 kB of Core C code, and 12 kB in strings. For comparison, the raw machine code when compiled for Intel x86 processors is 18 kB in size. The Core C version of the program contains 1,710 variables, of which 60 are global, most of which are string constants. The `qmail-smtp` program contains 119 reachable functions – that is, functions that are either reachable from the `main()` function or whose addresses have been taken.

All components of the static analysis are written in Haskell (about 20 kLOC) with the exception of the *treewriter* extension to GCC, which is mostly automatically generated C code, and the TVPI domain (*TVPI library* in the figure), which is comprised of 10 kLOC of C++ code. While the analysis of the small examples such as the string-copying loop from Chap. 11 terminates in less than a second, the analysis of `qmail-smtp` takes in excess of an hour without actually processing the whole program. The incomplete coverage is due to precision losses in the analyser. Specifically, arrays containing pointers are approximated in our analysis by an unbounded value and a zero points-to vector, corresponding to the points-to set {NULL}. Dereferencing such a pointer triggers a warning about accessing a value. The erroneous circumstance where an array element is dereferenced although it contains a value is removed, thereby stopping the analysis at that point since the points-to set contains no further l-values and the following statements are thus deemed unreachable. Before discussing how to improve the precision of the analysis such that the whole program is analysed, we comment on practical implementation issues and inherent limitations of the analysis.

14.2 Managing Abstract Domains

While the larger part of the code base of the analyser is concerned with the management of work lists, abstract states, transfer functions, and the handling of fields that make up memory regions, the bottleneck of the analysis remains the domain operations of the polyhedral domain. Thus, it is prudent to avoid unnecessary domain operations whenever possible or to replace costly operations with cheaper ones. In this context, observe that a reoccurring task is to check for correctness of a memory access; that is, to check if a warning must be emitted. Conceptually, these checks have been presented so far by refining an input state N to a state $N' = N \sqcap_N \{0 \leq o < s\}$, where the invariant that an access offset o lies within the bounds of a memory region of s bytes is expressed using the meet operator. If this invariant holds in N, then the constraint $0 \leq o < s$ is redundant in N and N' is no smaller than N; that is, $N \sqsubseteq_N N'$. Implementing this strategy directly requires a copy of the original domain N, the calculation of the intersection, and an entailment check. By enhancing the meet operation, it is possible to implement

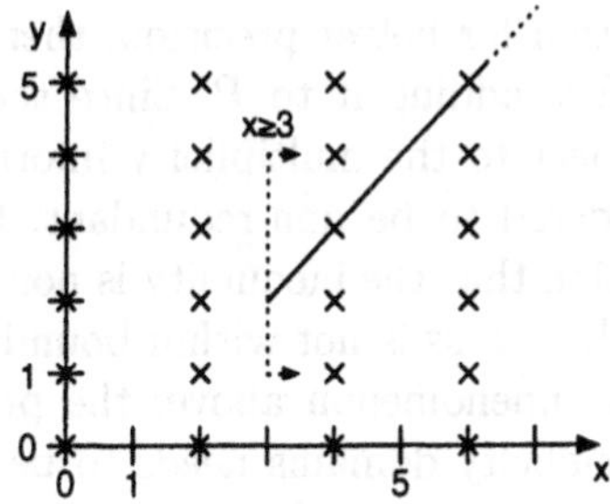

Fig. 14.2. Intersecting a domain with a redundant constraint. The crosses mark feasible points if x is a multiple of two.

a cheaper strategy that does not require a copy of the domain or an entailment check. Specifically, the meet operation is refined to distinguish three possible outcomes:

unsatisfiable: adding the constraint rendered the domain unsatisfiable
changed: the domain has changed but is still satisfiable
redundant: the constraint was redundant and the domain remains unchanged

The last outcome permits a cheaper implementation of the correctness checks. Rather than copying the current state and using an entailment check to determine if the constraint that expresses the bound check is redundant, it is sufficient to merely add the constraint and to check if it was redundant. If the program can be proved to be correct, all bound checks will be redundant, so enhancing the meet operation is a major performance improvement.

One subtlety that arises in this approach is that it is not always easy to determine if a constraint is indeed redundant. Ideally, a constraint is flagged as redundant if adding it to the inequality system leaves the state space represented by the new system unchanged. However, in the context of the integral TVPI domain, the inequalities and the multiplicity information constitute only an approximation to the enclosed integral set of points. Thus, a new constraint that is added to a domain might change the representation of the state space without changing the set of feasible integral points. We have observed this phenomenon when the multiplicity domain and the polyhedral domain are not implemented as a reduced product – that is, if both domains are used separately. For example, suppose that the inequalities $x = y + 1$ and $y \geq 2$ are added to an initially empty system. Furthermore, suppose the multiplicity is given by $M(y) = 0$ and $M(x) = 1$. The system $P = [\![x = y + 1, y \geq 2]\!]$ is shown in Fig. 14.2, where crosses mark those vertices that adhere to the multiplicity information. It can be seen that the lower bounds of x and y are not fully tightened since x must be a multiple of two. Now consider adding the redundant inequality $x \geq 3$. Even though the inequality is redundant with respect to P, the implementation of the meet operation will incorporate the multiplicity information for better precision, thereby tightening the new inequality to $x \geq 4$ before adding it to P. Since the bounds of P are not fully tightened with respect to the multiplicity information M, the new inequality $x \geq 4$ is considered to be non-redundant. If $x \geq 3$ arises from a bound check, the deduction that the inequality is non-redundant leads to the incorrect warning that the access is not within bounds.

In order to avoid the phenomenon above, the propagation between the polyhedral and the multiplicity domains needs to be improved. This goal is best served by avoiding the need for explicit propagation of information, which is achieved by implementing both domains as a reduced product. However, this new numeric domain still represents an approximation to the set of contained integral points in that a rational TVPI system is used to track integral points. Thus, adding constraints that are redundant with respect to the contained integral points may still change the representation of the domain and may thereby incorrectly trigger a warning. Tightening each polyhedron around the contained integral points would guarantee that a redundant constraint is flagged as such. However, as pointed out in Chap. 9 on integral TVPI polyhedra, ensuring that a polyhedron is always tightened to a $\mathbb{Z}$-polyhedron is too costly. Hence, flagging redundant constraints as non-redundant cannot be completely avoided.

Note that these problems are not necessarily linked to the way invariants are checked: Even when performing a bounds check by copying a reference domain and using an entailment test to check for a change, the same problems regarding different representations of state spaces may occur. In particular, intersecting the copy of the original domain with a redundant inequality might change the representation of the copy, even though the set of contained integral points does not change. Thus, the test if the original domain is entailed in the copy might fail since the representation of the copy has changed with respect to the representation of the original.

The next sections elaborate on how the number of domain operations can be reduced by using different iteration strategies. While an efficient bound check reduces the cost of evaluating a transfer function, a good iteration strategy may reduce the number of basic blocks that need to be evaluated. As such, the potential for a speedup is much larger.

14.3 Calculating Fixpoints

A reoccurring challenge in the analysis of programs is the inference of loop invariants. The presence of loops requires that the possible state space of a program be calculated as a fixpoint. The idea is to store an abstract state for each basic block that is valid at the beginning of that block. For an efficient analysis, the fixpoint calculation is based on a chaotic iteration strategy [34] in which the analysis executes some basic block whose input state has not

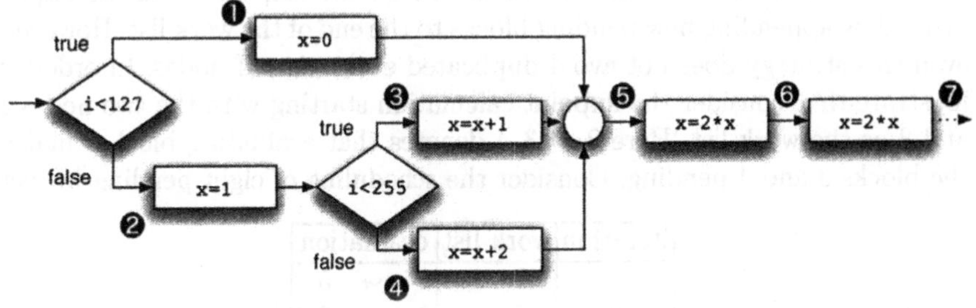

Fig. 14.3. Evaluating conditionals without repeated analysis of following blocks.

yet been propagated until a fixpoint is reached. In particular, analyses often employ a work list of basic blocks that are not yet stable; that is, blocks that have not been evaluated with their current input state. In each analysis step, one basic block is taken off the work list, its semantics is evaluated, and the new state is propagated along the outgoing control-flow edges. Each basic block that thereby receives a new input state is added to the work list. The process is repeated with the next basic block from the work list until the list is empty, at which point all basic blocks are stable and a fixpoint is reached.

14.3.1 Scheduling of Code without Loops

The challenge of an efficient fixpoint engine lies in finding a scheduling of basic blocks that minimises the number of times each basic block is evaluated. In order to illustrate the problems involved, consider the control-flow graph of two simple conditionals depicted in Fig. 14.3. Assume that the previous basic block has led to the evaluation of the condition `i<127`, in which both branches received a larger state. Given scheduling based on a work list, the two pending basic blocks are stored as the list 1,2. A simple FIFO propagation strategy consists of removing the first element from the work list, namely 1, evaluating the corresponding basic block, and adding any new pending basic blocks at the beginning of the list. In the context of the example, this corresponds to evaluating block 1, which will make block 5 pending, and it is hence prepended to the list. Thus, block 5 is executed next, making block 6 pending, and this is then executed. As soon as the evaluation of a basic block does not result in any new pending basic blocks, basic block 2 will be evaluated. As before, the state will be propagated to blockd 3, 5, 6, and so on. Note that before the evaluation of block 5, the output state of block 3 is joined with that of block 1 and evaluation continues with the joined state. Since the join includes the state space resulting from the evaluation of block 1, the previous evaluation of the blocks 5, 6, etc., is made obsolete. A depth-first iteration strategy is thus unsuitable for implementing an efficient analyser.

As an alternative, consider a breadth-first traversal, which can be implemented by appending new pending blocks to the end of the work list. However, even this strategy does not avoid duplicated evaluation of nodes. In order to illustrate this, consider the fixpoint calculation starting with the two nodes 1 and 2 as the work list. Here $2 \rightsquigarrow 3, 4$ denotes that evaluating block 2 makes the blocks 3 and 4 pending. Consider the scheduling of eight pending blocks:

iteration	work list	evaluation
1	1,2	1 $\rightsquigarrow$ 5
2	2,5	2 $\rightsquigarrow$ 3,4
3	5,3,4	5 $\rightsquigarrow$ 6
4	3,4,6	3 $\rightsquigarrow$ 5
5	4,6,5	4 $\rightsquigarrow$ 5
6	6,5	6 $\rightsquigarrow$ 7
7	5,7	5 $\rightsquigarrow$ 6
8	7,6	7 $\rightsquigarrow$ ...

Iteration 7 of the fixpoint calculation is noteworthy because the sequence 5,6,7 is analysed a second time. In order to ensure a minimum of repeated evaluations, we impose the restriction that a basic block may not be evaluated until all incoming nodes have fired – that is, updated the state of that block. Specifically, each block on the work list is associated with a set of edges that still need to fire before the evaluation of that block proceeds. If this set is empty and an incoming edge updates the state, the set is updated to all incoming edges except the one that just fired. Calculating a fixpoint using this strategy results in the following iterations, where the work list consists of tuples containing the basic block number and the set of incoming edges that have not yet fired. The first iterations now schedule the following blocks:

iteration	work list	evaluation
1	$\langle 1, \emptyset \rangle, \langle 2, \emptyset \rangle$	1 $\rightsquigarrow$ 5
2	$\langle 2, \emptyset \rangle, \langle 5, \{3, 4\} \rangle$	2 $\rightsquigarrow$ 3,4
3	$\langle 5, \{3, 4\} \rangle, \langle 3, \emptyset \rangle, \langle 4, \emptyset \rangle$	3 $\rightsquigarrow$ 5
4	$\langle 5, \{4\} \rangle, \langle 4, \emptyset \rangle$	4 $\rightsquigarrow$ 5
5	$\langle 5, \emptyset \rangle$	5 $\rightsquigarrow$ 6
6	$\langle 6, \emptyset \rangle$	6 $\rightsquigarrow$ 7
7	$\langle 7, \emptyset \rangle$	7 $\rightsquigarrow$ ...

The strategy above works well if the evaluation of each block leads to a new state for the following block. If the evaluation of a block results in a smaller or equal state space, the computation should stop since a fixpoint along this path has been reached. However, simply removing a block from the work list whose input state is stable will prevent later blocks from running. For instance, if the evaluation of block 3 in the example does not lead to a bigger state, the incoming edge 3 of block 5 will never fire, and hence the new state of block 5 is not propagated.

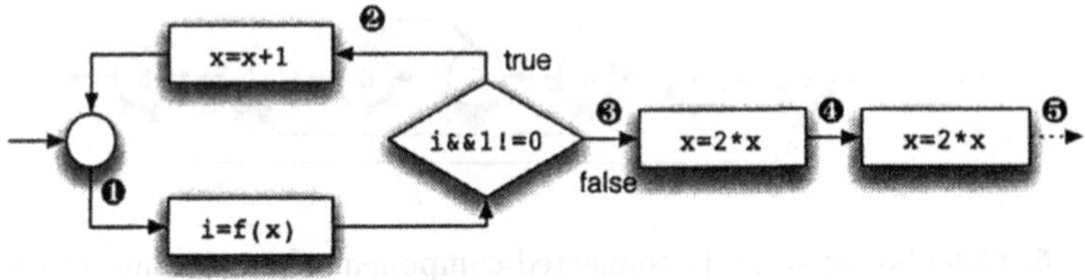

Fig. 14.4. Evaluating loops may update the following basic block several times.

One solution to this problem is merely to skip the actual evaluation of the basic block whenever the input state is already stable. Specifically, when a block with a stable input state is scheduled, all blocks along the outgoing edges are enqueued onto the work list as if the block had been evaluated. This strategy ensures that all incoming edges of a block will eventually fire. However, in the presence of loops and function calls, additional mechanisms are necessary.

14.3.2 Scheduling in the Presence of Loops and Function Calls

The simple approach of waiting for all incoming edges to fire before a basic block is scheduled only works for straight line code. In particular, tracking which incoming edges have already fired in the context of a loop creates a basic block in the work list with a pending edge. For instance, consider the loop in Fig. 14.4, which is entered at block 1 and thus block 1 is added to the work list with the edge from block 2 as pending. Since the back edge from block 2 will not fire until the loop itself is evaluated, block 1 is stalled in the work list. Given a non-empty but stalled work list, block 1 could be run. However, the same problem occurs after block 2 is evaluated, namely block 1 is put onto the work list and is stalled since only one incoming edge has fired.

Another loop-related scheduling problem is the repeated propagation out of the loop. For instance, the exit condition that tests the lower bit of `i` in Fig. 14.4 may propagate a changed state via its false branch several times during the fixpoint computation of the loop. Each time the condition is evaluated, block 3 is added to the work list and, since it has no other incoming edges that must fire, it can be evaluated immediately. As a consequence, the blocks 3,4,5 are evaluated as many times as the loop conditional is evaluated. This problem is exacerbated when analysing a sequence of several loops, where the evaluation of one loop leads to the repeated evaluation of the following loops.

Both problems, namely scheduling stalled basic blocks that constitute the heads of loops and the repeated scheduling of blocks that lie outside of a loop, can be circumvented in a reducible control-flow graph by adding additional rules for edges that enter or exit loops. However, the control-flow graphs of programs may not always be reducible. Similar problems arise due to function

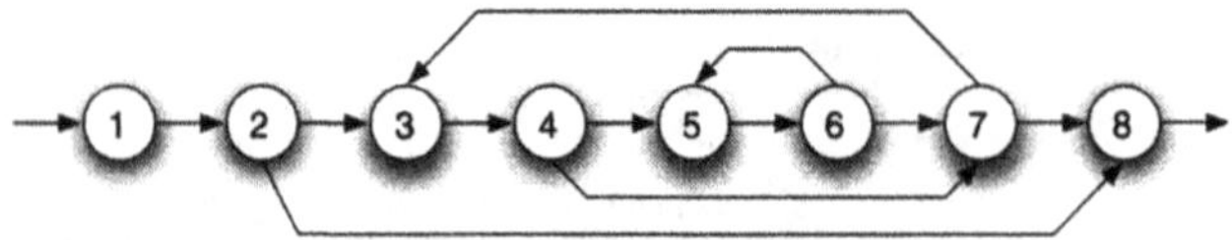

Fig. 14.5. Calculating strongly connected components from a control-flow graph.

calls. Specifically, functions may be called through a pointer, which implies that several functions are scheduled at once if the pointer contains more than one l-value. Upon the return of each function, the result will be propagated along the outgoing edges of the call site. Thus, the outgoing edges will fire once for each function that was called, with the effect that the following blocks might be executed several times. A reliable workaround that delays the scheduling of the blocks at the call site is much harder to devise since, for example, a function might call `exit()` and thereby not propagate any state.

14.3.3 Deriving an Iteration Strategy from Topology

One solution to the problem of calculating fixpoints of imperative programs was given by Bourdoncle [34]. He devised a variant of the classic Tarjan algorithm [178] to calculate strongly connected components (SCCs) even in the presence of irreducible control-flow graphs. His algorithm calculates a so-called hierarchical ordering – that is, a permutation of the set of basic blocks in which (possibly nested) SCCs are identified by parentheses. For example, a hierarchical ordering of the control-flow graph in Fig. 14.5 is the following:

$$1 \quad 2 \quad (\underline{3} \quad 4 \quad (\underline{5} \quad 6) \quad 7) \quad 8$$

This sequence is to be interpreted as two SCCs, where the SCC containing blocks 5 and 6 is contained within the SCC that is composed of blocks 3 to 7. Given this hierarchical ordering, one strategy for scheduling the computation is to evaluate the basic blocks starting from block 1 and to perform a fixpoint computation for each SCC encountered. In particular, the underlined blocks form the heads of the SCCs and are used as widening points. In the example, the blocks 1...6 are executed before a fixpoint for the SCC 5,6 is calculated using block 5 as a widening point. Once blocks 5 and 6 have stabilised, block 7 is evaluated and a fixpoint calculation of the SCC 3...7 commences. As part of this fixpoint calculation, several fixpoint calculations of the sub-SCC 5,6 may be performed. In general, the number of blocks that are evaluated with this strategy can grow exponentially with the nesting depth of SCCs.

The benefit of a scheduling strategy that is based on the topology of the control-flow graph is that the work list can be replaced by a simple index into the hierarchical ordering. Specifically, an index of the form $i_0.i_1 \ldots i_n$ identifies a basic block in an SCC of nesting depth n. Here, i_k denotes the block number at that level. Each sub-SCC has its own number. For instance, the control-flow graph in Fig. 14.5 has the following indices:

block	1	2	3	4	5	6	7	8
index	1	2	3.1	3.2	3.3.1	3.3.2	3.4	4

Instead of a work list, the analyser stores a single index that indicates that all basic blocks with a lexicographically smaller index are stable. This index is called the current index. Whenever a basic block is updated, for instance through a function return, the lexicographically smaller index of the current and the newly updated index becomes the new current index. After a block is processed, the index is advanced to the next index in the total order of the blocks unless the input state of a block with a smaller index was updated. This scheduling strategy is robust with respect to loops with several entry and exit nodes and with respect to function calls through pointers. In fact, a call stack of functions can be implemented easily by keeping a list containing triples consisting of a function, its call stack, and the current index within that function. By evaluating blocks in those functions that have the deepest call stacks, leaf functions are evaluated to completion before the evaluation of the caller is continued. In particular, if a function call through a pointer invokes several functions, their call stacks are all of the same depth and are thus evaluated concurrently. Only after the evaluation of all callees is completed does the evaluation at the call site continue.

As described in Sect. 14.3.1, it is beneficial to store a flag with each basic block that indicates whether the input state of that block needs propagating. If the current index refers to a block for which the flag states that evaluation is not necessary, the current index is merely advanced.

14.3.4 Related Work

Bourdoncle was the first to devise a chaotic iteration strategy from the topology of the control-flow graph rather than using a work list. He proposed two different strategies for computing a fixpoint: iterative and recursive. The iterative strategy repeatedly evaluates all basic blocks in the outermost SCC until a fixpoint is reached in the outermost and all contained SCCs. The recursive strategy was described in the previous section; it calculates a fixpoint for every inner SCC before continuing to the enclosing SCC. Bourdoncle observes that the worst-case complexity of the recursive iteration strategy is better than that of the iterative iteration strategy and that this result usually carries over to the actual implementation. Howe and King [103] compare different iteration strategies and point out that using Bourdoncle's algorithm to calculate SCCs is relatively expensive and may also perform significantly worse that simpler iteration strategies. Due to the complexity of the proposed polyhedral analysis, the overhead of running Bourdoncle's algorithm is not an issue. However, it might be possible to find fixpoints faster using a different iteration strategy, in particular in the presence of extrapolation strategies such as widening with landmarks.

Other work on iteration strategies for analysing imperative programs includes that of Burke, who describes how to construct iteration strategies by sorting the nodes in a graph topologically [37]. Loops that are reducible are replaced by a virtual edge connecting the head of the loop with the exit nodes. This virtual edge is then associated with a fixpoint of the transfer function of the loop. This transfer function is calculated by a traversal of the loop body in topological order without considering back edges. The disadvantage of this approach is that it requires a reducible control-flow graph and some non-trivial transformations on it and no flexibility of altering the iteration strategy.

Another aspect of calculating fixpoints efficiently lies in the way stability is detected. Once the SCCs of a graph are calculated, it is sufficient to check for entailment at the head of each SCC. In cases where the join operation is more expensive than an entailment check, a more efficient analysis may be possible by checking entailment at the basic block level. That is, when a new state is propagated along the outgoing edges to other basic blocks, it is first checked if this new state is already contained in the current input state of that block. If it is, no join needs to be calculated and the flag marking the basic block as pending is not set. This strategy can improve efficiency if the loop invariant is tested within the loop rather than at the head of the SCC. A surprising instance of a fixpoint calculation that stabilises in the loop body can be witnessed in the context of string buffer analysis. Consider the fixpoint calculation in Fig. 12.6 on p. 224, and specifically state $T_{13} = R_{13} \sqcap_N \{c \geq 1\}$. Suppose now that no integer tightening is performed so that $T_{13} = \{0 \leq i, 1 \leq c \leq 255, 255i + c \leq 2550\}$; that is, the upper bound of i is $\frac{2549}{255} \approx 0.996$. Given that $i \leq 9$ in the previous iteration, the loop is assumed to be unstable and a new iteration is calculated. However, when the equation for R defined on p. 222 is evaluated, the state is divided into three states, in which $i \leq 9$, $i = 10$, and $i \geq 11$. While the latter two intersections are empty, the inequality $i \leq 9$ discards the state where i takes on the values $9 < i < 0.997$. Thus, the resulting state for R is no larger than in the previous iteration and a fixpoint is reached. Thus, a fixpoint can be reached in the midst of a loop body.

Detecting stability at the level of individual basic blocks is only beneficial if the analysis can efficiently determine which other basic block still needs evaluation. For instance, using a flag for each basic block that indicates if that block is stable has the drawback that the flags of all blocks in the hierarchical order need to be examined until an unstable block is found. However, this test is cheap and in any case better than evaluating a basic block unnecessarily. Other ways of avoiding unnecessary evaluation were proposed by Jones in the context of attribute grammars [107]. However, the underlying iteration strategy resembles Gauss-Seidel iteration [55,58], and it is not clear how his technique maps to chaotic iteration with widening points. An orthogonal approach is taken by Le Charlier and Van Hentenryck [41], in which equations are evaluated in a demand-driven way, although the premature evaluation of basic blocks that later have to be evaluated again cannot be avoided either.

14.4 Limitations of the String Buffer Analysis

Tracking the first NUL position of string buffers is an abstraction that works well for many string-manipulating programs. However, since information about other NUL positions in the buffer is lost, there are evidently some programs with coding practices that are correct but that cannot be proven as such. Furthermore, our analysis is limited by the expressiveness of the TVPI domain, which can only track linear relations involving two variables. This section demonstrates how these limitations can lead to an imprecise analysis. While the examples presented are manufactured to elude the capability of the analyser, they are realistic enough to be found in standard C programs.

14.4.1 Weaknesses of Tracking First NUL Positions

String buffers that use a NUL character to indicate the end of a string can be constructed in many ways, some of which elude the expressiveness of the analysis presented. Consider the following C program that zeros the buffer pointed to by `s` and then writes the buffer character by character:

```
char *s = malloc(10);
char t[4];
memset(s,0,10);
s[0]='o'; s[1]='k';
strcpy(t, s);
```

Suppose that s_n denotes the NUL position of the dynamically allocated buffer. The call to `memset` will zero the whole memory region pointed to by `s` such that the first NUL position is at $s_n = 0$. After the first write, the analysis can only deduce $1 \leq s_n \leq 10$ since the first NUL (if it exists) must occur to the right of the character `'o'`. Likewise, after the second write, the analysis infers $2 \leq s_n \leq 10$. Copying the content of the pointer `s` to the four character buffer `t` is safe if $s_n \leq 3$. However, $2 \leq s_n \leq 10$ does not imply $s_n \leq 3$, and therefore the call to `strcpy` generates a spurious warning. In contrast, inserting `s[2]='a'; s[3]='y';` in front of the call to `strcpy` will correctly signal that `t` is accessed out-of-bounds in every execution of the program since the new NUL position $4 \leq s_n \leq 10$ implies that $s_n \leq 3$ cannot be satisfied. In this case, a definite error can be reported rather than a warning. On the contrary, writing the same four characters to the same positions in reverse order only leads to a warning: $s_n = 0$ holds true until `s[0]` is written, and this then updates the NUL position to $1 \leq s_n \leq 10$.

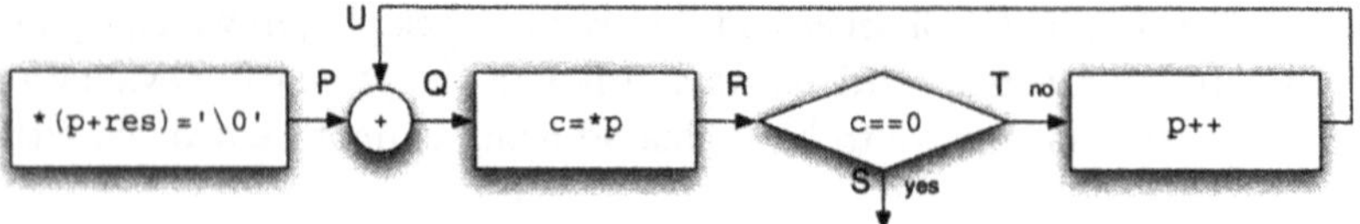

Fig. 14.6. The control-flow graph of the example that searches for the first NUL position in the buffer pointed-to by p.

In general, writing the zero character before writing the actual content of a buffer and writing several zero characters into a buffer cannot be adequately tracked by the analysis. Additional abstractions similar to the string buffer analysis are necessary to prove these examples correct. For instance, tracking the number of NUL characters towards the end of a buffer could infer that the s buffer is still NUL-terminated when its contents are copied. No such abstraction is currently implemented in our analyser since the need for such an abstraction has not arisen from the program under test. Note, however, that an abstraction that states that a whole range of elements only contains zero bytes is important for dealing with arrays of pointers, where this information is needed to express that pointers have no offset (in contrast to an arbitrary offset). The lack of such an abstraction in the analysis presented prohibits the verification of the example in Chap. 1. A more detailed explanation of this phenomenon is given in Sect. 14.5.

The next section details the limitation of the string buffer analysis that arises through the use of the TVPI domain.

14.4.2 Handling Symbolic NUL Positions

Mixing the convention of explicitly sized string buffers with buffers that are implicitly terminated by a NUL character can lead to subtle coding errors. In particular, strings received over the network by an attacker might include a deliberate NUL character to trigger a faulty behaviour in the program. The following example reads a string from the standard input into a fixed-sized buffer and then checks if the read string contains a NUL character, thereby defying the above-mentioned attack.

```
ssize_t res;
char buf[2048];
char c;
char* p=buf;
res = read(0, p, 2047);
if (res==-1) return 1;
*(p+res)='\0';
while (*p) p++;
if (p-buf<res)
  printf("The␣input␣contains␣a␣NUL␣character.\n");
```

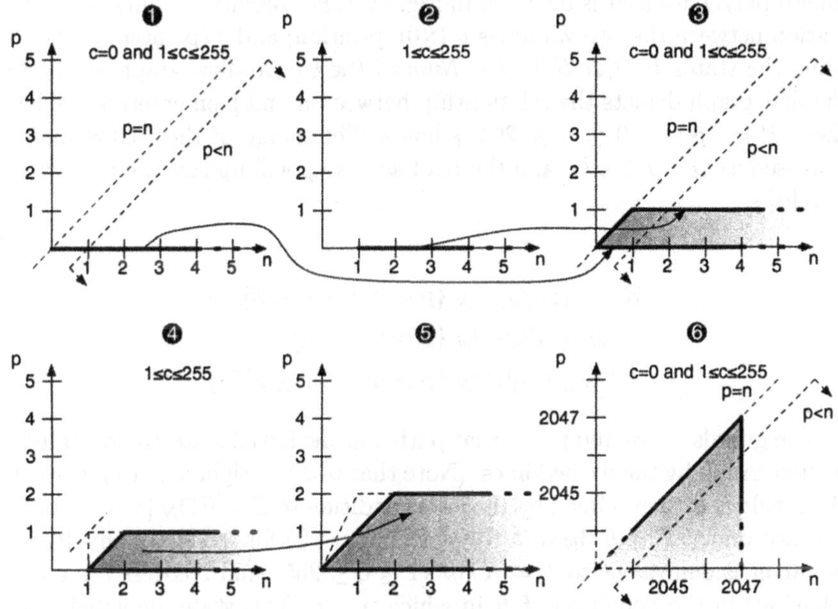

Fig. 14.7. Calculating the length of a string buffer in which the NUL position may be in a range of locations. In order to infer that the loop terminates, the NUL position of the buffer has to be always larger than the current pointer offset.

In this code snippet, the while loop calculates the length of the string by advancing the pointer to the first NUL position. The loop terminates at the latest when the offset of p is equal to `res` since the statement `*(p+res)='\0'` ensures that a NUL character exists at this index. However, the loop may exit earlier if the read data contain a NUL character. We shall discuss the fixpoint calculation of the loop using the control-flow graph shown in Fig. 14.6.

Let n denote the NUL position of `buf`, let p represent the offset of the pointer `p`, and let c denote the character that is read by the access to `*p`. Note that the call to the `read` function takes an upper limit on the number of characters that should be read (here 2047) and returns the number of characters actually read in `res` or -1 if an error occurs. Thus, since the error state is removed by `if (res==-1)`, the assignment `*(p+res)='\0'` always accesses `buf` within bounds and sets the abstract NUL position to `res`. Thus, on entering the loop, the first NUL position n is at indices less than or equal to `res` such that the state P satisfies $0 \leq n \leq 2047$. Since a NUL position exists somewhere within the buffer, the analysis should be able to prove that the **while**-loop never accesses `buf` out-of-bounds.

Unfortunately, the TVPI domain is not expressive enough to infer this. Before we discuss the loss of precision, consider the fixpoint computation using general polyhedra that is depicted in Fig. 14.7. For brevity, we only depict the relation between the two variables n (NUL position) and p (pointer offset) and define the states $P,Q,R,S,T,U \in Num$ of the control-flow graph on-the-fly. The first graph depicts the relationship between n and p on entering the loop where $P = \{p = 0, 0 \leq n \leq 2047\}$ holds. The merge of the control flow is expressed as $Q = P \sqcup_N U$, and the read access *p will update the character c as follows:

$$\begin{aligned} R = & \ (\exists_c(Q) \sqcap_N \{p < n, 1 \leq c \leq 255\}) \\ & \sqcup_N (\exists_c(Q) \sqcap_N \{p = n, c = 0\}) \\ & \sqcup_N (\exists_c(Q) \sqcap_N \{p > n, 0 \leq c \leq 255\}) \end{aligned}$$

The guards $p < n$ and $p = n$ that partition the initial state are indicated in the first graph by the dashed lines. (Note that $p < n$ is tightened to $p \leq n-1$.) The result R is then refined by the loop condition to $S = R \sqcap_N \{c = 0\}$ for the loop exit and to $T = R \sqcap_N \{c > 0\} \sqcup_N R \sqcap_N \{c < 0\}$ for the loop body. In fact, the latter is equivalent to $T = R \sqcap_N \{1 \leq c \leq 255\}$ and recovers the second behaviour in the definition of R in which $p < n$. This state, depicted in the second graph of Fig. 14.7, is transposed by one unit by the loop body p++ to $U = T \triangleright p := p+1$ before being joined with the state P from graph one. The result is a new state for Q, as depicted in graph three, which is again used to evaluate the read access *p, resulting in a new state R in which $c = 0$ is related to $p = n$ and $1 \leq c \leq 255$ is related to $p < n$. Intersecting the state R with the loop invariant $1 \leq c \leq 255$ yields a new state for T that corresponds to the region delineated by a bold line in graph four. This state is again translated by one unit and joined with P, yielding the state in graph five. A fixpoint of the loop is reached when p is incremented to coincide with the maximum NUL position $n = 2047$ as shown in graph six. This state is partitioned a final time by the read access *p. This time, the result of intersecting the state R with $1 \leq c \leq 255$ and translating it by one unit results in a state T that is entailed by the join Q of the previous iteration. Thus, using general polyhedra, a fixpoint of the **while** loop is detected where $p \leq n$ holds in Q, and hence $p \leq 2047$ and *p is within the bounds of buf.

The important observation in this example is that the read access sets c according to the relative position of p and n and thereby relates the three variables p, n, and c. Specifically, the abstract domain must be able to express that $p \leq n$ (from $p = n$) when $c = 0$ and that $p + 1 \leq n$ (from $p < n$) when $1 \leq c \leq 255$. These two behaviours are combined by the first join operation in the definition of R, which calculates a single inequality relating p, n, and c. This inequality can be derived manually as follows: For $c = 0$, the first inequality is equivalent to $p \leq n + fc$ for any $f \in \mathbb{N}$. The same inequality must hold for $p+1 \leq n+fc$ if $1 \leq c \leq 255$, for which $f = \frac{1}{255}$ is a solution. To ensure $f \in \mathbb{N}$, the inequality is multiplied by 255 to give $255p + 255 \leq 255n + c$. This inequality has three non-zero coefficients and is therefore not generated during the join of the TVPI domain. Using the techniques presented in Sect. 8.2.3, the three TVPI inequalities that approximate the above inequality $255p - 255n - c \leq -255$ are $p \leq n$ (for $minExp(-c, Q) = 255$), $255p - c \leq -255$ (for $minExp(-n, Q) = 0$), and $-255n - c \leq -255$ (for $minExp(p, Q) = 0$). Applying the loop invariant by calculating $T = R \sqcap_N \{1 \leq c \leq 255\}$ and performing integer tightening simplifies the three inequalities to $p \leq n$, $p \leq 1$, and $-n \leq -1$, respectively. These inequalities are indicated by the thin dotted boundaries in graph four. Thus, the intersection with $1 \leq c \leq 255$ cannot recover the relational information $p < n$. Translating this approximated state and joining it with P results in the additional state indicated by the thin dashed line in graph five. From this point onwards, the third behaviour of the read access is enabled such that the intersection with $1 \leq c \leq 255$ does not imply $p < n$ anymore. As a result, no fixpoint is detected when p is incremented to the maximum value of n and the verification of the loop fails.

When the join in a weakly relational domain discards valuable information with respect to a certain variable, the program can be analysed with several domains at once, one for each polyhedron that needs to remain separate. The challenge is to decide if and when to split the domain into two. For the task of retaining the necessary precision in the example above, this decision is easy since the two domains in which $c = 0$ and $1 \leq c \leq 255$ should be kept apart. Thus, the analysis can be reformulated in terms of pairs of polyhedra in which all polyhedral operations are applied to each element of the pair. The only exception is the read access, in which the state $\langle Q_1, Q_2 \rangle$ is joined and split according to the access position:

$$\begin{aligned}\langle R_1, R_2 \rangle = \langle\, & \exists_c((Q_1 \sqcap_N \{p < n\}) \sqcup_N (Q_2 \sqcap_N \{p < n\})) \sqcap_N \{1 \leq c \leq 255\}, \\ & (\exists_c((Q_1 \sqcap_N \{p = n\}) \sqcup_N (Q_2 \sqcap_N \{p = n\})) \sqcup_N \{c = 0\}) \sqcup_N \\ & (\exists_c((Q_1 \sqcap_N \{p = n\}) \sqcup_N (Q_2 \sqcap_N \{p > n\})) \sqcap_N \{0 \leq c \leq 255\}) \,\rangle\end{aligned}$$

In the definition above, the two incoming polyhedra are partitioned separately using the access position before they are joined and c is updated. In particular, in the context of the example above, the states $Q_2 \sqcap_N \{p < n\}$ and $Q_1 \sqcap_N \{p = n\}$ will always be empty, which indicates that the two behaviours, namely accessing in front of and at the NUL position, remain distinct. We chose to express the third behaviour, namely an access past the first NUL position, in the second polyhedron under the assumption that $p > n$ is not satisfied in a loop whose termination requires the existence of a NUL position. Thus, the states where $c = 0$ and where $1 \leq c \leq 255$ stay separate in the states Q_1, R_1 and Q_2, R_2, respectively.

Interestingly, splitting the control-flow path of the loop depending on the value of the polyhedral variable c is equivalent to using a Boolean flag since c can only assume two states, namely $c = 0$ and $1 \leq c \leq 255$. While these two states can be distinguished by a general polyhedral domain, the TVPI domain is not expressive enough. Specifically, the TVPI domain can only express that the range of a variable changes with the truth value of a Boolean flag. As shown in this example, string buffer analysis requires distinguishing the relationships $p \leq n - 1$ and $p \leq n$ using a Boolean flag.

So far, a split of the control flow using two TVPI domains instead of one has not been implemented as it involves a major redesign of the interface between the abstract transfer functions and the domain. In particular, since the analysis should be fully automatic, a heuristic is needed that introduces split points and join points. Up to now, the need for more than three variables has only arisen when analysing string buffers that have a non-constant NUL position. Since the relationships that the character c should distinguish are known, it might be easier to monitor the value of c and to add the inequality $q \leq n - 1$ whenever c is restricted to zero. Such a refinement is known as a propagation rule in the context of constraint handling rules [76].

14.5 Proposed Future Refinements

One deficiency of the analysis presented is its inability to argue about the contents of arrays. Ignoring individual array elements is acceptable if arrays contain pure data that have little impact on the overall execution of the program. However, in the case where an array contains pointers, a fatal precision loss occurs that impacts on the usability of the analysis. Recall that, whenever the elements of an array are written within a loop, no new fields are added to the array such that the contents of the array must be assumed to be arbitrary. If a program – like **`qmail-smtp`** – uses a hash table that contains pointers, an array holding the hash table will be cleared and afterwards filled with pointers. Since no fields of the array are modelled, the written points-to sets and the pointer offsets are lost. As a consequence, whenever an element from the hash table is retrieved, the analysis approximates the read value with a temporary polyhedral variable that is restricted to the range of an unsigned 32-bit value. Hence, when dereferencing this value as a pointer, the analyser will issue a warning that a value is dereferenced. After removing this erroneous assumption, the remaining state is empty since the points-to set of the read value is empty. Thus, while the approximation to the values of the array elements is sound, it prevents the analyser from completing the analysis of the program. This precision loss is one reason why we were not able to obtain a full analysis of the `qmail-smtp` program. Future work therefore has to focus on finding an abstraction to array elements that retains enough information on pointers. In particular, it is necessary to infer that all elements of an array are zero (to model the fact that the pointers stored in the array have a zero offset) and to summarise the possible l-values that are stored in the different elements of the array in a single set of l-values. Such an abstraction would enable the analysis of a hash table as well as a full analysis of the introductory example in Chap. 1 where an array of pointers to command-line arguments is passed to the `main` function of the program.

15

Conclusion and Outlook

Conclusion

In this book we formally, yet concisely, defined a value-range analysis of C programs. Moreover, all aspects of the analysis were described, ranging from the concrete and abstract semantics including the abstraction relation, over an efficient polyhedral domain, up to necessary implementation details.

With respect to the analysis of C programs, we defined an intermediate language called Core C that reduces general C programs to operations on memory. In fact, Core C is closer to assembler than C but retains information on boundaries of variables and assignments of structures, thereby enabling a simpler analysis and more precise warnings. A novelty in presenting an analysis for full C is a concise abstraction relation that relates the bits of integer variables with polyhedral variables. Furthermore, the abstraction allows for linear relationships when accessing the same memory regions with different sizes. In order to attain a fully automated analysis of C, we proposed to infer relevant fields of C program variables at analysis time.

With respect to the second strand of this book, namely the research into efficient polyhedral domains, we presented the TVPI domain. The TVPI domain constitutes the first efficient relational domain that allows for arbitrary coefficients in inequalities. The latter is a requirement for expressing the analysis of overlapping fields and NUL positions in string buffers. The implementation of the TVPI domain also addresses the problem of excessively growing coefficients during fixpoint calculations on polyhedra by partially tightening a TVPI domain around the contained integral point set.

As a third strand, we proposed a new abstraction for tracking NUL positions in string buffers, an acceleration technique called widening with landmarks that overcomes deficiencies in the common widening/narrowing approach, and a combination of points-to analysis with numeric analysis.

We hope that the analysis presented will form a reference for future developments, be it as an extension to the analysis in the form of additional

abstractions (which can be added analogously to the tracking of NUL positions) or using different approaches to modelling low-level memory accesses and other low-level operations. In the long run, we expect static analyses of memory management to mature and to grow into commonplace tools that are part of every development process. Ideally, verifying correct memory management will become a requirement in future software development since it constitutes a measurable quality of software.
We conclude this book with some possible extensions to the presented analysis.

Outlook

In addition to the refinement of modelling arrays that was proposed in Chap. 14, the analysis can be extended and improved in several directions, some of which are outlined below.

Replacing the Polyhedral Domain While the TVPI domain offers operations that run in polynomial time, new insights into the manipulation of general polyhedra [169] might yield an implementation that outperforms the TVPI domain in practice. However, general polyhedra require techniques for integral tightening to address the excessive growth of coefficients that can occur during fixpoint calculations.

Analysing Assembler instead of Core C While the conciseness of our intermediate language Core C is key to the definition of the semantics and the analysis itself, an analysis of the executable may be more practical, as the source code of a program and the linked-in libraries is not required. Furthermore, the correctness of an analysis at the assembler level can be assessed with respect to the semantics of each machine instruction, whereas our approach assumes that the translation from GCC's internal structure to Core C is correct and that the translation of the intermediate structure to the machine code is correct. Analysing machine code can therefore improve the confidence in the analysis results. Future work should assess the feasibility of this approach and determine if the ample set of assembler instructions can be translated into a small number of primitives. Static analysis on assembler has already been used to aid in inspecting executables [15] for security vulnerabilities. In this context, it has been observed that debugging information might be absent or deliberately misleading. However, debugging information could still be exploited to make warnings raised by the analysis intelligible to the user, a problem that can severely restrict the usefulness of analysing assembler. In this case, and if the analysis of assembler is too slow, a hybrid approach is possible. For instance, Rival proposes to infer the invariants on the source code and check that they still hold on the assembler level by using debugging information [150]. A similar approach is used within the CIL framework [137], which inserts range checks into C programs. Harren and Necula augmented this program transformation to check the validity of its output in the executable [94].

Better Analysis of Dynamically Allocated Memory The analysis thus far can only prove programs correct that allocate and free simple buffers, in particular, recursive data structures are handled inadequately. New domains are necessary to express the shape of data structures in memory. Interesting candidates for domains that are able to model the recursive nature of data structures are, for instance, regular expressions [70] or domains based on separation logic [40]. Challenging research tasks are therefore the definition of efficient domain operations and the combination of these domains with the polyhedral domain.

Analysing Floating-Point Arithmetic Calculations using floating-point arithmetic can be expressed as operations on two polyhedral variables that represent the mantissa and the exponent [165]. Rounding modes of the IEEE 754 Standard [105] can be implemented using approaches similar to those of right shifts on integers, as shown in Fig. 3.7 on p. 61. A practical implementation is required to assess the practicality of this approach.

Context-Sensitive Analysis For an analysis to scale to a large code basis, a context-sensitive analysis seems to be an inevitable requirement. Future work needs to address how a context-sensitive analysis can be achieved using polyhedra. In general this is not possible as polyhedra are not meet distributive. This implies that an input-output relationship of a function expressed as a polyhedron cannot be refined at each call site by merely intersecting with the ranges of the arguments without loosing precision.

Analysis of Concurrent Programs Through the use of libraries, C programs can exploit the multi-tasking capabilities of the underlying operating system. The analysis of such concurrent programs poses the problem of identifying those parts of the program that may run in parallel and to simulate all possible interleaved executions of these program parts.

A

Core C Example

The following Core C program is the result of translating the program in Fig. 1.2. The program contains line annotations that are not part of the grammar presented in Chap. 2. Furthermore, the variable declarations of the `main` function are split into input, output, and local variables to facilitate the removal of variables at function exit. Variable names of the form `t#X` are temporary variables that are mostly created as anonymous temporaries within the compiler but partly stem from the translation into Core C. The call to `printf` is automatically translated as primitive, as the definition is not available.

Note that this is the "incorrect" version in which the signed character `t#3` in block `b105` is sign-extended to the signed integer `t#4`. The corrected version contains an extra cast statement, `(uint8) t#3.1 = (int8) t#3`, and `t#3.1` is then assigned to `t#4`.

```
var t#12 : 11;
  t#12  = "'%c'␣:␣%i\n";
function main
input   argc : 4, argv : 4;
output  #result : 4;
local   i : 4, str : 4, dist : 1024, t#0 : 4,
    t#1 : 4, t#2 : 4, t#3 : 1, t#4 : 4,
    t#5 : 4, t#6 : 4, t#7 : 4, t#8 : 4,
    t#9 : 1, i.0 : 4, t#10 : 4, t#11 : 4,
    t#13 : 4;
b20
  #line=9
  if (int32) argc!=2 then jump b41
  jump b55
b41
  (int32) t#0 = 1;
  jump b358
b55
```

```
  #line=10
  (uint32) t#1 = argv+4;
  (uint32) str = *t#1;
  #line=11
  t#2  = &dist;
  memset((uint32) t#2, (int32) 0, (uint32) 1024)
  jump b197
b105
  #line=14
  (int8) t#3 = *str;
  (int32) t#4 = (int8) t#3;
  t#6  = &dist;
  (int32) t#6 = 4*t#4+t#6;
  (int32) t#5 = *t#6;
  (int32) t#7 = t#5+1;
  t#8  = &dist;
  (int32) t#8 = 4*t#4+t#8;
  (int32) *t#8 = t#7;
  #line=15
  (uint32) str = str+1;
  jump b197
b197
  #line=13
  (int8) t#9 = *str;
  if (int8) t#9!=0 then jump b105
  jump b224
b224
  #line=18
  (int32) i = 32;
  jump b322
b241
  #line=19
  (int32) i.0 = i;
  t#11  = &dist;
  (int32) t#11 = 4*i.0+t#11;
  (int32) t#10 = *t#11;
  t#13  = &t#12;
  printf((uint32) t#13, (int32) i, (int32) t#10)
  #line=18
  (int32) i = i+1;
  jump b322
b322
  if (int32) i<=127 then jump b241
  jump b342
b342
```

```
  #line=21
  (int32) t#0 = 0;
  jump b358
b358
  (int32) #result = t#0;
  return
```

References

1. A. V. Aho, R. Sethi, and J. D. Ullman. *Compilers: Principles, Techniques, and Tools*. Addison-Wesley, 1986.
2. S. G. Akl and G. T. Toussaint. A fast convex hull algorithm. *Information Processing Letters*, 7(5):219–222, 1978.
3. L. O. Andersen. *Program Analysis and Specialization for the C Programming Language*. PhD thesis, DIKU, University of Copenhagen, May 1994.
4. K. R. Anderson. A Reevaluation of an Efficient Algorithm for Determining the Convex Hull of a Finite Planar Set. *Information Processing Letters*, 7(1):53–55, 1978.
5. A. M. Andrews. Another efficient algorithm for convex hulls in two dimensions. *Information Processing Letters*, 9(5):216–219, 1979.
6. D. Avis and K. Fukuda. A Pivoting Algorithm for Convex Hulls and Vertex Enumeration of Arrangements and Polyhedra. *Discrete & Computational Geometry*, 8:295–313, 1992.
7. R. Bagnara. *Data-Flow Analysis for Constraint Logic-Based Languages*. PhD thesis, Università di Pisa, Dipartimento di Informatica, Pisa, Italy, 1997.
8. R. Bagnara, K. Dobson, P. M. Hill, M. Mundell, and E. Zaffanella. Grids: A Domain for Analyzing the Distribution of Numerical Values. In A. King, editor, *Logic-Based Program Synthesis and Transformation*, volume 4407 of *LNCS*, pages 219–235, Kogens Lyngby, Denmark, August 2007. Springer.
9. R. Bagnara, P. M. Hill, E. Mazzi, and E. Zaffanella. Widening Operators for Weakly-Relational Numeric Abstractions. In C. Hankin and I. Siveroni, editors, *Static Analysis Symposium*, volume 3672 of *LNCS*, pages 3–18, London, UK, September 2005. Springer.
10. R. Bagnara, P. M. Hill, E. Ricci, and E. Zaffanella. Precise Widening Operators for Convex Polyhedra. *Science of Computer Programming*, 58(1–2):28–56, 2005.
11. R. Bagnara, P. M. Hill, and E. Zaffanella. Widening Operators for Powerset Domains. In B. Steffen and G. Levi, editors, *Verification, Model Checking and Abstract Interpretation*, volume 2937 of *LNCS*, pages 135–148, Venice, Italy, January 2004. Springer.
12. R. Bagnara, P. M. Hill, and E. Zaffanella. Not Necessarily Closed Convex Polyhedra and the Double Description Method. *Formal Aspects of Computing*, 17(2):222–257, 2005.

13. R. Bagnara, P. M. Hill, and E. Zaffanella. An Improved Tight Closure Algorithm for Integer Octagonal Constraints. In F. Logozzo, D. Peled, and L. D. Zuck, editors, *Verification, Model Checking and Abstract Interpretation*, volume 4905 of *LNCS*, pages 8–21, San Francisco, California, USA, January 2008. Springer.
14. R. Bagnara, E. Ricci, E. Zaffanella, and P. M. Hill. Possibly Not Closed Convex Polyhedra and the Parma Polyhedra Library. In M. V. Hermenegildo and G. Puebla, editors, *Static Analysis Symposium*, volume 2477 of *LNCS*, pages 213–229, Madrid, Spain, September 2002. Springer.
15. G. Balakrishnan, G. Grurian, T. Reps, and T. Teitelbaum. CodeSurfer/x86 – A Platform for Analyzing x86 Executables. In *Compiler Construction*, volume 3443 of *LNCS*, pages 250–254, Edinburgh, Scotland, April 2005. Springer. Tool-Demonstration Paper.
16. V. Balasundaram and K. Kennedy. A Technique for Summarizing Data Access and Its Use in Parallelism Enhancing Transformations. In *Programming Language Design and Implementation*, pages 41–53, Snowbird, Utah, USA, June 1989. ACM.
17. T. Ball, E. Bounimova, B. Cook, V. Levin, J. Lichtenberg, C. McGarvey, B. Ondrusek, S. Rajamani, and A. Ustuner. Thorough Static Analysis of Device Drivers. In Y. Berbers and W. Zwaenepoel, editors, *European Systems Conference*, pages 73–85, Leuven, Belgium, April 2006. ACM.
18. T. Ball, B. Cook, V. Levin, and S. K. Rajamani. SLAM and Static Driver Verifier: Technology Transfer of Formal Methods inside Microsoft. Technical report, Microsoft Research, January 2004.
19. T. Ball and S. K. Rajamani. Bebop: A Symbolic Model Checker for Boolean Programs. In K. Havelund, J. Penix, and W. Visser, editors, *SPIN Workshop on Model Checking and Software Verification*, volume 1885 of *LNCS*, pages 113–130, Stanford, California, USA, August 2000. Springer.
20. T. Ball and S. K. Rajamani. Automatically Validating Temporal Safety Properties of Interfaces. In M. B. Dwyer, editor, *SPIN Workshop on Model Checking of Software*, volume 2057 of *LNCS*, pages 103–122, Toronto, Canada, May 2001. Springer.
21. F. Banterle and R. Giacobazzi. A Fast Implementation of the Octagon Abstract Domain on Graphics Hardware. In H. R. Nielson and G. Filé, editors, *Static Analysis Symposium*, volume 4634 of *LNCS*, pages 315–332, Kongens Lyngby, Denmark, August 2007. Springer.
22. A. Baratloo, N. Singh, and T. Tsai. Transparent Run-Time Defense Against Stack-Smashing Attacks. In *USENIX Annual Technical Conference*, pages 251–262, San Diego, California, USA, June 2000. USENIX Association.
23. G. Behrmann, P. Bouyer, K. G. Larsen, and R. Pelánek. Lower and Upper Bounds in Zone Based Abstractions of Timed Automata. In K. Jensen and A. Podelski, editors, *Tools and Algorithms for the Construction and Analysis of Systems*, volume 2988 of *LNCS*, pages 312–326, Barcelona, Spain, March 2004. Springer.
24. F. Benoy, A. King, and F. Mesnard. Computing Convex Hulls with a Linear Solver. *Theory and Practice of Logic Programming*, 5(1-2):259–271, 2005.
25. P. M. Benoy. *Polyhedral Domains for Abstract Interpretation in Logic Programming*. PhD thesis, Computing Lab., University of Kent, Canterbury, UK, January 2002.

26. M. Berndl, O. Lhoták, F. Qian, L. Hendren, and N. Umanee. Points-to analysis using BDDs. In *Programming Language Design and Implementation*, pages 103–114, San Diego, California, USA, June 2003. ACM.
27. J. Bertrand. NewPolka. `http://www.irisa.fr/prive/bjeannet/newpolka.html`, 2005.
28. F. Besson, T. P. Jensen, and J.-P. Talpin. Polyhedral Analysis for Synchronous Languages. In A. Cortesi and G. Filé, editors, *Static Analysis Symposium*, volume 1694 of *LNCS*, pages 51–68, Venice, Italy, September 1999. Springer.
29. G. Birkhoff. *Lattice Theory*, volume XXV. Colloquium Publications, American Mathematical Society, Providence, Rhode Island, USA, 3rd edition, 1967.
30. B. Blanchet, P. Cousot, R. Cousot, J. Feret, L. Mauborgne, A. Miné, D. Monniaux, and X. Rival. Design and Implementation of a Special-Purpose Static Program Analyzer for Safety-Critical Real-Time Embedded Software. In T. Æ. Mogensen, D. A. Schmidt, and I. H. Sudborough, editors, *The Essence of Computation: Complexity, Analysis, Transformation. Essays Dedicated to Neil D. Jones*, volume 2566 of *LNCS*, pages 85–108. Springer, 2002.
31. B. Blanchet, P. Cousot, R. Cousot, J. Feret, L. Mauborgne, A. Miné, D. Monniaux, and X. Rival. A Static Analyzer for Large Safety-Critical Software. In *Programming Language Design and Implementation*, San Diego, California, USA, June 2003. ACM.
32. H.-J. Boehm and M. Weiser. Garbage Collection in an Uncooperative Environment. *Software Practice and Experience*, 18(9):807–820, 1988.
33. F. Bourdoncle. Abstract Debugging of Higher-Order Imperative Languages. In *Programming Language Design and Implementation*, pages 46–55, Albuquerque, New Mexico, USA, June 1993. ACM.
34. F. Bourdoncle. Efficient Chaotic Iteration Strategies with Widenings. In D. Bjørner, M. Broy, and I. V. Pottosin, editors, *Formal Methods in Programming and Their Applications*, volume 735 of *LNCS*, pages 128–141, Novosibirsk, Russia, June 1993. Springer.
35. O. Bournez, O. Maler, and A. Pnueli. Orthogonal Polyhedra: Representation and Computation. In F. W. Vaandrager and J. H. van Schuppen, editors, *Hybrid Systems: Computation and Control*, volume 1569 of *LNCS*, pages 46–60, Berg en Dal, The Netherlands, March 1999. Springer.
36. H. Brönnimann, J. Iacono, J. Katajainen, P. Morin, J. Morrison, and G. T. Toussaint. In-Place Planar Convex Hull Algorithms. In *Latin American Symposium on Theoretical Informatics*, volume 2286 of *LNCS*, pages 494–507, Cancun, Mexico, April 2002. Springer.
37. M. Burke. An Interval-Based Approach to Exhaustive and Incremental Interprocedural Data-Flow Analysis. *ACM Transactions on Programming Languages and Systems*, 12(3):341–395, 1990.
38. W. R. Bush, J. D. Pincus, and D. J. Sielaff. A static analyzer for finding dynamic programming errors. *Software Practice and Experience*, 30(7):775–802, 2000.
39. V. T. Chakaravarthy and S. Horwitz. On the non-approximability of points-to analysis. *Acta Informatica*, 38(8):587–598, July 2002.
40. B.-Y. E. Chang, X. Rival, and G. C. Necula. Shape Analysis with Structural Invariant Checkers. In H. R. Nielson and G. Filé, editors, *Static Analysis Symposium*, volume 4634 of *LNCS*, pages 384–401, Kongens Lyngby, Denmark, August 2007. Springer.

41. B. Le Charlier and P. Van Hentenryck. Experimental Evaluation of a Generic Abstract Interpretation Algorithm for PROLOG. *ACM Transactions on Programming Languages and Systems*, 16(1):35–101, 1994.
42. D. Chase, M. Wegman, and F. Zadeck. Analysis of Pointers and Structures. In *Programming Language Design and Implementation*, pages 296–310, White Plains, New York, USA, June 1990. ACM.
43. B. Cheng and W. W. Hwu. Modular Interprocedural Pointer Analysis Using Access Paths: Design, Implementation, and Evaluation. *ACM SIGPLAN Notices*, 35(5):57–69, 2000.
44. N. V. Chernikova. Algorithm for Discovering the Set of All Solutions of a Linear Programming Problem. *USSR Computational Mathematics and Mathematical Physics*, 8(6):282–293, 1968.
45. B. Chess. Improving Computer Security Using Extended Static Checking. In *IEEE Symposium on Security and Privacy*, page 160, Berkeley, California, USA, 2002. IEEE Computer Society.
46. J.-D. Choi, M. Burke, and P. Carini. Efficient Flow-Sensitive Interprocedural Computation of Pointer-Induced Aliases and Side Effects. In *Principles of Programming Languages*, pages 232–245, Charleston, South Carolina, USA, January 1993. ACM.
47. R. Clariso and J. Cortadella. The Octahedron Abstract Domain. In R. Giacobazzi, editor, *Static Analysis Symposium*, volume 3148 of *LNCS*, Verona, Italy, August 2004. Springer.
48. E. M. Clarke, O. Grumberg, S. Jha, Y. Lu, and H. Veith. Progress on the State Explosion Problem in Model Checking. In R. Wilhelm, editor, *Informatics*, volume 2000 of *LNCS*, pages 176–194, Dagstuhl, Germany, 2001. Springer.
49. E. M. Clarke, O. Grumberg, and D. Peled. *Model Checking*. MIT Press, December 1999.
50. M. Codish, A. Mulkers, M. Bruynooghe, M. García de la Banda, and M. Hermenegildo. Improving Abstract Interpretations by Combining Domains. *ACM Transactions on Programming Languages and Systems*, 17(1):28–44, 1995.
51. Joint Technical Committee. International Standard ISO/IEC of C 98/99, 1999.
52. B. Cook, A. Podelski, and A. Rybalchenko. Abstraction Refinement for Termination. In C. Hankin and I. Siveroni, editors, *Static Analysis Symposium*, volume 3672 of *LNCS*, pages 87–101, London, UK, September 2005. Springer.
53. D. C. Cooper. Theorem proving in arithmetic without manipulation. *Machine Intelligence*, 7:91–99, 1972.
54. T. H. Cormen, C. Stein, R. L. Rivest, and C. E. Leiserson. *Introduction to Algorithms*. McGraw-Hill, 2001.
55. P. Cousot and R. Cousot. Static Determination of Dynamic Properties of Programs. In B. Robinet, editor, *International Symposium on Programming*, pages 106–130, Paris, France, April 1976.
56. P. Cousot and R. Cousot. Abstract Interpretation: A Unified Lattice Model for Static Analysis of Programs by Construction or Approximation of Fixpoints. In *Principles of Programming Languages*, pages 238–252, Los Angeles, California, USA, January 1977. ACM.
57. P. Cousot and R. Cousot. Systematic Design of Program Analysis Frameworks. In *Principles of Programming Languages*, pages 269–282, San Antonio, Texas, USA, January 1979. ACM.

58. P. Cousot and R. Cousot. Abstract Interpretation and Application to Logic Programs. *Journal of Logic Programming*, 13(2–3):103–179, 1992.
59. P. Cousot and R. Cousot. Comparing the Galois Connection and Widening/-Narrowing Approaches to Abstract Interpretation. In M. Bruynooghe and M. Wirsing, editors, *Programming Language Implementation and Logic Programming*, volume 631 of *LNCS*, pages 269–295, Leuven, Belgium, August 1992. Springer.
60. P. Cousot, R. Cousot, J. Feret, L. Mauborgne, A. Miné, D. Monniaux, and X. Rival. The ASTRÉE Analyzer. In M. Sagiv, editor, *European Symposium on Programming*, pages 21–30, Edinburgh, Scotland, April 2005. Springer.
61. P. Cousot, R. Cousot, J. Feret, A. Miné, L. Mauborgne, D. Monniaux, and X. Rival. Combination of Abstractions in the ASTRÉE Static Analyzer. In M. Okada and I. Satoh, editors, *Asian Computing Science Conference*, volume 4435 of *LNCS*, pages 272–300, Tokyo, Japan, December 2006. Springer.
62. P. Cousot and N. Halbwachs. Automatic Discovery of Linear Constraints among Variables of a Program. In *Principles of Programming Languages*, pages 84–97, Tucson, Arizona, USA, January 1978. ACM.
63. C. Cowan, C. Pu, D. Maier, H. Hinton, J. Walpole, P. Bakke, S. Beattie, A. Grier, P. Wagle, and Q. Zhang. Stackguard: Automatic Adaptive Detection and Prevention of Buffer-Overflow Attacks. In *USENIX Security Symposium*, pages 63–78. USENIX Association, 1998.
64. C. Cowan, P. Wagle, C. Pu, S. Beattie, and J. Walpole. Buffer Overflows: Attacks and Defenses for the Vulnerability of the Decade. In *Information Survivability Conference and Exposition*, volume II, pages 154–163. IEEE Computer Society, 1998.
65. M. Das. Unification-based pointer analysis with directional assignments. *ACM SIGPLAN Notices*, 35(5):35–46, 2000.
66. M. Das, B. Liblit, M. Fähndrich, and J. Rehof. Estimating the Impact of Scalable Pointer Analysis on Optimization. *LNCS*, 2126:260–279, July 2001.
67. H. Davenport. *The Higher Arithmetic*. Cambridge University Press, 7th edition, 1952.
68. A. M. Day. The implementation of a 2D convex hull algorithm using perturbation. *Computer Graphics Forum*, 9(4):309–316, 1990.
69. D. Delmas and J. Souyris. Astrée: From Research to Industry. In H. R. Nielson and G. Filé, editors, *Static Analysis Symposium*, volume 4634 of *LNCS*, pages 437–451, Kogens Lyngby, Denmark, August 2007. Springer.
70. A. Deutsch. Interprocedural May-Alias Analysis for Pointers: Beyond k-limiting. In *Programming Language Design and Implementation*, pages 230–241, Orlando, Florida, USA, June 1994. ACM.
71. N. Dor, M. Rodeh, and M. Sagiv. Cleanness Checking of String Manipulations in C Programs via Integer Analysis. In P. Cousot, editor, *Static Analysis Symposium*, volume 2126 of *LNCS*, pages 194–212, Paris, France, July 2001. Springer.
72. N. Dor, M. Rodeh, and M. Sagiv. CSSV: Towards a Realistic Tool for Statically Detecting All Buffer Overflows in C. In R. Gupta, editor, *Programming Language Design and Implementation*, pages 155–167, San Diego, California, USA, June 2003. ACM.
73. J. Elgaard, A. Møller, and M. I. Schwartzbach. Compile-Time Debugging of C Programs Working on Trees. In G. Smolka, editor, *European Symposium on*

Programming, volume 1782 of *LNCS*, pages 119–135, Berlin, Germany, March 2000. Springer.
74. M. Emami, R. Ghiya, and L. Hendren. Context-Sensitive Interprocedural Analysis in the Presence of Function Pointers. In *Programming Language Design and Implementation*, pages 242–256, Orlando, Florida, USA, June 1994. ACM.
75. D. Evans, J. Guttag, J. Horning, and Y. M. Tan. LCLint: a Tool for Using Specifications to Check Code. In *Symposium on Foundations of Software Engineering*, pages 87–96, San Diego, California, USA, December 1994. ACM.
76. T. Frühwirth. Theory and Practice of Constraint Handling Rules. *Journal of Logic Programming, Special Issue on Constraint Logic Programming*, 37(1-3): 95–138, October 1998.
77. A. Ghosh, T. O'Connor, and G. McGraw. An Automated Approach for Identifying Potential Vulnerabilities in Software. In *IEEE Symposium on Security and Privacy*, pages 104–114, Oakland, California, USA, May 1998. IEEE Computer Society.
78. R. Giacobazzi and E. Quintarelli. Incompleteness, Counterexamples and Refinements in Abstract Model-Checking. In P. Cousot, editor, *Static Analysis Symposium*, volume 2126 of *LNCS*, pages 356–373, Paris, France, July 2001. Springer.
79. R. Giacobazzi, F. Ranzato, and F. Scozzari. Making abstract interpretations complete. *Journal of the ACM*, 47(2):361–416, 2000.
80. D. Gopan, F. DiMaio, N. Dor, T. Reps, and M. Sagiv. Numeric Domains with Summarized Dimensions. In K. Jensen and A. Podelski, editors, *Tools and Algorithms for the Construction and Analysis of Systems*, Barcelona, Spain, March 2004.
81. D. Gopan and T. Reps. Lookahead Widening. In T. Ball and R. B. Jones, editors, *Computer-Aided Verification*, volume 4144 of *LNCS*, Seattle, Washington, USA, August 2006. Springer.
82. D. Gopan and T. W. Reps. Guided Static Analysis. In H. R. Nielson and G. Filé, editors, *Static Analysis Symposium*, volume 4634 of *LNCS*, pages 349–365, Kogens Lyngby, Denmark, August 2007. Springer.
83. D. Gopan and T. W. Reps. Low-Level Library Analysis and Summarization. In W. Damm and H. Hermanns, editors, *Computer Aided Verification*, volume 4590 of *LNCS*, pages 68–81, Berlin, Germany, July 2007. Springer.
84. R. L. Graham. An Efficient Algorithm for Determining the Convex Hull of a Finite Planar Set. *Information Processing Letters*, 1(4):132–133, 1972.
85. P. Granger. Static Analysis of Arithmetic Congruences. *International Journal of Computer Mathematics*, 30:165–199, 1989.
86. P. Granger. Static Analysis of Linear Congruence Equalities among Variables of a Program. In S. Abramsky and T. S. E. Maibaum, editors, *Theory and Practice of Software Development*, volume 493 of *LNCS*, pages 169–192, Brighton, UK, April 1991. Springer.
87. P. Granger. Static Analyses of Congruence Properties on Rational Numbers (Extended Abstract). In *Symposium on Static Analysis*, pages 278–292, Paris, France, September 1997. Springer.
88. D. Gries and I. Stojmenović. A note on Graham's convex hull algorithm. *Information Processing Letters*, 25(5):323–327, 1987.
89. B. S. Gulavani and S. K. Rajamani. Counterexample Driven Refinement for Abstract Interpretation. In Holger Hermanns and Jens Palsberg, editors, *Tools*

and Algorithms for the Construction and Analysis of Systems, volume 3920 of *LNCS*, pages 474–488, Vienna, Austria, March 2006. Springer.

90. B. Hackett, M. Das, D. Wang, and Z. Yang. Modular Checking for Buffer Overflows in the Large. In *International Conference on Software Engineering*, pages 232–241, Shanghai, China, July 2006. ACM.
91. N. Halbwachs. *Détermination Automatique de Relations Linéaires Vérifiées par les Variables d'un Programme.* Thèse de 3ème cicle d'informatique, Université scientifique et médicale de Grenoble, Grenoble, France, March 1979.
92. N. Halbwachs, Y.-E. Proy, and P. Raymond. Verification of Linear Hybrid Systems by Means of Convex Approximations. In B. Le Charlier, editor, *Static Analysis Symposium*, Namur, Belgium, September 1994. Springer.
93. N. Halbwachs, Y.-E. Proy, and P. Roumanoff. Verification of Real-Time Systems using Linear Relation Analysis. *Formal Methods in System Design*, 11(2):157–185, August 1997.
94. M. Harren and G.C. Necula. Using Dependent Types to Certify the Safety of Assembly Code. In C. Hankin and I. Siveroni, editors, *Static Analysis Symposium*, LNCS, pages 155–170, London, UK, September 2005. Springer.
95. W. H. Harrison. Compiler Analysis of the Value Ranges for Variables. *Transactions on Software Engineering*, 3(3):243–250, May 1977.
96. W. Harvey. Computing Two-Dimensional Integer Hulls. *SIAM Journal on Computing*, 28(6):2285–2299, 1999.
97. W. Harvey and P. J. Stuckey. A Unit Two Variable per Inequality Integer Constraint Solver for Constraint Logic Programming. *Australian Computer Science Communications*, 19(1):102–111, 1997.
98. E. Haugh and M. Bishop. Testing C Programs for Buffer Overflow Vulnerabilities. In M. Tripunitara, editor, *Network and Distributed System Security*, San Diego, California, USA, February 2003. Internet Society (ISOC).
99. N. Heintze and O. Tardieu. Ultra-fast Aliasing Analysis using CLA: A Million Lines of C Code in a Second. In *Programming Language Design and Implementation*, pages 254–263, Snowbirth, Utah, USA, June 2001. ACM.
100. M. Hind and A. Pioli. Which Pointer Analysis Should I Use? In *International Symposium on Software Testing and Analysis*, pages 113–123, Portland, Oregon, USA, August 2000. ACM.
101. D. S. Hochbaum. Monotonizing linear programs with up to two nonzeroes per column. *Operations Research Letters*, 32(1):49–58, 2003.
102. D. S. Hochbaum and J. Naor. Simple and Fast Algorithms for Linear and Integer Programs with Two Variables per Inequality. *SIAM Journal on Computing*, 23(6):1179–1192, 1994.
103. J. M. Howe and A. King. Efficient Groundness Analysis in Prolog. *Theory and Practice of Logic Programming*, 3(1):95–124, January 2003.
104. J.-L. Imbert. Fourier's Elimination: Which to Choose? In *Principles and Practice of Constraint Programming*, pages 117–129, 1993.
105. Institute of Electrical and Electronics Engineers, Inc, New York, USA. *IEEE Standard for Binary Floating-Point Arithmetic*, 1985.
106. J. Jaffar, M. J. Maher, P. J. Stuckey, and R. H. C. Yap. Beyond Finite Domains. In A. Borning, editor, *Principles and Practice of Constraint Programming*, volume 874 of *LNCS*, pages 86–94, Rosario, Orcas Island, Washington, USA, May 1994. Springer.
107. L. G. Jones. Efficient Evaluation of Circular Attribute Grammars. *ACM Transactions on Programming Languages and Systems*, 12(3):429–462, 1990.

108. R. W. M. Jones and P. H. J. Kelly. Backwards-Compatible Bounds Checking for Arrays and Pointers in C Programs. In M. Kamkar, editor, *Automated and Algorithmic Debugging*, pages 13–26, Linköping, Sweden, May 1997.
109. M. Karr. On affine relationships among variables of a program. *Acta Informatica*, 6(2):133–151, 1976.
110. A. King and L. Lu. A Backward Analysis for Constraint Logic Programs. *Theory and Practice of Logic Programming*, page 32, July 2002.
111. A. King and J. C. Martin. Control Generation by Program Transformation. *Fundamenta Informaticae*, 69(1–2):179–218, 2006.
112. D. E. Knuth. *The Art of Computer Programming: Fundamental Algorithms*, volume 1. Addison-Wesley, 2nd edition, 1973.
113. J. Koplowitz and D. Jouppi. A more efficient convex hull algorithm. *Information Processing Letters*, 7(1):56–57, 1978.
114. J. C. Lagarias. The Computational Complexity of Simultaneous Diophantine Approximation Problems. *SIAM Journal on Computing*, 14(1):196–209, 1985.
115. D. Larochelle and D. Evans. Statically Detecting likely Buffer Overflow Vulnerabilities. In *USENIX Security Symposium*, Washington DC, USA, August 2001. USENIX Association.
116. D. Larochelle and D. Evans. Improving Security Using Extensible Lightweight Static Analysis. *IEEE Software*, 19(1):42–51, 2002.
117. H. Le Verge. A Note on Chernikova's Algorithm. Technical Report 1662, Campus Universitaire de Beaulieu, Institut de Recherche en Informatique, Beaulieu, France, 1992.
118. D. Liang and M. J. Harrold. Efficient Computation of Parameterized Pointer Information for Interprocedural Analyses. In P. Cousot, editor, *Static Analysis Symposium*, volume 2126 of *LNCS*, pages 279–298, Paris, France, July 2001. Springer.
119. V. Loechner. PolyLib. `http://icps.u-strasbg.fr/polylib/`, 2005.
120. L. Lu and A. King. Determinacy Inference for Logic Programs. In S. Sagiv, editor, *European Symposium on Programming*, volume 3444 of *LNCS*, pages 108–123, Edinbourgh, UK, April 2005. Springer.
121. M. J. Maher. Abduction of Linear Arithmetic Constraints. In M. Gabbrielli and G. Gupta, editors, *International Conference on Logic Programming*, volume 3668 of *LNCS*, pages 174–188, Sitges, Spain, October 2005. Springer.
122. K. Marriott. Frameworks for Abstract Interpretation. *Acta Informatica*, 30(2):103–129, 1993.
123. L. Mauborgne and X. Rival. Trace Partitioning in Abstract Interpretation Based Static Analyzers. In M. Sagiv, editor, *European Symposium on Programming*, volume 3444 of *LNCS*, pages 5–20, Edinburgh, UK, April 2005. Springer.
124. K. Mehlhorn. *Sorting and Searching*, volume 1 of *ETACS Monographs*. Springer, 1984.
125. F. Mesnard and R. Bagnara. cTI: a Constraint-Based Termination Inference Tool for ISO-Prolog. *Theory and Practice of Logic Programming*, 5(1-2):243–257, 2005.
126. B. Miller, L. Fredrikson, and B. So. An Empirical Study of the Reliability of UNIX Utilities. *Communications of the ACM*, 33(12):32–44, 1990.
127. A. Miné. A New Numerical Abstract Domain Based on Difference-Bound Matrices. In O. Danvy and A. Filinski, editors, *Programs as Data Objects*, volume 2053 of *LNCS*, pages 155–172, Aarhus, Denmark, May 2001. Springer.

128. A. Miné. The Octagon Abstract Domain. In *Conference on Reverse Engineering*, pages 310–319, Stuttgart, Germany, October 2001. IEEE Computer Society.
129. A. Miné. A Few Graph-Based Relational Numerical Abstract Domains. In M. V. Hermenegildo and G. Puebla, editors, *Static Analysis Symposium*, volume 2477 of *LNCS*, Madrid, Spain, September 2002. Springer.
130. A. Miné. The Octagon Abstract Domain. *Higher-Order and Symbolic Computation*, 19:31–100, 2006.
131. A. Miné. Symbolic Methods to Enhance the Precision of Numerical Abstract Domains. In E. A. Emerson and K. S. Namjoshi, editors, *Verification, Model Checking and Abstract Interpretation*, volume 3855 of *LNCS*, pages 348–363, Charleston, South Carolina, USA, January 2006. Springer.
132. T. S. Motzkin, H. Raiffa, G. L. Thompson, and R. M. Thrall. The Double Description Method. In H. W. Kuhn and A. W. Tucker, editors, *Contributions to the Theory of Games*, number 28 in Annals of Mathematics Study. Princeton University Press, 1953.
133. M. Müller-Olm and H. Seidl. A Note on Karr's Algorithm. In J. Díaz, J. Karhumäki, A. Lepistö, and D. Sannella, editors, *Automata, Languages and Programming*, volume 3142, pages 1016–1028, July 2004.
134. M. Müller-Olm and H. Seidl. Precise Interprocedural Analysis through Linear Algebra. In *Principles of Programming Languages*, pages 330–341. ACM, January 2004.
135. M. Müller-Olm and H. Seidl. Analysis of Modular Arithmetic. In S. Sagiv, editor, *European Symposium on Programming*, volume 3444 of *LNCS*, pages 46–60, Edinburgh, UK, April 2005. Springer.
136. G. C. Necula, J. Condit, M. Harren, S. McPeak, and W. Weimer. CCured: Type-Safe Retrofitting of Legacy Software. *ACM Transactions on Programming Languages and Systems*, 25(3):1–50, 2005.
137. G. C. Necula, S. McPeak, S. P. Rahul, and W. Weimer. CIL: Intermediate Language and Tools for Analysis and Transformation of C Programs. In R. N. Horspool, editor, *Compiler Construction*, volume 2304 of *LNCS*, Grenoble, France, April 2002. Springer.
138. C. G. Nelson. An $n^{\log(n)}$ Algorithm for the Two-Variable-Per-Constraint Linear Programming Satisfiability Problem. Technical Report STAN-CS-78-689, Stanford University, November 1978.
139. E. M. Nystrom, H. S. Kim, and W. W. Hwu. Bottom-up and Top-down Context-Sensitive Summary-Based Pointer Analysis. In R. Giacobazzi, editor, *Static Analysis Symposium*, volume 3148 of *LNCS*, pages 165–180, Verona, Italy, August 2004. Springer.
140. E. M. Nystrom, H. S. Kim, and W. W. Hwu. Importance of Heap Specialization in Pointer Analysis. In C. Flanagan and A. Zeller, editors, *Program Analysis for Software Tools and Engineering*, Washington DC, USA, June 2004. ACM.
141. Aleph One. Smashing the Stack for Fun and Profit. *Phrack Magazine*, 7(49), 1996. `www.phrack.org`.
142. H. Pande and B. Ryder. Data-flow-based Virtual Function Resolution. In R. Cousot and D. A. Schmidt, editors, *Static Analysis Symposium*, volume 1145 of *LNCS*, pages 238–254, Aachen, Germany, September 1996. Springer.
143. N. Papaspyrou. *A Formal Semantics for the C Programming Language*. PhD thesis, National Technical University of Athens, 1998.

144. D. J. Pearce. *Some directed graph algorithms and their application to pointer analysis.* PhD thesis, Imperial College of Science, Technology and Medicine, University of London, London, UK, February 2005.
145. V. R. Pratt. Two Easy Theories Whose Combination is Hard, September 1977. `boole.stanford.edu/pub/sefnp.pdf`.
146. F. P. Preparata and S. J. Hong. Convex Hulls of Finite Sets of Points in Two and Three Dimensions. *Communications of the ACM*, 20(2):87–93, 1977.
147. F. P. Preparata and M. I. Shamos. *Computational Geometry.* Texts and Monographs in Computer Science. Springer, 1985.
148. W. Pugh. The Omega test: a fast and practical integer programming algorithm for dependence analysis. *Communications of the ACM*, 8:102–114, 1992.
149. F. Ranzato. Closures on CPOs form complete lattices. *Information and Computation*, 152(2):236–249, 1999.
150. X. Rival. Abstract Interpretation-Based Certification of Assembly Code. In L. D. Zuck, P. C. Attie, A. Cortesi, and S. Mukhopadhyay, editors, *Verification, Model Checking and Abstract Interpretation*, volume 2575 of *LNCS*, pages 41–55, New York, New York, USA, January 2003. Springer.
151. X. Rival. Understanding the Origin of Alarms in Astrée. In C. Hankin and I. Siveroni, editors, *Static Analysis Symposium*, volume 3672 of *LNCS*, pages 303–319, London, UK, September 2005. Springer.
152. M. Sagiv, T. Reps, and R. Wilhelm. Parametric Shape Analysis via 3-Valued Logic. *ACM Transactions on Programming Languages and Systems*, 24(3):217–298, 2002.
153. S. Sankaranarayanan, M. Colón, H. B. Sipma, and Z. Manna. Efficient Strongly Relational Polyhedral Analysis. In E. A. Emerson and K. S. Namjoshi, editors, *Verification, Model Checking and Abstract Interpretation*, LNCS, pages 111–125, Charleston, South Carolina, USA, January 2006. Springer.
154. S. Sankaranarayanan, F. Ivancic, and A. Gupta. Program Analysis Using Symbolic Ranges. In H. R. Nielson and G. Filé, editors, *Static Analysis Symposium*, volume 4634 of *LNCS*, pages 366–383, Kogens Lyngby, Denmark, August 2007. Springer.
155. B. Schlich, M. Rohrbach, M. Weber, and S. Kowalewski. Model Checking Software for Microcontrollers. Technical Report AIB-2006-11, RWTH Aachen, August 2006.
156. D. A. Schmidt. Comparing Completeness Properties of Static Analyses and Their Logics. In N. Kobayashi, editor, *Asian Symposium on Programming Languages and Systems*, volume 4279 of *LNCS*, pages 183–199, Sydney, Australia, November 2006. Springer.
157. A. Schrijver. *Theory of Linear and Integer Programming.* John Wiley & Sons, 1998.
158. R. Sedgewick. *Algorithms in C.* Addison-Wesley, 1988.
159. D. Seeley. A Tour of the Worm. In *Winter Usenix Conference*, San Diego, California, USA, 1989. USENIX Association.
160. R. Seidel. Convex Hull Computations. In J. E. Goodman and J. O'Rourke, editors, *Handbook of Discrete and Computational Geometry*, pages 361–376. CRC Press, 1997.
161. R. Shaham, H. Kolodner, and M. Sagiv. Automatic Removal of Array Memory Leaks in Java. In D. A. Watt, editor, *Compiler Construction*, volume 1781 of *LNCS*, pages 50–66, Berlin, Germany, March 2000. Springer.

162. U. Shankar, K. Talwar, J. S. Foster, and D. Wagner. Detecting Format String Vulnerabilities With Type Qualifiers. In *USENIX Security Symposium*. USENIX Association, 2001.
163. R. Shostak. On the SUP-INF method for proving Presburger formulas. *Journal of the ACM*, 24(4):529–543, 1977.
164. R. Shostak. Deciding Linear Inequalities by Computing Loop Residues. *Journal of the ACM*, 28(4):769–779, 1981.
165. A. Simon. Relational Analysis of Floating-Point Arithmetic. In *Workshop on Numerical and Symbolic Abstract Domains*, Paris, France, 2005.
166. A. Simon. Splitting the Control Flow with Boolean Flags. In M. Alpuente and G. Vidal, editors, *Static Analysis Symposium*, volume 5079 of *LNCS*, pages 315–330, Valencia, Spain, July 2008. Springer.
167. A. Simon and A. King. Analyzing String Buffers in C. In H. Kirchner and C. Ringeissen, editors, *Algebraic Methodology and Software Technology*, volume 2422 of *LNCS*, pages 365–379, Reunion Island, France, September 2002. Springer.
168. A. Simon and A. King. Convex Hull of Planar H-Polyhedra. *International Journal of Computer Mathematics*, 81(4):259–271, March 2004.
169. A. Simon and A. King. Exploiting Sparsity in Polyhedral Analysis. In C. Hankin and I. Siveroni, editors, *Static Analysis Symposium*, volume 3672 of *LNCS*, pages 336–351, London, UK, September 2005. Springer.
170. A. Simon and A. King. Widening Polyhedra with Landmarks. In N. Kobayashi, editor, *Asian Symposium on Programming Languages and Systems*, volume 4279 of *LNCS*, pages 166–182, Sydney, Australia, November 2006. Springer.
171. A. Simon and A. King. Taming the Wrapping of Integer Arithmetic. In G. File and H. R. Nielson, editors, *Static Analysis Symposium*, volume 4634 of *LNCS*, pages 121–136, Kongens Lyngby, Denmark, August 2007. Springer.
172. A. Simon, A. King, and J. M. Howe. Two Variables per Linear Inequality as an Abstract Domain. In M. Leuschel, editor, *Logic-Based Program Synthesis and Transformation*, volume 2664 of *LNCS*, pages 71–89, Madrid, Spain, September 2003. Springer.
173. B. Snow. Panel Discussion on the Future of Security. In *IEEE Symposium on Security and Privacy*. IEEE Computer Society, 1999.
174. A. Sotirov. Reverse engineering techniques to find security bugs: A case study of the ANI. Google tech talk, May 2007. `http://tinyurl.com/yw99c8`.
175. B. Steensgaard. Point-to analysis by Type Inference of Programs with Structures and Unions. In T. Gyimothy, editor, *Compiler Construction*, volume 1060 of *LNCS*, Linköping, Sweden, April 1996. Springer.
176. B. Steensgaard. Points-to Analysis in Almost Linear Time. In *Principles of Progamming Languages*, pages 32–41, St. Petersburg Beach, Florida, USA, January 1996. ACM.
177. Z. Su and D. Wagner. A Class of Polynomially Solvable Range Constraints for Interval Analysis without Widenings. *Theoretical Computer Science*, 345(1):122–138, 2005.
178. R. Tarjan. Depth-first search and linear graph algorithms. *SIAM Journal on Computing*, 1:146–160, 1972.
179. G. T. Toussaint and D. Avis. On a convex hull algorithm for polygons and its application to triangulation problems. *Pattern Recognition Letters*, 15(1):23–29, 1982.

180. G. T. Toussaint and H. El Gindy. A counterexample to an algorithm for computing monotone hulls of simple polygons. *Pattern Recognition Letters*, 1:219–222, 1983.
181. A. M. Turing. On computable numbers: With an application to the Entscheidungsproblem. In *Proceedings of the London Mathematical Society*, volume 42, pages 230–265, 1936.
182. A. Venet and G. Brat. Precise and Efficient Static Array Bound Checking for Large Embedded C Programs. In *Programming Language Design and Implementation*, pages 231–242, Washington DC, USA, June 2004. ACM.
183. J. Viega, J. T. Bloch, T. Kohno, and G. McGraw. ITS4: A Static Vulnerability Scanner for C and C++ Code. In *Computer Security Applications Conference*, New Orleans, Louisiana, USA, December 2000. IEEE Computer Society.
184. D. Wagner. *Static analysis and computer security: New techniques for software assurance*. PhD thesis, University of California at Berkeley, December 2000.
185. D. Wagner, J. S. Foster, E. A. Brewer, and A. Aiken. A First Step Towards Detection of Buffer Overrun Vulnerabilities. In *Network and Distributed System Security Symposium*, San Diego, California, USA, 2000. Internet Society (ISOC).
186. K. D. Wayne. A polynomial combinatorial algorithm for generalized minimum cost flow. In *Theory of Computing*, pages 11–18, Atlanta, Georgia, USA, May 1999. ACM.
187. N. Weaver and V. Paxson. A Worst-Case Worm. In *Workshop on Economics and Information Security*, Minneapolis, Minnesota, USA, May 2004.
188. J. Whaley and M. S. Lam. Cloning-Based Context-Sensitive Pointer Alias Analysis Using Binary Decision Diagrams. In *Programming Language Design and Implementation*, Washington DC, USA, June 2004. ACM.
189. Y. Xie, A. Chou, and D. Engler. Archer: Using Symbolic, Path-sensitive Analysis to Detect Memory Access Errors. *Foundations of Software Engineering*, pages 327–336, 2003.

Index